Biomaterials Science and Technology

Fundamentals and Developments

Biomaterials Science and Technology

Fundamentals and Developments

Yaser Dahman
Professor of Chemical Engineering
Ryerson University
Toronto, Canada

CRC Press
Taylor & Francis Group
Boca Raton London New York

CRC Press is an imprint of the
Taylor & Francis Group, an **informa** business

CRC Press
Taylor & Francis Group
6000 Broken Sound Parkway NW, Suite 300
Boca Raton, FL 33487-2742

Printed on acid-free paper

International Standard Book Number-13: 978-1-138-61147-4 (Hardback)

Library of Congress Cataloging-in-Publication Data

Names: Dahman, Yaser, author.
Title: Biomaterials Science and Technology Fundamentals and Developments / Yaser Dahman.
Description: Boca Raton : Taylor & Francis, 2019. | Includes bibliographical references.
Identifiers: LCCN 2018044697| ISBN 9781138611474 (hardback : alk. paper) | ISBN 9780429465345 (ebook)
Subjects: | MESH: Biocompatible Materials | Biocompatible Materials--therapeutic use | Nanotechnology
Classification: LCC R856 | NLM QT 37 | DDC 610.28--dc23
LC record available at https://lccn.loc.gov/2018044697

Visit the Taylor & Francis Web site at
http://www.taylorandfrancis.com

and the CRC Press Web site at
http://www.crcpress.com

Contents

Preface

This book presents a broad scope of the entire field of biomaterial science and technology, focusing on theoretical background, in addition to the most recent advances and applications of biomaterials. It reviews thoroughly the fabrication and properties of different classes of biomaterials, such as bioinert, bioactive, and bioresorbable, in addition to their biocompatibility. This includes the different classes of biomaterials derived from metals, ceramics, polymers, biopolymers, in addition to biocomposites. It further details traditional and recent techniques and methods that are utilized to characterize major properties of biomaterials. This includes surface and bulk properties, in addition to biocompatibility, mechanical, thermal, and biodegradability properties. The book details modifications of biomaterials in order to tailor properties and thus accommodate different applications in the Biomedical Engineering fields. It also summarizes the nanotechnology approach to the field of biomaterials. The last section reviews traditional and emerging applications of the different classes of biomaterials together with defining sets of tailored properties accordingly. Major applications are in the emerging fields of regenerative medicine as soft and hard tissues scaffolds, 3D printing such as bioinks, and in drug delivery applications.

This book targets students in advanced undergraduate and graduate levels in majors related to fields of Chemical Engineering, Materials Engineering and Science, Biomedical Engineering, Bioengineering, and Life Sciences. It mainly assists in understanding major concepts of fabrication, modification, and possible applications of different classes of biomaterials that include metals, ceramics, polymers, biopolymers, in addition to biocomposites. Additionally, the book is intended for professionals who are interested in recent advances in the emerging field of biomaterials.

Topics presented in the book cover a wide range of biomaterial science. Readers will learn about biomaterial classes of metals, ceramics, polymers, biopolymers, in addition to biocomposites. The book offers a wide scope of topics that cover introductory and theoretical materials that deal with major classes of biomaterials' fabrications, modifications, and applications. It also presents an up-to-date state of the literature on recent advances in biomaterials sciences in terms of fabrications and applications.

Chapter 1 discusses the general properties of biomaterials, as well as the methods used to characterize them. The chapter also discusses how researchers analyse biomaterials to determine their suitability for specific applications. Chapter 2 provides a review of the recent advances in biocompatibility with a study mostly focusing on orthopedic applications involving metal. Chapter 3 provides a review of polymeric-based biomaterials while examining past and current research in the field. Chapter 4 presents a review of ceramic-based biopolymers with a greater focus on their history, synthesis, characteristics, and applications of four specific types of bioceramics. Chapter 5 discusses the use of biomaterial composites as implants along with the characteristics and properties of several different biomaterial composites. Chapter 6 presents the applications of biopolymers with a large focus on the biopolymer Chitosan. Chapter 7 talks about the surface modification of biopolymers in order to elicit a positive response from the body. Three major surface modification methods will be discussed in greater detail. Chapter 8 is dedicated to the nano-approach. The chapter talks about the use of nanotechnology in conjunction with biomaterials to enhance the features of the biomaterials, as well as a focus on using nanotechnology for other biomaterial applications such as improving drug delivery systems, which will also be discussed in more detail in Chapter 9. Chapter 10 presents information about tissue engineering and regenerative medicine as alternative solutions to treating different diseases rather than only alleviating pain. The next two chapters review past and current methods of tissue replacement in the body for treatment; Chapter 11 focuses on hard tissue replacement, while Chapter 12 focuses on soft tissue replacement. With the current advances in technology, Chapter 13 discusses the use of 3D printing and bioinks in the creation of biomaterials with a greater emphasis on background information and current applications. Continuing the trend of advanced technology, Chapter 14 discusses the application of smart polymers as biomaterials. The chapter provides background information about smart polymers, their current applications, and their future prospects.

Acknowledgements

Development of this book owes much to the efforts of several former graduate students in the Department of Chemical Engineering, Ryerson University, Toronto, Canada. Their individual chapter contributions are as follows:

Patricia Lilian Torena Perez (Chapter 1), Ahmed Chaudhry (Chapter 2), Luqmaan Moolla (Chapter 3), Daniela Herrera (Chapter 4), Umme Sharmeen Hyder (Chapter 5), Chad Gorman Groves (Chapter 6), Stefan Balko (Chapter 7), Bhavik Vyas (Chapter 8), Mohamed Khattab (Chapter 9), Filip Jelic (Chapter 10), Ragvendra Pratap Singh (Chapter 11), Zaid Dwaik (Chapter 12), Thivjhan Kanaganavagam (Chapter 13), Amirhossein Kouhpour (Chapter 14).

Author

Dr. Yaser Dahman is a faculty member at the rank of "Professor" in the department of Chemical Engineering at Ryerson University, in Toronto, Canada. He obtained his Ph.D. in Chemical and Biochemical Engineering in addition to an MBA from Ryerson University. He commissioned a world-class research laboratory named "Nanocomposites and Biomaterials Engineering Laboratory", which was homed at Ryerson University. Later, this research facility became part of the "Center of Green Research Technology" that hosted all of research projects. Dr Dahman has supervised more than 40 graduate students' thesis projects and published several peer-reviewed articles, book chapters, and conference proceedings.

He has research focusing on the design, fabrication, and characterization of polymeric based biomaterials. His area of expertise includes developing chemical pathways for surface and bulk modification of biomaterials to tailor their characteristics. He collaborates with a number of surgeons from the bone fracture clinic at Mount Sinai Hospital in downtown Toronto to fabricate and produce new medical devices. He is also a fiber engineer who is specialized in producing nanostructured biocomposites for application in the healthcare industry. Recently, he published a book entitled "Nanotechnology and Functional Materials for Engineers - Elsevier".

His research interest has evolved into utilizing renewable and sustainable resources of agro-industrial wastes, algal biomass, and HEMP (a form of cannabis plant grown for its fibres) in advanced aerobic and anaerobic Simultaneous Saccharification and Fermentation (SSF) and Separate Hydrolysis and Fermentation (SHF) to produce green biomaterials, green biodegradable plastics, and green biobutanol and bioethanol. Different pretreatment methods were intensively tested and utilized at different conditions for the conversion of agricultural cellulose and hemicellulose to monomeric sugars essential to conducting the green fermentation reactions. Biocatalysis were also broadly utilized to catalyse saccharification reactions.

Another research project resulted in developing novel classes of bioreactors with unique designs that exhibit improved hydrodynamics of mixing characteristics and operating conditions. This class of bioreactors includes internally and externally recirculated airlift bioreactors. Least invasive techniques using Electrical Resistance Tomography were utilized to visualize the hydrodynamics of mixing on the micro and macro scales inside the bioreactors through 2D and 3D tomograms.

1 General Properties and Characterization Methods of Biomaterials

1.1 INTRODUCTION

For decades, biomaterials have been successfully used in medical applications to improve both the quality and the length of life of many people. Their wide ranges of functions vary from drug delivery systems to replacing, augmenting, or repairing tissues, organs, or functions of the body. But these functions were only achieved after many different attempts, research works, and historical developments made over the centuries.

The introduction of non-biological materials into the human body was found to happen far back in prehistory. Dental implants are traceable to early Egyptians and South Central American cultures, while the earliest operations for the restoration of missing parts were found to be about 600 BC (Saini et al., 2015). The earliest evidence of the application of metal for surgical procedures is from 1565, which was not successful due to infections after implantation (Bhat, 2005). The poor understanding of biocompatibility and sterilization explains why the majority of implants developed before 1950 had low chances to succeed. In 1891, the first hip replacement was performed by Theodore Gluck, a German surgeon, with no success (Ratner et al., 2004). It was only during the nineteenth century that huge implant advancements were achieved, making possible surgical procedures – such as heart valve implantation – which previously were considered to be impossible. Among the many modern implant developments made throughout the twentieth century are artificial kidneys, hemodialyzer, artificial heart, hip and knee prostheses, screws and plates for fracture fixation, intraocular lenses and corneal replacements, pacemakers, etc.

The performance and success of every implant and medical device used to accomplish specific functions in the human body depend upon the properties and biocompatibility of the biomaterials chosen for their construction. Herein lies the importance of establishing a process that permits a complete characterization of the biomaterial and understanding of the biological responses attained when implanted in a living system. Many methods from medical and engineering disciplines are used to characterize biomaterials, which provide elemental information allowing determination of whether the device is suitable for its intended biomedical function and if it meets the regulatory standards.

by Yaser Dahman and Patricia Lilian Torena Perez

This work aims to provide a comprehensive review of the main characteristics of biomaterials used in current biomedical practices and the specific methodologies employed to characterize their properties and biological responses within living systems. Chemical, physical, mechanical, surface, and biological properties are revised, and an analysis is made of the properties that are usually desired in biomaterials. The most common characterization techniques based on medicine and physical, engineering and biological sciences are then discussed, complemented by an insight into how researchers integrate these different techniques to thoroughly analyse biomaterials and determine their suitability for the intended biomedical application. Finally, recent studies carried out in the field of biomaterials are overviewed in order to provide insight on the future perspectives and efforts directed towards the improvement of current biomaterial characterization practices.

1.2 PROPERTIES OF BIOMATERIALS

The different properties of biomaterials have been shown to have an important influence in their dynamic interactions with the biological surroundings when used as medical implants and organ or tissue replacements. The performance and success of each one of these implants depend upon the properties and biocompatibility of the biomaterial used to construct it. Thus, when designing any medical device or implant, it is critical to evaluate the material properties within the context of its future biomedical application.

In general, the most important properties associated with biomaterials are classified into chemical, physical, mechanical, and biological in relation to the bulk and surface of the material. These are summarized in Figure 1.1 and explained in detail below.

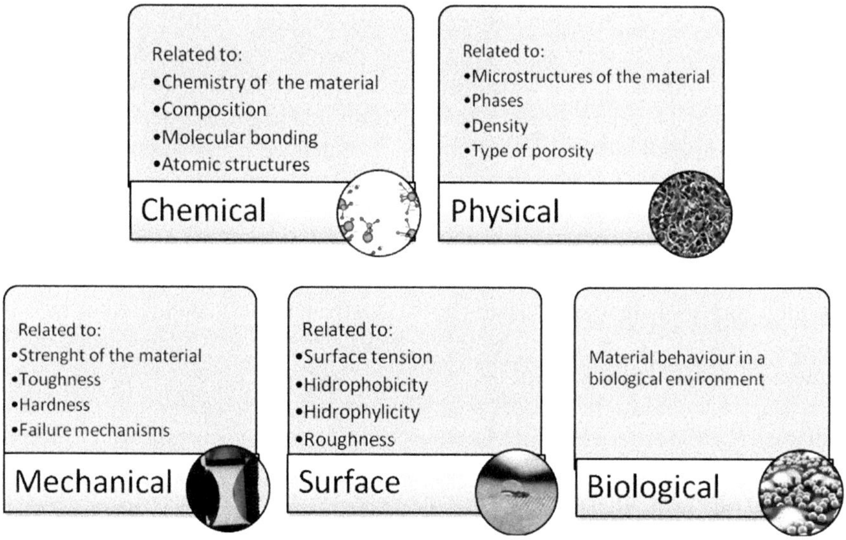

FIGURE 1.1 Scheme of biomaterials properties.

1.2.1 Chemical Properties

When talking about the chemistry of biomaterials, one usually refers to the molecular bonds and atomic structures. Since solids have constituent atoms being held together by strong interatomic forces, the nature of these forces (i.e. bonding) will govern the properties of every solid material. Unique intermolecular forces that result in different electrical and thermal conductivities govern different types of bonding. There are basically three types of strong or primary bonds: ionic, covalent, and metallic. Metallic bonds have free valence electrons, which allow high electrical and heat conductivity; ionic bonds have electrons that are tightly held together by the ions, which prevent heat and electrical conductivity; covalent bonds share valence electrons with adjacent atoms, which causes a poor electrical and thermal conductivity (Bhat, 2005). Examples of materials with these types of bonds are metals (metallic bond), ceramics (ionic bond), and polymers (covalent bond). In addition to strong bonds, there are weak or secondary bonds, which also influence the properties of the materials. The most relevant ones are van der Waals and hydrogen bonding since they have strengths that are about 3 to 10% of a primary carbon–carbon covalent bond (Ratner et al., 2004). In van der Waals bonds, electrons are not evenly distributed among the ions capable of forming dipoles, while in hydrogen bonds, hydrogen atoms are covalently bonded to an electronegative atom which turns into a positive ion (Bhat, 2005). For instance, polymers possess two bonding types: covalent, between atoms in a molecule, and van der Waals, between molecules; therefore, its properties are influenced by the individual intermolecular forces from each bond type. The strength order and schematic representation of these bonds are shown in Figure 1.2.

Furthermore, each bond type exhibits different ranges of bonding energy (Table 1.1), which is the amount of energy required to separate the bonded atoms. The magnitude of bonding energy influences certain material properties, such as thermal: heat of fusion, conductivity, heat capacity, and melting temperature. The higher the bonding energy, the more external energy is required to separate the atoms, which traduces to, e.g. higher melting temperatures. The bonding energy and melting temperature of several substances commonly used in biomaterials are listed in Table 1.1.

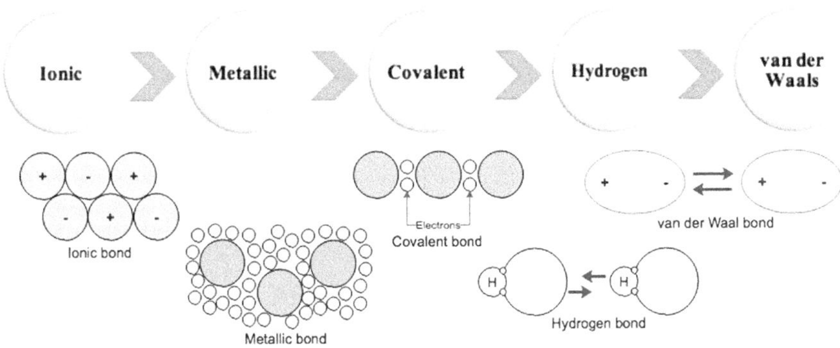

FIGURE 1.2 Strength order of bonding types and their schematic representations.

TABLE 1.1
Bonding Energies and Melting Temperatures for Various Substances Used in Biomaterials

Bonding Type	Bonding Energy Range[a] (kJ/mol)	Substance	Bonding Energy (kJ/mol)	Melting Temperature (°C)
Ionic	700–4,000	MgO	1,000	2,800
Covalent	200–1,000	Si	450	1,410
		C (diamond)	713	>3,550
Metallic		Hg	68	−39
	40–850	Al	324	660
		Fe	406	1,538
		W	849	3,410
Hydrogen	10–40	NH_3	35	−78
Van der Waals	≤	Cl_2	31	−101

Source: Callister and Rethwisch (2006).
[a]Some values taken from *Science Encyclopedia*, available at http://science.jrank.org/pages/984/Bond-Energy.html.

On the other hand, atomic structures refer to the arrangement of atoms or ions in any solid material, which will define if the solid is crystalline or amorphous. Crystalline solids are those whose atoms or ions are closely packed and arranged in a three-dimensional lattice following a particular repeated pattern – also called a unit cell – whereas amorphous solids lack this systematic and regular arrangement (Bhat, 2005; Callister and Rethwisch, 2006). Examples of biomaterials showing a crystalline structure are metals and ceramics, while many polymers and resins show amorphous structures. Usually, materials used for medical implants are made of two or more elements. Hence, the atomic structure of these biomaterials will depend on the constitutive elements and the relative size of their atoms. Figure 1.3 illustrates the unit cell of hydroxyapatite (HAp), a versatile biomaterial used in many applications such as bone grafting and tissue engineering (Zilm et al., 2016).

1.2.2 Physical Properties

The physical properties of materials are related to their internal microstructure, which is the way that grains are tightly packed and firmly bound together in most crystalline solids, usually metals and ceramics. The microstructure of materials includes the different phases present in the structure; size, shape, and spatial distribution of these phases or grains, and possible crystal imperfections such as point and line defects. For instance, the microstructure of the Si_3N_4–bioglass composite obtained by scanning electron microscopy (SEM) analysis (Figure 1.4), after CF_4 plasma etching, reveals a glassy phase – grey – that surrounds the α-Si_3N_4 grains (Amaral et al., 2002). All the microstructural features will influence other properties

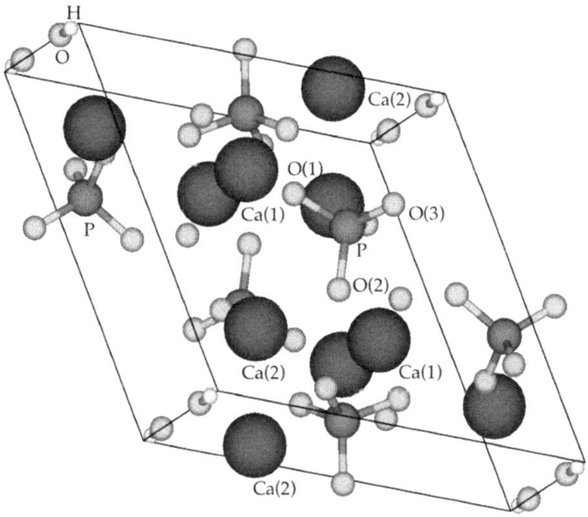

FIGURE 1.3 Unit cell of hydroxyapatite (Zilm et al., 2016).

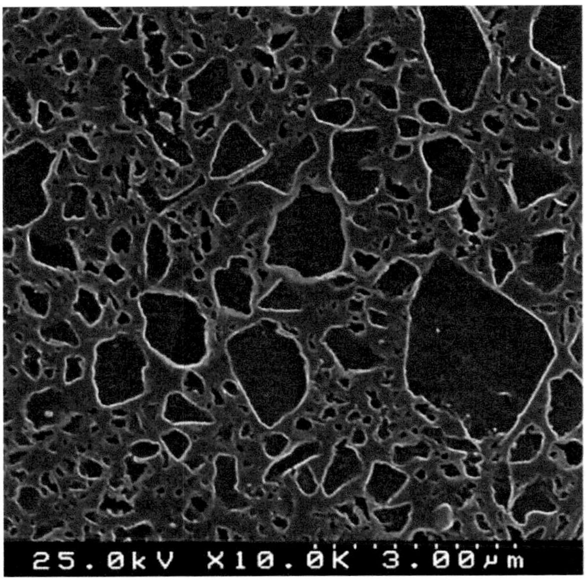

FIGURE 1.4 Microstructure of Si_3N_4–bioglass composite by SEM analysis after CF_4 etching (Amaral et al., 2002).

of the biomaterial, e.g. mechanical properties. Table 1.2 lists some examples of microstructures of several titanium alloys used in biomedical applications, depending on different thermomechanical treatments.

There are certain substances that can show two or more phases in solid state, which is called allotropy. Iron is an example of this phenomenon, exhibiting three

TABLE 1.2
Microstructure of Titanium Alloys Used in Biomedical Applications

Alloy	Microstructure
Pure Ti	(ASTM F67-89)
Ti-6Al-4V ELI (ASTM F136-84, F620-87)	α+β type
Ti-6Al-4V (ASTM F1108-88)	α+β type
Ti-6Al-7Nb (ASTM F1295-92, ISO5832-11)	α+β type (Swiss)
Ti-5Al-2.5Fe (ISO5832-10)	α+β type (Germany)
Ti-5Al-3Mo-4Zr	α+β type (Japan)
Ti-15Sn-4Nb-2Ta-0.2Pd	α+β type (Japan)
Ti-15Zr-4Nb-2Ta-0.2Pd	α+β type (Japan)
Ti-13Nb-13Zr (ASTM F1713-96)	near β type (U.S.), Low modulus
Ti-12Mo-6Zr-2Fe (ASTM F1813-97)	β type (U.S.), Low modulus
Ti-15Mo	β type (U.S.), Low modulus
Ti-16Nb-10Hf	β type (U.S.), Low modulus
Ti-15Mo-5Zr-3Al	β type (Japan), Low modulus
Ti-15Mo-2.8Nb-0.2Si-0.26O	β type (U.S.), Low modulus
Ti-35Nb-7Zr-5Ta	β type (U.S.), Low modulus
Ti-29Nb-13Ta-4.6Zr	β type (Japan), Low modulus
Ti-40Ta, Ti-50Ta	β type (U.S.), High corrosion resistance

Source: Hussein et al. (2015).

different crystal structures: 'bcc' below 916°C, 'fcc' between 916–1389°C, and 'bcc' above 1389°C (Bhat, 2005). Moreover, in a binary or multicomponent system, different phases at different compositions can be observed. Such is the case of Ti-6Al-4V alloy listed in Table 1.2 (used for bone screws and joint replacement), which can acquire multiple microstructures along with geometrical configurations of α and β phases according to the specific thermomechanical treatment, as is shown in detail in the phase diagram from Figure 1.5 (Boyer, Welsch, and Collings, 1994).

Aside from the phases and their spatial distribution in biomaterials, there are also possible crystal imperfections that need to be examined. These imperfections can have effects on other properties, depending on their level of concentration. Even when this concentration is low, these defects have a significant effect on optical and electrical properties, and a dominant effect on mechanical properties (Bhat, 2006). For example, in the case of metals, crystalline defects have an impact on their ductility. The imperfections are classified into point defect and linear defect, depending on geometry and dimension of the defect. Point defect is associated with vacancies and self-interstitials; the former refers to vacant lattice sites from which an atom is missing, while the latter refers to an atom that occupies a small space that normally is not occupied (Callister and Rethwisch, 2006). On the other hand, linear defects are related to edge or screw dislocations occurring on the planes of atoms.

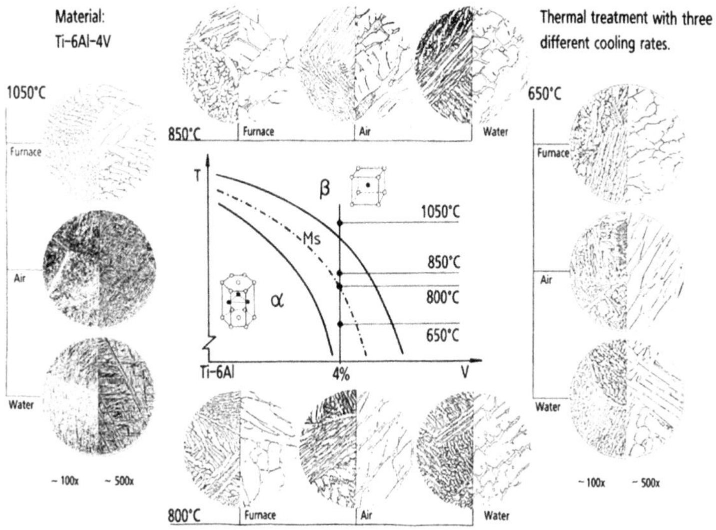

FIGURE 1.5 Phase diagram of Ti-6Al-6V alloy for different thermomechanical treatments (Ducato et al., 2013).

Furthermore, there are defects related to the chemical nature of materials, which are called impurities; impurities can also modify the properties, whether they are there by accident or have been added intentionally.

1.2.3 Mechanical Properties

Ductility, strength, and hardness are some of the common mechanical properties studied in biomaterials. These properties are related to the material's ability to bear static loads. Among the most fundamental mechanical properties of materials are: elastic modulus, yield strength, yield strain, ductility, fracture strength, resilience, and toughness; they are obtained from a stress–strain curve calculated from the measured load and deformation throughout a mechanical loading test. Of these, strength is considered one of the most relevant properties for biomedical applications, since medical devices must meet certain mechanical loading requirements; it is defined as the stress required to break a material into two pieces (Park, 2008). Likewise, the elastic modulus or Young's modulus is an important property; it indicates the stiffness of the material and its tendency to deform along an axis with opposing forces. Hardness or toughness, on the other hand, is a measure of the resistance of a material to localized plastic deformation; e.g. a biomaterial is very hard if it can withstand high stresses while suffering considerable plastic deformation, whereas if it only withstands low stresses but suffers high deformations, it is much less hard (Bhat, 2005). Some of these mechanical properties of common implant biomaterials and tissues are presented in Table 1.3, and compared with those of human hard and soft tissues.

In this context, it is also important to determine the ductility of materials. Ductility indicates the degree to which a structure will have plastic deformation upon rupture.

TABLE 1.3
Mechanical Properties of Implant Biomaterials and Tissues

Material	Elastic Modulus (GPa)	Tensile strength (MPa)	Reference
Alumina	350–410	1,000–10,000	Amaral et al. (2002), Ratner et al. (2004)
Co-Cr Alloys	195–253	655–1,896	Hussein, Mohammed, and Al-Aqeeli (2015)
Tantalum[a]	190	517	Park and Lakes (2007)
Titanium	110	760	Hussein et al. (2015)
PMMA[b]	3	35–50	Park (2008), Ratner et al. (2004)
Bioglass	35	200	Park (2008)
Ti-6Al-4V	116	965–1,103	Hussein et al. (2015)
Hydroxyapatites	80–120	40	Amaral et al. (2002), Chen et al. (2012)
Stainless Steel	190	480–1,351	Black (1988), Hussein et al. (2015)
Platinum	100–105	550–860	Black (1988)
Synthetic rubber	4	10–12	Black (1988)
Polyethylene[c]	0.6–1.8	23–40	Ratner et al. (2004)
Cancellous bone	4.6–15	51–193	Cowin, Buskirt, and Ashman (1987)
Cortical bone	15–30	70–150	Cowin et al. (1987), Ratner et al. (2004)
Tendon/ligament	65–2,500 MPa	30–300	Silver (1994)

[a]Tantalum (F560) cold worked.
[b]PMMA: poly(methyl methacrylate).
[c]High density polyethylene.

Hence, if it experiences scanty or no plastic deformation upon rupture, it means the material is brittle; and conversely, if it possesses a high degree of plastic deformation upon rupture, this means the material is ductile. Ductility is expressed as the percentage of elongation; as a general rule of thumb, when this percentage is less than about 5%, the material is considered brittle (Callister and Rethwisch, 2006).

Biomaterials exhibit different typical mechanical behaviours depending on their type, e.g. ceramic, metal, polymer or composite, as is observed in Table 1.3. For instance, the differences in elastic modulus and strength are a result of different atomic bonding. As discussed beforehand, whether an atomic bond is ionic, metallic, or covalent, a certain amount of energy is required to break the bond apart; therefore, the stronger the bond, the more energy is required to break it, resulting in a higher material strength or stiffness. Ceramics, such as hydroxyapatite, alumina, and bioglass, usually exhibit high stiffness and the highest elastic modulus; their strong covalent or ionic bonding is produced by the high electronegativity difference between their constitutive metal and non-metal ions. However, they are brittle and fracture before any plastic deformation can occur during tensile load. This brittle fracture process involves the creation and propagation of cracks through the cross-section of material, and the ability of the ceramic material to resist fracture when these cracks are present is measured in terms of fracture toughness (Callister and Rethwisch, 2006). Similarly, metal alloys such as CoCr and titanium alloys typically show a high elastic modulus and the highest tension strength, whereas pure metals, e.g. titanium and stainless steel, usually show lower strength. They are ductile and their metallic bonding possesses a small electronegativity difference between atoms. However, there are other metals with relatively poor mechanical properties such as tantalum, which is limited to a few specialized applications like wire sutures used by plastic surgeons and neurosurgeons (Park and Lakes, 2007). Polymers, like polyethylene and poly(methyl methacrylate) PMMA, are the lowest in both strength and elastic modulus. As mentioned previously, they have typically covalent and van der Waals forces between atoms and molecules, respectively; the higher influence of van der Waals forces, which are weak by nature, result in low strength and Young's modulus. Finally, since composites (e.g. Si_3N_4–bioglass) contain two or more different materials or phases, their mechanical behaviours depend on the inclusions and individual mechanical properties. Composites combine the advantageous properties of each material; however, their strength depends on the brittleness or ductility of the substances and the matrix. Typical tensile stress–strain curves for pure poly(glycerol sebacate) (PGS), an elastomer used for engineered soft tissues, and the PGS/bioglass composites are shown in Figure 1.6. This composite exhibits an enhanced Young's modulus and increased mechanical strength and strain at rupture.

1.2.4 Surface Properties

The surface properties are among the most important biomaterial properties since the surface chemistry will determine how the material of the implanted device interacts with the surrounding tissues or fluids. The atoms that reside at the surface have a distinctive organization and reactivity and control most of the biological reactions such as protein adsorption, cell adhesion, cell growth, etc. (Ratner et al., 2004). Many parameters allow description of the surface properties: chemical composition, surface energy, surface tension, and roughness, among others. Surface energy denotes how unsatisfied are the bonds at the surface, which creates a force field that attracts atoms into the bulk. In other words, when there is an additional energy at the surface of the material, adsorption occurs; atoms and molecules are adsorbed onto the surface so that its thermodynamic stability is achieved. For instance, ceramics and metals have large surface energies whereas polymers have far smaller surface energies (Bhat, 2005). Another important parameter is surface tension; in biomaterials science, the quantitative measurement of the contact angles has been used to determine the surface tension, as an indicator of the wettability, hydrophobicity, and hydrophilicity of biomaterials. These contact angles are observed when a drop of liquid is sitting on a solid surface and depending on the type of interfacial free energies between the two substances, the liquid may remain on the surface as a droplet or it may spread out all over it. This phenomenon represents a balance between the cohesive

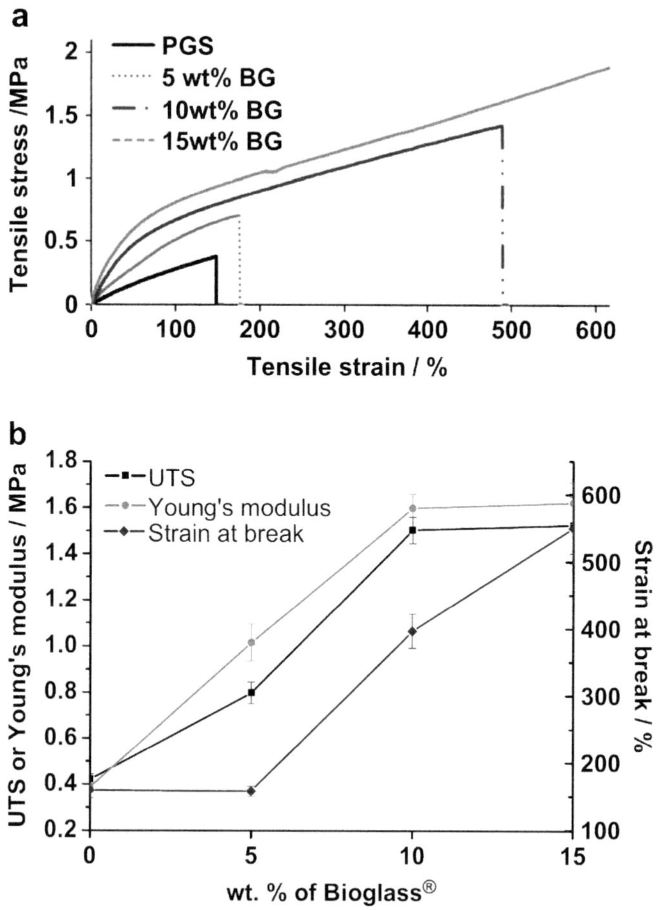

FIGURE 1.6 Typical tensile stress–strain curves of pure PGS and PGS composites of 5, 10, or 15 wt.% bioglass (a). Ultimate tensile strength (UTS), Young's modulus, and strain at break of pure PGS and PGS/bioglass composites (Liang et al., 2010).

forces of the liquid molecules and the adhesive forces by which these molecules are attracted to the surface of the solid. The equilibrium established between these forces (Figure 1.7) is described by a Young–Dupre equation (Gillis, 2006):

$$\gamma_{s/g} = \gamma_{s/l} + \gamma_{l/g} \cos\theta$$

where $\gamma_{s/g}$ is the interfacial free energy between the solid and gas, $\gamma_{s/l}$ between the solid and liquid, $\gamma_{l/g}$ between the liquid and gas, and θ the contact angle.

The wetting characteristic is generalized as complete wetting when the contact angle is $\theta_{=}\theta$; partial wetting when the contact angle is $0 < \theta \leq 90°$ and non-wetting when $\theta > 90°$ The different values of contact angle of a liquid with different values of $\gamma_{l/g}$ an be plotted $\cos\theta$ vs $\gamma_{l/g}$ to find the critical surface tension of the material; this critical

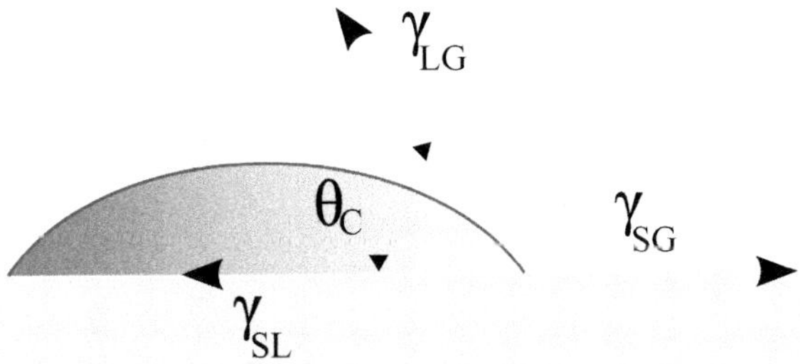

FIGURE 1.7 Contact angle (Gillis, 2006).

point is defined as the value of $\gamma_{l/g}$ at which $\cos\theta = 1$ and the more hydrophobic the biomaterial surface, the lower its critical surface tension (Bhat, 2006). Hydrophobic surfaces have characteristics such as high contact angle, poor adhesiveness, poor wettability, and low solid surface free energy, whereas hydrophilic surfaces have low contact angle, good adhesiveness, good wettability, and high solid surface free energy.

Lastly, the surface roughness of a biomaterial refers to the small irregularities on its surface. These irregularities and their alterations may affect the response of cells and tissues by increasing or decreasing the surface area of an implant, which can improve or worsen cell attachment, cell growth or protein adsorption (Saini et al., 2015). Implant surfaces can be classified by their roughness as minimally rough (0.5–1 m), intermediately rough (1–2 m), and rough (2–3 m) (Rompen et al., 2006).

1.2.5 Biological Properties

Biological properties of biomaterials are related to their behaviour in a biological environment. Since biomaterials are intended to perform certain functions within a living system, it is imperative that their biological properties are studied in order to determine the biocompatibility. Biocompatibility is defined as the ability of the biomaterial or medical device to perform its intended function with an appropriate and favourable host response in a particular biological environment. Depending on these responses, the material can be classified as toxic, bioinert, bioactive, or bioresorbable. Bioinert materials are those that are both nontoxic and biologically inactive, like most metallic implants, e.g. titanium alloys, zirconium, and aluminum oxides; however, they usually produce some response in the adjacent tissues that leads to fibrous encapsulation of the implant. Bioactive materials are nontoxic but biologically active, hence the material and host tissue interact and form interfacial bonds; most calcium phosphate ceramics like hydroxyapatite and bioactive glasses such as bioglass are some examples (Bandyopadhyay and Bose, 2013). A bioresorbable material is, in contrast, nontoxic and completely reabsorbed and replaced by the host tissue. Biodegradable polymers such as poly(glycolic acid), block copolymers like PLGA-PEG, bioactive glasses, calcium sulfate, tricalcium phosphate, all undergo bioresorption. Lastly, toxic materials cause death to surrounding tissue. Table 1.4 lists

TABLE 1.4
Biological Properties of Synthetic Biomaterials Used in Bone Tissue Engineering

Biomaterial	Biological Properties
Calcium phosphates Hydroxyapatite (HAp) Tricalcium phosphate (TCP)	Excellent biocompatibility Cell activity support Biodegradable
Na-containing silicate bioactive glasses	Excellent biocompatibility Cell activity support Biodegradable
Borate bioactive glasses	Biodegradable Risk of toxicity due to the release of borate ions
Bioactive glass ceramics	Excellent biocompatibility Cell activity support Biodegradable
Bulk biodegradable polymers Poly(lactic acid) (PLA) Poly(glycolic acid) (PGA) Poly(lactic-co-glycolicacid) (PLGA) Poly(propylene fumarate) (PPF) Poly(polyol sebacate) (PPS)	Good biocompatibility Biodegradable Bioresorbable
Poly(ortho esters) Poly(anhydrides) Poly(phosphazene)	Good biocompatibility Not completely replaced by new bone tissue
Composites (Containing bioactive phases)	Excellent biocompatibility Cell activity support Biodegradable

Source: Chen et al. (2012).

the biological properties of synthetic biomaterials used in bone tissue engineering (Chen, Zhu, and Thouas, 2012).

Biocompatibility depends on many factors such as material degradation, protein deposition, and encapsulation; but for medical implants, it depends to a large extent on the corrosion resistance and cytotoxicity of corrosive products. Corrosion and corrosion resistance relate to the loss of metallic ions from a metal surface to the biological surrounding and can be classified as crevice, pitting, galvanic, and electrochemical (Saini et al., 2015).

1.2.6 Desired Properties of Biomaterials

Different material systems possess generic advantageous properties that make them more suitable for certain applications. For instance, metallic materials are widely known for their high ductility, strength, and toughness; their excellent mechanical properties, added to a good corrosion resistance, make them an ideal material for bone and joint replacement and bone fracture fixation. Ceramic materials stand out for their hardness and low wear rate; while they are not as resistant to fracture as metals, they have the highest compression strength, which is why ceramics are widely used in load-bearing implants. Moreover, ceramics are resistant to microbial attack, bioresorbable, and bioactive, hence extensively used in bone tissue engineering and implant coatings for bioactivity improvement. Polymers may not be as strong as metals and ceramic but they are the most widely used materials in health care due to their versatility and other valuable properties. Polymeric materials can be either flexible or rigid, biodegradable or bioinert, strong or fragile, chemically resistant, etc., depending on their intended function. Their multiple uses range from sutures, blood vessels, and soft tissues, to facial and orthopedic prostheses; polymer structures with a strong interaction with water are used as hydrogels since their zero or low interfacial tension with the surroundings reduce protein adsorption and cell adhesion. Finally, composite materials have myriad properties that depend on their constituents and individual properties. Usually, composites are designed to either combine the best of two different materials or to enhance the properties of one of them. For example, bioactive glasses have poor mechanical strength and incompatibility with bone, which limits their application in load-bearing applications; however, these materials are combined with polymers to form composites with improved mechanical properties and biocompatibility that can be used in bone reparations (Chen et al., 2012). Other applications of composites are found to be in dental resins and bone cement. Some of the advantageous properties of common materials used in biomedical applications are shown in Table 1.5.

1.3 CHARACTERIZATION OF BIOMATERIALS

Biomaterials are used to construct biomedical devices to perform many specific functions in the human body; selecting the most appropriate material for these functions is a complex process that depends on particular mechanical, chemical, structural, and biological requirements. During this process, it is essential to characterize the material in order to reveal its properties and verify if these requirements are met. Many disciplines are involved in the characterization of biomaterials; physical and engineering sciences focus on the physical, mechanical, and surface properties, whereas biological sciences and medicine focus on biological properties and biocompatibility. The integration of different techniques based on these sciences allows researchers to thoroughly analyse the properties of biomaterials and determine their suitability for the intended biomedical application. A schematic outline of biomaterial characterization is shown in Figure 1.8; the first level of testing is achieved when basic material properties (i.e. physical, chemical, mechanical, and surface) are determined; only after a second level of testing, which includes biological in-vitro and in-vivo studies, can devices be considered for medical trials and manufacturing.

TABLE 1.5
Properties of Common Biomaterials Used in Medical Devices

Materials	Properties	Biomedical Applications medical Applications
	Polymers	
Poly(ethylene)	Excellent electrical insulation Excellent chemical resistance High elastic modulus and tensile yield strength High hardness Good biocompatibility	Orthopedic implants, hip joint and knee joint prostheses, tubes for various catheters, bags, ear parts, esophagus segments; facial implants, dentures, and finger joints
Poly(vinyl chloride)	Excellent resistance to abrasion Dimensional stability High chemical resistance to acids, alkalis, oils, fats, alcohols, and aliphatic hydrocarbons Good biocompatibility	Flexible or semi-flexible medical tubes, catheter, ear parts, facial prosthesis, heart components, inner tubes, components of dialysis installation, temporary blood storage Devices
Poly(methyl methacrylate) (PMMA)	Exceptional transparency High elastic modulus and compressive strength Very hard and brittle Excellent biocompatibility Optical properties Excellent weathering properties	Intraocular lenses, dentures, bone cement, fixation of articular prostheses
Poly(ethylene terephthalate) (PET)	Good transparency Good resistance to traction and tearing Resistance to oils, fats, organic solvents Good biocompatibility	Vascular, laryngeal, esophageal prostheses, surgical sutures, knitted vascular prosthesis, heart valves, artificial vascular graft
Poly(amide) (Nylon)	Very good mechanical properties Excellent fiber-forming ability High degree of crystallinity Resistance to abrasion and breaking Stability to shock and fatigue Good thermal properties Good chemical resistance	Intracardiac catheters, urethral sound, surgical suture, dialysis devices components, heart mitral valves, tracheal tubes, bones, and joints
Poly(urethane) (PU)	High resistance to breaking Very high elastic modulus at compression Resistance to abrasion Elongation to breaking Good biocompatibility	Biodegradable sutures, implant coating, adhesives, dental materials, blood pumps, artificial heart, and skin

(Continued)

TABLE 1.5 (CONTINUED)
Properties of Common Biomaterials Used in Medical Devices

Materials	Properties	Biomedical Applications medical Applications
Poly(hydroxyethyl methacrylate) (PHEMA)	Good tensile strength Semipermeable Hydrophilic Good biocompatibility Low or zero interfacial tension	Hydrogel for synthetic articular cartilage in reconstructive joint surgery
Stainless steel (type 316 and 316L)	Resistant to corrosion (for temporary devices) Nonmagnetic Uniform microstructure, fine grains (wrought alloy) Low mechanical strength (annealed)	Orthopedic implants, joint replacement, mandibular staple bone plates, heart valves, Mayfield and Scwhartz clips, bone screws, hip nails
Cobalt–Chromium alloys	Large grains and low mechanical properties (cast alloy) Uniform microstructure with fine grains (wrought alloy) Highest mechanical strength when cold worked High degree of corrosion resistance to seawater	Dental implants, artificial joints, bone ingrowth, knee and hip prosthesis, heart valves, fracture plates
Titanium-based alloys: Ti-6Al-4V	Exceptional strength-to-weight ratio Good strength, toughness, ductility, and fatigue Low elastic modulus, flexibility Good biocompatibility	Bone screws, partial and total hip, knee, elbow, jaw, finger, and shoulder replacement joints, dental implants
Platinum and other noble metals (Pd, Rh, and Ir)	Highly resistant to corrosion Poor mechanical properties Low threshold potential	Pacemaker tips
Alumina	High hardness Excellent corrosion resistance Good biocompatibility High wear resistance Good mechanical strength Small grain size High elastic modulus	Dental implants, load-bearing hip prosthesis, hip and knee joints, tibia plate, femur shaft, shoulders, radius, vertebra, leg lengthening spacer, ankle joint prostheses, reconstructive maxillofacial surgery, bone tissue engineering
Zirconia (Yttria stabilized)	High hardness Excellent biocompatibility Good wear resistance High fracture strength and toughness Fine grain size and well-controlled microstructure	Orthopedic prosthesis, shoulder, hip joint heads, total coating in dental implants

(*Continued*)

TABLE 1.5 (CONTINUED)
Properties of Common Biomaterials Used in Medical Devices

Materials	Properties	Biomedical Applications medical Applications
Hydroxyapatite	High elastic modulus Excellent biocompatibility High stiffness, compression and tensile strength, and fracture toughness (when high crystalline) Biodegradable	Bone tissue engineering (scaffolds, coating for tissue attachment materials), drug delivery (anticancer drugs), stem cell differentiation
Glass-ceramics (bioglass)	Fine grained-structure Excellent mechanical and thermal properties High brittleness Resistance to scratching and abrasion Excellent biocompatibility Bioactive and bioresorbable	Fillers for bone cement, dental restorative composite, coating material, stem cell differentiation
	Composites	
Poly(glycerol)/45S5 bioglass	Good biocompatibility High elastic modulus High elongation at break Biodegradable	Soft tissue engineering
TCP/PLA	Good biocompatibility Biodegradable Highly porous	Bone tissue engineering
Bioglass/ poly(glycerol sebacate)	Good biocompatibility Biodegradable Highly porous (>90%)	Bone tissue engineering

1.3.1 Physical and Chemical Characterization

The physical and chemical properties of biomaterials, as explained beforehand, are closely related to their microstructure. In order to study the features of the microstructure, including crystal imperfections and phases/grains in the structure, it is essential to carry out morphology, crystallography, and chemical composition analyses. Morphology analysis allows the finding of details on (1) porosity; (2) crystal imperfections; and (3) dimension, form, and spatial distribution of the grains or phases. Crystallography analysis reveals the phases existent in the structure and the atomic packing within these phases (Bandyopadhyay and Bose, 2013). Lastly, chemical composition analysis is simply the detection and quantification of the different chemical elements that constitute the material. Many techniques based in physical sciences are employed to perform these analyses; the most

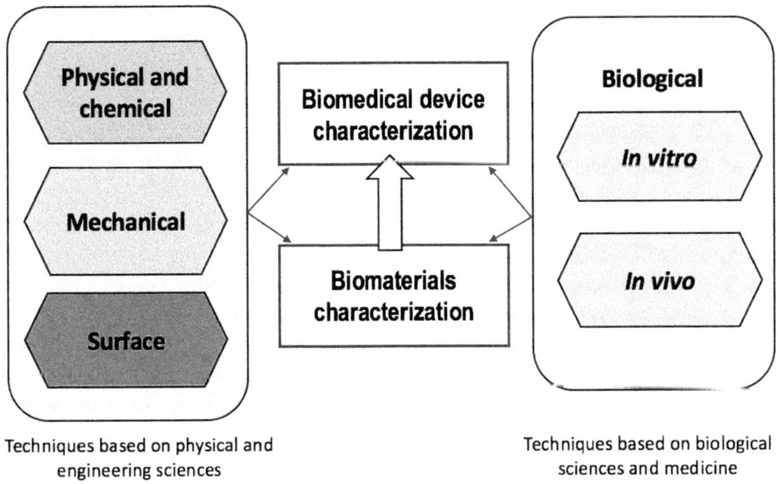

FIGURE 1.8 Schematic outline of biomaterials characterization.

TABLE 1.6
Common Techniques for Physical/Chemical Characterization of Biomaterials and Their Capabilities

Method	Crystallography	Morphology	Chemical
Optical microscopy			
Transmission electron microscopy (**TEM**)			
Scanning electron microscopy (**SEM**)			
Scanning tunneling microscopy (**STM**)			
Atomic force microscopy (AFM)			
X-ray diffraction (**XRD**)			
Fourier transform infrared (**FTIR**)			
Mercury intrusion porosimetry (MIP)			
Gas adsorption			

common ones are described in Table 1.6 along with their capabilities for different analyses.

The principle of imaging in all these microscopic and spectroscopic techniques is to interact with a specimen with a form of probe – such as X-rays, visible light, and excited electrons – to collect and analyse the scattered signal. Figure 1.9 shows

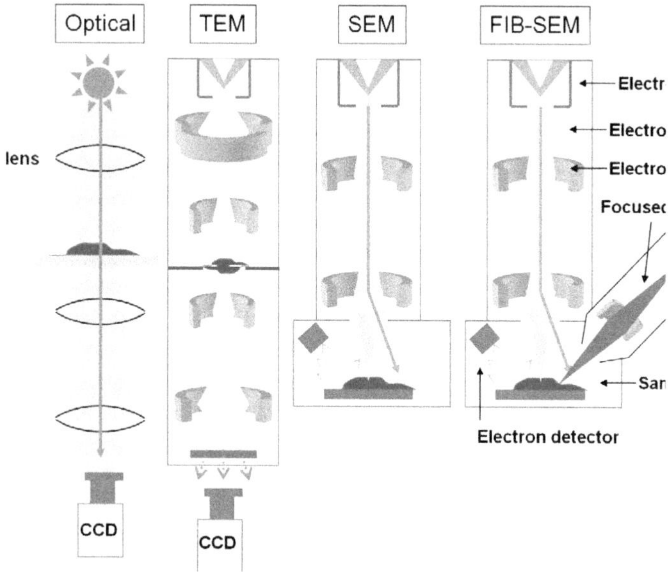

FIGURE 1.9 Principles of imaging by optical microscopy compared with electron microscopy techniques: transmission electron microscopy (TEM), scanning electron microscopy (SEM), and focused ion beam SEM (FIB-SEM) (Dudkiewicz et al., 2011).

the principles of some of these microscopic techniques. On the other hand, mercury infusion porosimetry (MIP) and gas adsorption are methods based on different principles. The MIP technique, according to the U.S. Pharmacopeial Convention (2011), measures the mercury volume intruded into a porous solid at a certain pressure following the simple principle that the pressure required to force a non-wetting liquid (mercury) to fill the pores of a specimen is inversely proportional to the diameter of the pores; while the gas adsorption technique measures the amount of adsorbed inert gas on solid surfaces.

A good physical/chemical characterization will integrate two or more of these techniques and their capabilities for different types of analysis, so that all the physical and chemical parameters of the sample can be fully investigated. For instance, a combination of SEM and TEM imaging techniques were used by Arkharova et al. (2016) to study the microstructure and porosity of a bacterial cellulose/hydroxyapatite (BC/HAp) composite (Figure 1.10).

1.3.2 Mechanical Characterization

The characterization of biomaterials' mechanical properties is elemental since every medical device has specific load-bearing requirements. The fundamental mechanical properties are determined from a stress–strain curve, which is obtained from the load measurements and deformations occurred during a mechanical loading test (Figure 1.11). In this test, different modes of forces can be applied to the specimen (Figure 1.12), and depending on the material class, several types of stress–strain

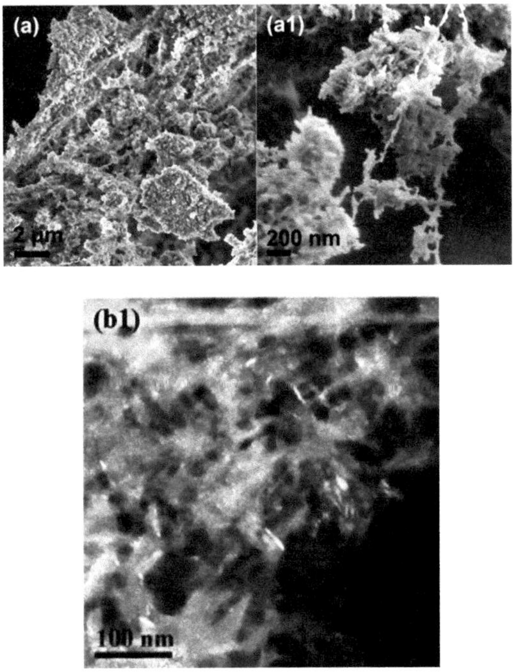

FIGURE 1.10 SEM images of BC/25%HAp (a, a1) and TEM image of BC/4%HAp (b1) (Arkharova et al., 2016).

curves can be obtained. Each type represents different physical behaviours that materials may exhibit, e.g. linear and nonlinear elastic deformation, perfect plastic deformation, and plastic deformation with discontinuities.

A schematic stress–strain curve depicting the fundamental properties is shown in Figure 1.13. Stress (σ) is defined as the force applied, in *N*, per cross-sectional area in m^2, and has units of N/m^2. Strain (ε) is defined as the change in length per original length (%). Thus, the stress–strain curve represents the continuous response of a specimen to the applied force. One of the most important properties found in this curve is the Young's modulus or elastic modulus, which corresponds to the slope of the plastic region. This parameter is based on Hooke's law, which states that strain increases proportionally to the applied stress (Callister and Rethwisch, 2006):

$$E = \frac{\sigma}{\varepsilon}$$

The elastic modulus is an indicator of the stiffness of the material; the higher the value of *E*, the harder it is to deform. The values of the cross-sectional area of the specimen are assumed to be constant during the loading test, which is why this curve is called engineering stress-curve. A true stress-curve can be constructed taking into account the changes which occurred in the specimen cross-sectional area throughout the test (Ratner et al., 2004). The area under the curve, up to the

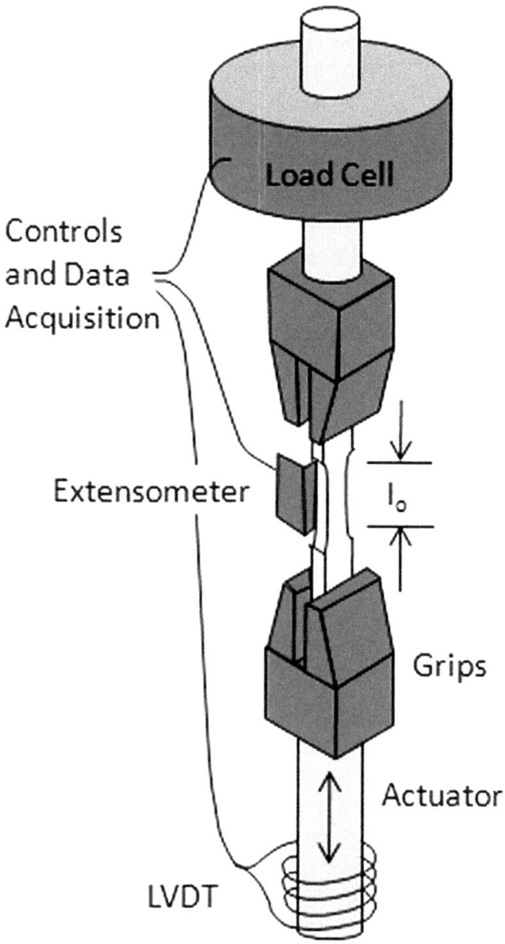

FIGURE 1.11 Setup for the application of load to a specimen.

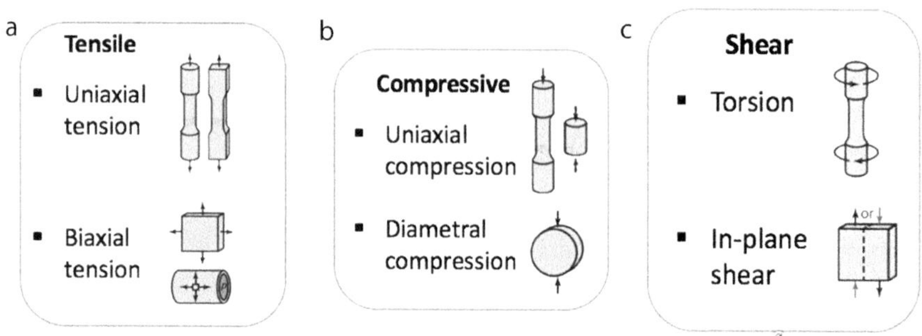

FIGURE 1.12 Modes of loading in mechanical testing.

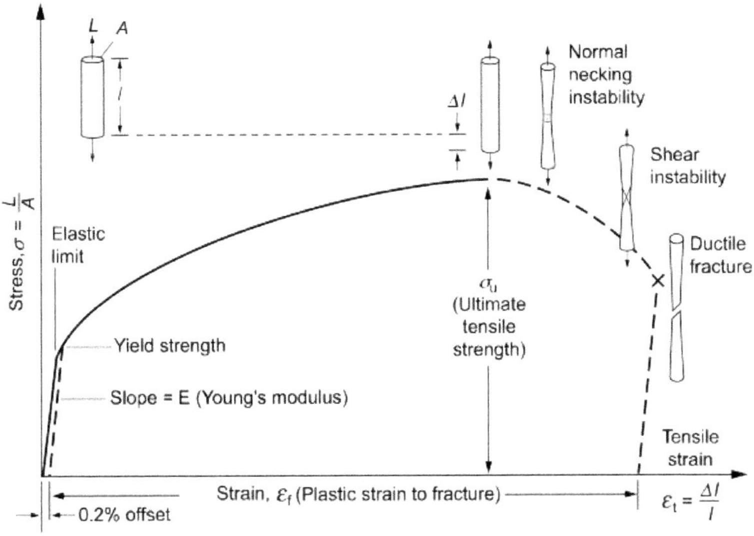

FIGURE 1.13 Schematic stress–strain curve for a specimen loaded in quasistatic uniaxial tension (Gray et al., 2005).

yield point, corresponds to the modulus of resilience, U_r. When it is up to the fracture point, it is called the modulus of toughness, U_t. This magnitude represents the amount of energy necessary to deform the material and it is usually reported in Newtons per meter, Nm. Both parameters, U_r and U_t, are indicators of the amount of energy absorbed in the plastic region and upon fracture, respectively.

Although loading mechanical testing provides detailed information regarding the mechanical behaviour of materials, its main drawback is that it is destructive to the sample. In recent years, efforts have been directed to find alternative mechanical testing methods that are non-destructive. Ultrasound techniques are found to be an attractive option, since they allow accurate measurement of elastic constants and stiffness coefficients through the use of ultrasonic wave propagation, without being destructive to the sample (Zhou et al., 2016).

1.3.3 Surface Characterization

When developing biomedical implant devices and new biomaterials it is key to study the surface properties and the possible interactions between the surface and the surroundings. Materials can undergo surface attack depending on the biological environment in which they are exposed, and the specific physicochemical conditions dictate the chances of incidence of this attack. Reactivity, adsorption, surface oxidation, surface chemical reactions, and wettability are just some of the main aspects that are investigated in surface characterization. The more of these parameters which are described, the more complete will be the surface analysis of the biomaterial. Many techniques are employed to this end, each one with different spatial resolutions and sensitivities. Table 1.7 presents some of the most common techniques for

TABLE 1.7
Techniques for Surface Characterization of Biomaterials

Method	Principle	Spatial Resolution	Analytical Sensitivity
Contact angles	Liquid wetting of surfaces is used to estimate the energy of surfaces	1 mm	Low or high depending on the chemistry
Scanning electron microscopy **(SEM)**	Secondary electron emission induced by a focused electron beam is spatially imaged	40 Å, typically	High, but not quantitative
Scanning force microscopy **(SFM)** **or** Atomic force microscopy **(AFM)**	A mechanical probe touches or feels the surface through precise movements	1 Å	Single atom
Electron spectroscopy for chemical analysis **(ESCA) or XPS**	X-rays induce the emission of electrons of characteristic energy	10–150 μm	0.1 at%
Secondary ion mass spectrometry **(SIMS)**	Ion bombardment sputters secondary ions from the surface	100 Å	Very high
Attenuated total reflection infrared **(AT-IR)**	IR radiation is adsorbed and excites molecular vibrations	10 μm	1 mol%
Surface Matrix-assisted Laser desorption ionization mass spectrometry **(MALDI-MS)**	Laser energy absorbing matrix creates ions and allows protonation (deprotonation) of the analyte molecules	>25 μm	Very high, but not quantitative. Suitable for DNA, proteins, and organic molecules
Raman spectroscopy	Sample is illuminated with monochromatic laser beam creating a scattered light, which is used to compare with inelastic scattering and create a Raman spectrum	1 μm	High; can be enhanced by using metals such as silver or gold (surface-enhanced Raman scattering SERS)

surface characterization along with their principles, spatial resolutions and analytical sensitivities.

X-ray photoelectron spectroscopy (XPS) is the most widely used technique for the surface chemistry analysis of solid biomaterials since it is very simple and flexible. Similarly, Raman spectroscopy is considered a great tool to determine surface chemistry since it is one of the most vibrational spectroscopy techniques. Moreover, secondary ion mass spectrometry (SIMS) is a very sensitive technique that is popular for semiconductor analysis of elemental and molecular contents. It can be static or dynamic; the former has excellent surface sensitivity and mass selectivity, whereas the latter allows depth profiling with a high resolution. Attenuated total reflection infrared (AT-IR), on the other hand, is a powerful tool used qualitatively for the characterization of the surface chemistry of opaque materials. Other techniques are used for surface topography such as SEM and atomic force microscopy (AFM). SEM is very useful for examining surface topography and obtaining two-dimensional topographical images of high spatial resolution of even <1 nm. AFM constitutes one of the essential tools for nanoscale surface imaging (Figure 1.14); not only it is less expensive and less time consuming, but also provides detailed three-dimensional maps rather than two-dimensional images. Another imaging technique called surface matrix-assisted laser desorption ionization mass spectrometry (MALDI-MS), which is relatively new, is used to analyse biomaterial surfaces and its main advantage is that it is suitable for fragile and easily fragmented molecules like polymers, as well as proteins and DNA. Nevertheless, it is not quantitative and requires similar analytical techniques to reach absolute quantification. Aside from these microscopic and spectroscopic techniques, contact angle measurement comprises a thermodynamic method for the description of surface wettability of biomaterials. The investigation of this parameter is crucial to the cell adhesion and spreading of materials. Knowing the surface energy of biomaterials is key to predict different interactions with the surroundings, especially adsorption.

1.3.4 Biological Characterization

This is perhaps the most important characterization type since it is a pre-requisite for clinical trial studies and manufacturing of any medical device. Here the main concern is to evaluate the biocompatibility through the biological responses of any living system to a biomaterial. Many parameters must be analysed to understand how the cells adhere to and colonize the surface as well as their growth, differentiation and migration processes. The biological characterization can be performed in two different ways: in-vitro and in-vivo. The difference resides in that for in-vitro procedures the biological environment is created in a Petri dish, while for in-vivo the biological environment is inside an animal or human body.

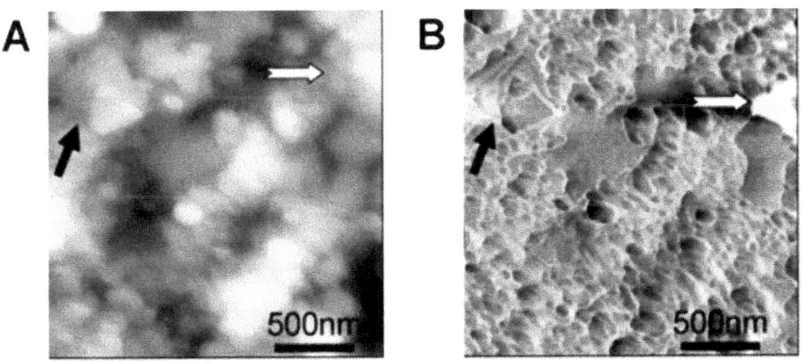

FIGURE 1.14 AFM height (a) and phase (b) images of polycarbonate dinol polyurethane (Ye and Zhao, 2010).

Moreover, in-vitro studies are based on controlled cell-biomaterial interactions, where cell biology and cell interaction to the biomaterial surface are examined. This procedure allows identification and understanding of the mechanisms in which these cells induce morphological and physiological changes. Among all the possible cell-biomaterial interactions, the ones occurring between bacteria and biomaterial are the most critical, since all biomaterials are prone to infections. To this end, investigators evaluate factors like initial bacterial adhesion to surface and biofilm formation (Figure 1.15). The bacterial biofilm or extra

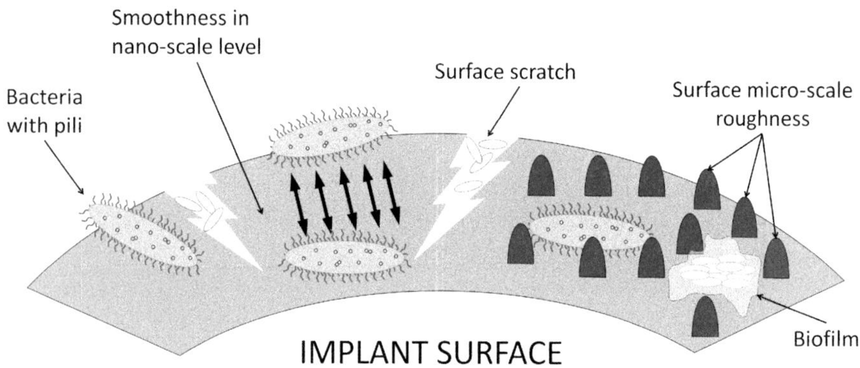

FIGURE 1.15 Bacterial adhesion to orthopedic implants (Gallo, Holinka, and Moucha, 2014).

TABLE 1.8
Techniques for Biological Characterization of Biomaterials

Evaluation Type/Test Methods	Effect/Indication
Toxicological in-vitro:	
• Cell culture toxicity/Mutagenicity	• Inhibition of growth or modification of cellular characteristic
• In-vivo:	• Tumour formation/DNA mutation
• Carcinogenicity	• Toxic nature
• Systemic toxicity	• Malformation of the fetus
• Teratogenicity	• Inflammation/irritation
• Intradermal and mucous membrane irritation test	
Blood compatibility in-vitro:	• Effect on hemorheological parameters
• Hemorheological assay	• Red cell rupture
• Hemolytic assay	• Blood clogging
• Platelet adhesion	• Blood clotting
• Clotting time	
Immunocompatibility	• Immunostimulation and rejection
Long-term implantation	• Retention of mechanical properties
Pyrogenicity	• Microbial contamination

polymeric matrix (EPM) is usually imaged and quantified by microscopic and spectroscopic techniques.

In-vivo characterization is about the understanding of cell-tissue-medical implant interactions within a living system. Myriad factors such as biological response, material degradation, and tissue regeneration are examined through different methods that have been standardized by prominent international and national organizations, summarized in Table 1.8. These traditional methods still rely on histology: the device is implanted into an animal and after a certain period of time, the animal is sacrificed, and the device is extracted for further analyses using all the physical, chemical, mechanical, and surface techniques described previously. Current efforts are being directed towards development of new measurement tools, engineering methods, and especially, non-invasive and real-time imaging tests.

1.4 RECENT RESEARCH IN BIOMATERIALS

In recent years, the fast-paced growth of the biomedical field has led scientists and engineers to increase their efforts into developing new and enhanced biomaterials and optimizing the current techniques used for their characterization, as well as exploring the utilization of new techniques. Many different studies and research works have been made to improve the traditional techniques – or to substitute histological techniques – by enhancing their resolution, versatility, and applicability. Other works have been dedicated to exploring novel techniques that allow for obtaining more in-depth details about the complex interactions between biological systems and medical implants. Moreover, numerous scientists have focused their efforts in finding non-destructive and advanced in-vitro and in-vivo characterization techniques.

For instance, many techniques used in the physicochemical characterization of a biomaterial's surface – such as FTIR, Raman, and contact angle measurements – do not allow observation of the influence of biological environments on the surface of the biomaterial. To solve this, a group of researchers studied the application of inverse liquid chromatography (ILC) to determine the surface properties of two ceramic biomaterials: Hydroxyapatite (HAp) and β-tricalcium phosphate (β-TCP) (Adamska, Kadlec, and Voelkel, 2016). The researchers applied this technique, commonly used for studying solid–liquid interactions, in order to predict the biocompatibility and stability of these two biomaterials under various conditions. They recreated conditions similar to those present in living organisms by using non-standard mobile phases such as simulated body fluid and successfully demonstrated the applicability of ILC for the surface characterization of these biomaterials. This finding shows the possibility to study physicochemical and biological interactions at the surface level of any biomaterial using multiple mobile phases that recreate different substances existing in biological systems.

Similarly, there are certain restraints associated with imaging techniques. Many of the imaging methods for evaluating and monitoring biomaterials in tissue engineering and regeneration, drug delivery, and cell therapies, present challenges when assessing their properties in tissue – such as hydrogels – including sample destruction and limitation to two-dimensional assessments. In 2015, Appel et al. worked to overcome these issues and investigated the applicability of the X-ray phase contrast (XPC) imaging technique for imaging a polyethylene glycol (PEG) hydrogel and soft tissue structure in a tissue engineering system (Appel et al., 2016). Their purpose was to use this technique on hydrogel implants to enable their imaging with high spatial resolution and without contrast agents. The obtained results were comparable to those from histology but also showed the three-dimensional imaging capability of XPC without the need for sectioning or staining the samples. This work enabled demonstration that this technique could be used for three-dimensional imaging of biomaterials used as soft tissues in tissue engineering.

Furthermore, the current methods available for mechanical characterization of biomaterials, which have been essential for studying mechanical properties in e.g. the development of drug delivery platforms and tissue engineering scaffolds, have also certain limitations. These methods – usually based on mechanical loading tests – involve direct mechanical measurements, which require animal sacrifice and sample destruction. To tackle these restraints, other techniques such as ultrasound elastography (UE) have been considered and explored recently. Although UE has been found to be a valuable tool since it is non-invasive and non-destructive, its capability to detect stiff samples is limited. In 2016, a group of researchers, Zhou et al., worked on assessing the capability of using strain imaging to characterize a biomaterial with a wide range of stiffness (47 kPa to 4 MPa). Their results showed a detection limit of strain imaging as high as 4 MPa, which means that this technique allows study not only of materials with moduli within the range of soft tissues but also those with a wider range of stiffness, while being non-invasive and non-destructive to the sample (Zhou et al., 2016).

In the same context, the non-invasive and non-destructive features of the ultrasound technique have made it an attractive biomedical imaging modality for many other applications such as longitudinal device characterization. It has a low cost, wide availability and good safety profile, in contrast to the standard methods that involve costly and invasive preclinical animal tests that are destructive to the sample. In 2016, Stukel et al. worked on demonstrating the utility of clinical ultrasound in longitudinal characterization of three-dimensional hydrogel matrices supporting cell growth. They developed a quantitative technique using a B-mode ultrasound to differentiate collagen content and fibroblast density within PEG hydrogels and validated it in an in-vitro phantom environment (Stukel et al., 2016). The results showed the feasibility of using clinical ultrasound for characterization of PEG-based hydrogels in a tissue-mimicking phantom and demonstrated that, with further validation, this approach can be used to non-destructively monitor in-vivo performance of implanted tissue engineering scaffolds over time.

On the other hand, several methods have been employed for understanding interactions between proteins, cell and material surfaces, which are critical when developing biomaterials; some of them are used for studying cell adhesion and include direct cell counting, centrifugation, and morphological observation. Although these techniques have been used for many years, they are time-consuming and may incur cell damage. Thus, other methods such as quartz crystal microbalance with dissipation monitoring (QCM-D) have been studied. Groups of researchers, like the one integrated by Kushiro, Lee, and Takai, studied QCM-D method and indicated that its main advantages are that, for protein analysis, it provides information about the viscoelastic properties of the adsorbed protein layer; and for cell studies, enables non-invasive, real-time *in situ* analyses and dynamic monitoring of the entire adhesion process. Therefore, they carried out a study to systematically analyse the influence of chemical functional groups on protein adsorption and consequent cell adhesion on various self-assembled monolayer (SAM) surfaces by means of QCM-D (Kushiro, Lee, and Takai, 2016). They were able to demonstrate that the characteristic features of the QCM-D plot – changes in frequency, Δf, vs changes in dissipation factor, ΔD – can dynamically and quantitatively reflect the state of the proteins and the cells on the surface and distinguish the different modes of interactions. Hence, this highly sensitive and quantitative technique can be applied for in-vitro characterization of biomaterial–protein–cell interactions.

Although many works and efforts have been made to tackle the limitations of different characterization methods, there is still a long road to go. The variety and complexity of the many possible interactions between biological systems and medical devices, along with the wide number of intended functions, requires extensive and exhaustive studies from scientists and engineers from all medical and engineering disciplines.

1.5 CONCLUSION

The field of biomedicine and its fast-paced growth have improved the quality life of many and will continue to do so thanks to the development and use of biomaterials and medical devices that are in a continuous upgrading and evolving process.

REFERENCES

Adamska, K., Kadlec, K., and Voelkel, A. (2016). Application of inverse liquid chromatography for surface characterization of biomaterials. *Chromatographia*, 79(7), 473–480.

Amaral, M., Lopes, M. A., Silva, R. F., and Santos, J. D. (2002). Densification route and mechanical properties of Si3N4-bioglass biocomposites. *Biomaterials*, 23(3), 857–862.

Appel, A. A., Larson, J. C., Jiang, B., Zhong, Z., Anastasio, M. A. and Brey, E. M. (2016). X-ray phase contrast allows three dimensional, quantitative imaging of hydrogel implants. *Annals of Biomedical Engineering*, 44(3), 773–781.

Arkharova, N. A., Suvorova, E. I., Severin, A. V., Khripunov, A. K., Krasheninnikov, S. V., and Klechkovskaya, V. V. (2016). SEM and TEM for structure and properties characterization of bacterial cellulose/hydroxyapatite composites: SEM and TEM of BC/HAP composites. *Scanning*, 38(6), 757–765.

Bandyopadhyay, A., and Bose, S. (2013). *Characterization of biomaterials*. Elsevier.

Bhat, S. V. (2006). *Biomaterials* (2nd ed.). Alpha Science International.

Black, J. (1988). *Orthopaedic Biomaterials in Research and Practice*. Churchill Livingstone.

Boyer, R., Welsch, G., and Collings, E. W. (1994). *Materials Properties Handbook: Titanium Alloys*. ASM International.

Callister, W. D. and Rethwisch, D. G. (2006). *Materials Science and Engineering: An Introduction* (7th ed.). J. Wiley & Sons.

Chen, Q., Zhu, C., and Thouas, G. A. (2012). Progress and challenges in biomaterials used for bone tissue engineering: Bioactive glasses and elastomeric composites. *Progress in Biomaterials*, 1(1), 1–22.

Cowin, S. C., Buskirt, W. C. V., and Ashman, R. B. (1987). *Handbook of Bioengineering*. McGraw Hill.

Dudkiewicz, A., Tiede, K., Loeschner, K., Jensen, E., Jensen, L. H. S., Wierzbicki, R., Boxall, A. B. A., and Molhave, K. (2011). Characterization of nanomaterials in food by electron microscopy. *Trends in Analytical Chemistry*, 30(1), 28–43.

Ducato, A. Fratini, L., La Cascia, M., and Mazzola, G. (2013) An automated visual inspection system for the classification of the phases of Ti-6Al-4V titanium alloy. In Wilson R., Hancock E., Bors A., Smith W. (Eds.). *Computer Analysis of Images and Patterns*. Lecture Notes in Computer Science, vol 8048. Springer.

Gillis, J. (2006). Contact angle. [Online Image]. Retrieved from https://commons.wikimedia.org/wiki/File:Contact_angle.svg.

Gray, G. T., Maudlin, P. J., Hull, L. M., Zuo, Q. K., and Chen, S. (2005). Predicting material strength, damage, and fracture the synergy between experiment and modeling. *Journal of Failure Analysis and Prevention*, 5(3), 7–17. doi:10.1361/15477020523725

Hussein, M. A., Mohammed, A. S., and Al-Aqeeli, N. (2015). Wear characteristics of metallic biomaterials: A review. *Materials*, *8*(5), 2749–2768.

Kushiro, K., Lee, C., and Takai, M. (2016). Simultaneous characterization of protein-material and cell-protein interactions using dynamic QCM-D analysis on SAM surfaces. *Biomaterials Science*, 4(6), 989–997.

Liang, S., Cook, W. D., Thouas, G. A., and Chen, Q. (2010). The mechanical characteristics and in vitro biocompatibility of poly(glycerol sebacate)-bioglass elastomeric composites. *Biomaterials*, 31(33), 8516–8529.

Park, J. B. (2008). *Bioceramics: Properties, characterizations, and applications*. Springer.

Park, J. B., and Lakes, R. S. (2007). *Biomaterials: An Introduction* (3rd ed.). Springer.

Ratner, B. D., Hoffman, A. S., Schoen, F. J., and Lemons, J. E. (2004). *Biomaterials Science: An Introduction to Materials in Medicine* (2nd ed.). Elsevier Academic Press.

Rompen, E., Domken, O., Degidi, M., Farias Pontes, A. E., and Piattelli, A. (2006). The effect of material characteristics, of surface topography and of implant components and connections on soft tissue integration: A literature review. *Clinical Oral Implants Research*, 17(S2), 55–67.

Saini, M., Singh, Y., Arora, P., Arora, V., and Jain, K. (2015). Implant biomaterials: A comprehensive review. *World Journal of Clinical Cases*, 3(1), 52.

Silver, F.H. 1994. *Biomaterials, Medical Devices and Tissue Engineering: An Integrated Approach*. Springer Science and Business Media.

Stukel, J. M., Goss, M., Zhou, H., Zhou, W., Willits, R. K., and Exner, A. A. (2016). Development of a high-throughput ultrasound technique for the analysis of tissue engineering constructs. *Annals of Biomedical Engineering*, 44(3), 793–802.

Ye, Z., and Zhao, X. (2010). Phase imaging atomic force microscopy in the characterization of biomaterials. *Journal of Microscopy*, 238(1), 27–35.

Zhou, H., Goss, M., Hernandez, C., Mansour, J. M., and Exner, A. (2016). Validation of ultrasound elastography imaging for nondestructive characterization of stiffer biomaterials. *Annals of Biomedical Engineering*, 44(5), 1515–1523.

Zilm, M. E., Chen, L., Sharma, V., McDannald, A., Jain, M., Ramprasad, R., and Wei, M. (2016). Hydroxyapatite substituted by transition metals: Experiment and theory. *Physical Chemistry Chemical Physics*, 18(24), 16457–16465.

2 Recent Advances in Biocompatibility

2.1 INTRODUCTION

2.1.1 BIOCOMPATIBILITY

The field of biocompatibility and biomaterials involves the collaboration between medicine and engineering along with other applied sciences (Ratner and Bryant, 2004). "Biocompatibility refers to the ability of a biomaterial to perform its desired function with respect to a medical therapy, without eliciting any undesirable local or systemic effects in the recipient or beneficiary of that therapy, but generating the most appropriate beneficial cellular or tissue response in that specific situation, and optimizing the clinically relevant performance of that therapy" (Williams, 2008). Biomaterials are now used in many different applications such as drug delivery, tissue engineering, degradable implantable systems, biotechnology, and nanotechnology. It is important to investigate and define biomaterial-tissue interactions in order to ensure no unnecessary harm comes to the patient while functioning within the body for a long period of time (Williams, 2014) (Figure 2.1).

2.1.2 BIOMATERIALS

Ideally, biomaterials should be non-toxic, non-irritant, and non-carcinogenic. The decision to choose a specific type of biomaterial is based on three major factors. The first factor is based on choosing the type of biomaterial. This is important since different applications require different outcomes (long term, short term, different sites of action) (Williams, 2008). Table 2.1 provides a list of commonly used biomaterials for long-term implantable devices. The second factor is based on biomaterial properties that result in improved interactions with the tissue and relating cells (Table 2.2). The final factor considers biomaterial-tissue interactions at the site of action. The final factor assumes that biomaterials perform differently under different conditions within the body. Therefore, influencing different immune responses such as:

Protein adsorption
Protein desorption
Macrophage activation
Inflammation chronic or acute
Fibrous tissue formation

by Yaser Dahman and Ahmed Chaudhry

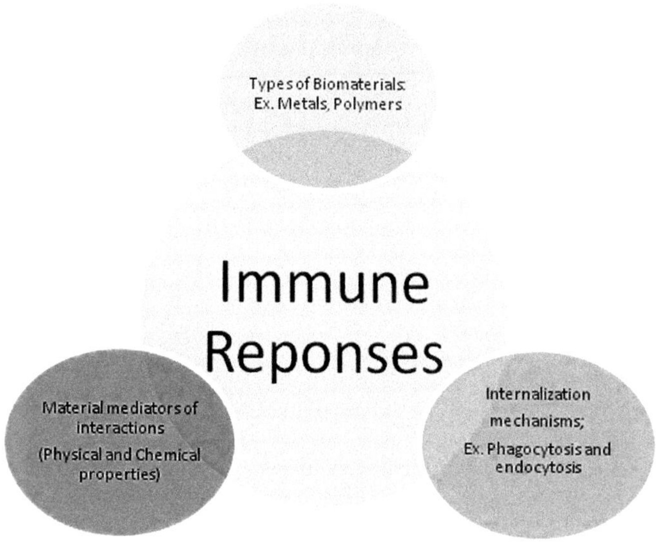

FIGURE 2.1 Illustrates factors leading to immune response in cellular response (Williams, 2014).

TABLE 2.1
List Commonly Used Biomaterial and Corresponding Applications

Biomaterial	Applications
Titanium	Dental implants Heart valves Fracture plates Spinal cages
Nitinol (Nickel Titanium)	Memory shape applications
Stainless steel	Intravascular stents Orthopedic implants
Carbon	Heart valves
PEEK: Polyether Ether Ketone	Spinal cages
PMMA: Poly(methyl methacrylate)	Bone cement
UHMW-polyethylene: Ultra High Molecular Weight polyethylene	Bearing surfaces
Silicone	Soft tissue augmentations

Source: Williams (2008).

Neutrophil activation
Platelet adhesion, activation, and aggregation
Delayed hypersensitivity
Clotting cascade activation
Tumour formation

TABLE 2.2
Displays Biomaterial Properties That Influence Immune Response

Bulk Properties	Surface Properties	Degradation Properties
Structure type (macro-, micro-, nano-)	Chemical composition	Metals: Ion accumulation Resulting in toxicity Ion release Parameters Corrosion
Crystalline	Topography	Polymers: Accumulation of degradation by products resulting in toxicity. Additives toxicity
Elasticity	Energy	Ceramics: Accumulation of degradation by products resulting in toxicity. Dissolution profile
Hydrophilic - Hydrophobic Interactions	Electrical potential	
Porosity		

Source: Williams (2008).

2.2 WOUND HEALING PROCESS

Normal wound healing processes are different from immune processes induced by implanted biomaterials. Normal wound healing processes occur when the tissue is injured/damaged resulting in acute inflammation and granulation formation (Ratner and Bryant, 2004). After injury, blood is rushed to the site of action. There, fibrinogen within the blood is used to stimulate platelet adhesion to produce blood clots (Ratner and Bryant, 2004). Neutrophils and monocytes are then released (received signals from growth factors and cytokines) into the site (Ratner and Bryant, 2004). Macrophages are then developed and used to engulf foreign materials and bacteria while at the same time recruiting fibroblast. The recruited fibroblasts are used to create a highly vascularized granulation tissue. Ideally, the granulation tissue should be replaced by extracellular matrix (ECM). However, granulation tissue usually is converted into scar tissue (Ratner and Bryant, 2004).

Specific protein adsorption occurs in normal wound healing process. However, after biomaterial implantation, non-specific protein adsorption occurs. The non-specific protein adsorption results in foreign body reactions (Ratner and Bryant, 2004). The foreign body reactions involve platelets and monocytes binding to the surface of the implant, which leads to increase cytokine levels in the implant site. The high cytokine levels influence pro-inflammatory reactions (Ratner and Bryant, 2004).

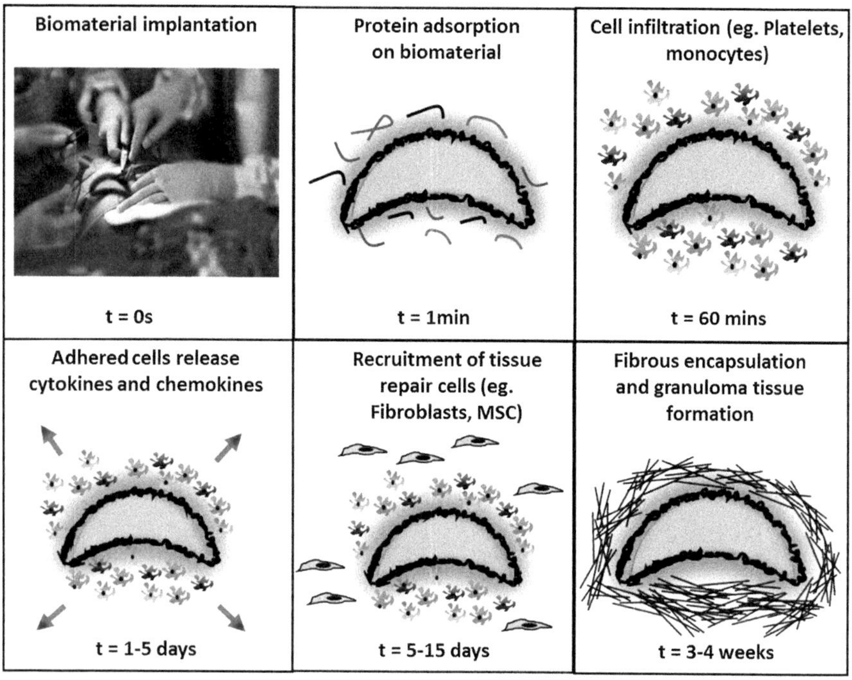

FIGURE 2.2 Illustrates the immune response activity after medical device implantation (Sridharan et al., 2015).

Macrophage sizes are influenced by the biomaterial, which leads to adverse events. The macrophages develop into multinucleated giant cells resulting in chronic inflammation. Total biocompatibility failure occurs when collagenous fibrous tissue (50–200 μm) separates the implant from healthy tissue. (Ratner and Bryant, 2004). Figure 2.2 displays the immune response in relation to implanted biomaterials (Sridharan et al., 2015).

2.3 LONG-TERM IMPLANTS

Implants tend to lose functionality over time. For example, it is estimated that between 2005 and 2030 that the total number of knee revision surgeries is going to increase by 607% (Gepreel and Niinomi, 2013). Implant failure is characterized as the need to revise surgery due to different factors caused by either the patient, surgeon, or by the biomaterial used (Zinar and Schmalzried, 2015). For example, infections resulting in revised surgery can be caused by the surgeon, the surgical team, the surgical equipment, or by the operating/recovery site (Zinar and Schmalzried, 2015). Patient health conditions can influence biomaterial performance in the body. For example, depression and renal dysfunction have shown to influence the biomaterial-tissue response (Zinar and Schmalzried, 2015). High life expectancy of patients is another factor relating to an increased number of revised surgeries (Gepreel and Niinomi, 2013).

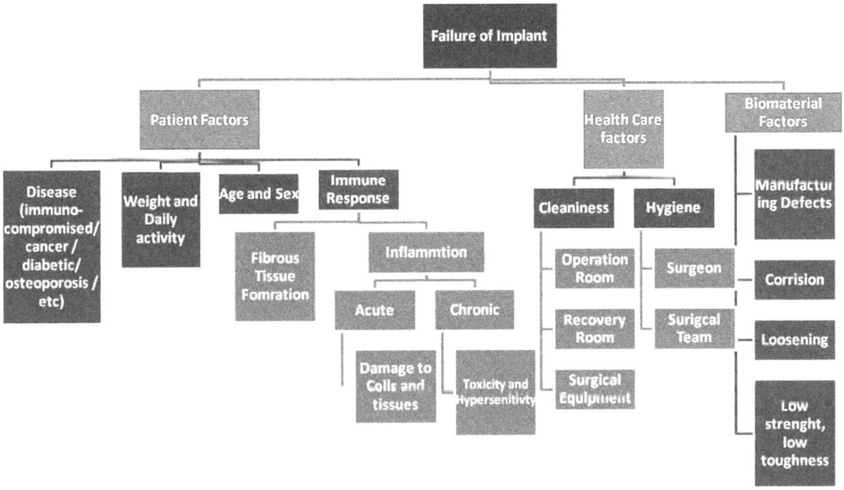

FIGURE 2.3 Represents different contributing factors that result in the failure of an implant (Gepreel and Niinomi, 2013).

Implant dislocation can result in surgical revision due to surgical technique and other patient factors including bone quality, daily activities (e.g. walking or running), and weight (Zinar and Schmalzried, 2015). Manufacturing defects such as design and biomaterials used can also influence the success/failure of implanted medical devices (Zinar and Schmalzried, 2015). Both patients and implant manufacturers suffer when biocompatibility between the implant and host tissue fails. The patients may experience tissue damage at the implant site. Revised surgery can result in tissue death while replacing the implant (Dyer, 2016). The manufacturer will suffer both a financial cost and their tarnished reputation at the same time (Dyer, 2016). Figure 2.3 summarizes implant failure factors. Research today focuses on developing biomaterials with long-term biocompatibility, which can be continuously functional in the body over time (Gepreel and Niinomi, 2013).

2.4 ORTHOPEDIC IMPLANTS FOR JOINT REPLACEMENT

By 2016, it was estimated that the orthopedic surgery industry would generate \$ 41.1 billion (Gibon et al., 2016). The purpose of orthopedic surgery is to improve the patient's quality of life by relieving pain and restoring function (Gibon et al., 2016). Currently, surgeons and bioengineers have access to different materials used in orthopedic surgeries (Gibon et al., 2016). The two most commonly used materials are metals and polymers (Gepreel and Niinomi, 2013). Metal implants can be either placed in the body as a single piece or in different pieces put together and used to replace femoral stem and acetabular shell. Polymers are used to replace the articular surface (acetabular side) and can be used as cement within the bone (Gibon et al., 2016). Biocompatibility can be achieved when both metals and polymers are implanted at the site of action and produce no adverse effects (Gepreel and Niinomi, 2013).

2.4.1 Metals

Titanium, chromium, and niobium are commonly used passive metals in orthopedic surgeries (Gibon et al., 2016). The main advantage of passive metals is the formation of the protective oxide layer (Chaturvedi, 2016). The passive oxide layer inhibits redox reaction between the metal surface and surrounding tissue (Gibon et al., 2016). The passive oxide layer is formed via passivation that occurs when oxygen interacts with the metal surface (Figure 2.4). Over time the passive layer may get damaged, however, it has the ability to re-establish the layer through reoxidation. When the rate of damage exceeds the rate of re-passivation, corrosion occurs (Gibon et al., 2016).

2.4.2 Metal Disadvantages

Corrosion is defined as the loss of metallic ions from the implant surface and transference to the surrounding environment (Saini et al., 2015). The human body is a hostile electrolytic environment that can influence the performance of metal-based implants. Factors such as the pH of bodily fluids and oxygen concentrations can influence corrosion (Chen and Thouas, 2015; Chaturvedi, 2016). Different parts of the human body have different pH values and pH fluctuates due to different conditions in the body such as surgery and the accumulation of other ions (Chen and Thouas, 2015).

Oxygen concentrations also influence corrosion (Chen and Thouas, 2015). The body only contains about 1/4 of the atmospheric pressure, resulting in low concentrations that can decrease the rate of passive oxide layer formation and increase the rate of corrosion (Chen and Thouas, 2015). The ideal metal should be able to withstand the hostile electrolytic environment and be able to function for an extended period of time (Chen and Thouas, 2015).

Corrosion results in health problems such as allergic reactions, fractures, local pain, swelling, and other cytotoxic responses thus reducing biocompatibility (Chaturvedi, 2016). Allergic reactions are caused by metal ions released from the implant surface. This results in immune system stimulation, which is carried out into two stages (Mahmood and Hammert, 2015). The first stage, known as the induction stage, involves antigen-specific T cells becoming sensitized. The next stage, known

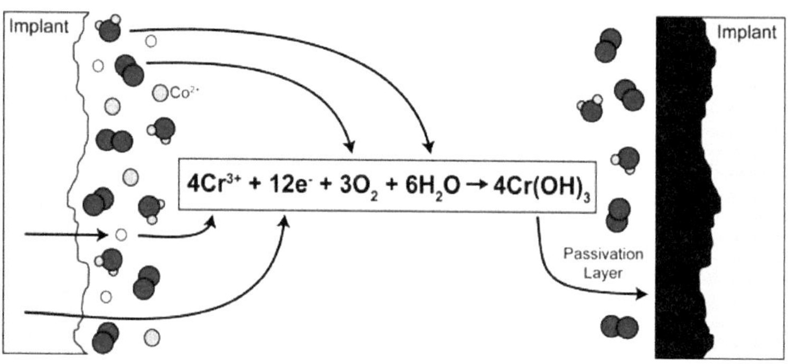

FIGURE 2.4 Displays the passivation oxide layer formation reaction of metal surfaces when placed in the body (Gibon et al., 2016).

as the effect phase, is where macrophages are released by stimulating the sensitized antigen-specific T cells. (Mahmood and Hammert, 2015).

Wear also influences corrosion and mainly occurs between metal-metal and metal-polymer articulations (Gibon et al., 2016). Wear processes start once the articulates are placed with the host. One metal will physically rub into the other surface until a modified area is formed. Once optimized, the wear rate will decrease until metal-metal articulations lose functionality (Gibon et al., 2016). Wear particles are associated with localized inflammation, tissue damage, and loosening, resulting in pain and loss of motion (Dapunt et al., 2014).

2.4.3 Stainless Steel

Stainless steel (316 L) is a cheap biomaterial and is regarded as a biomaterial with useful mechanical properties (elasticity and stiffness) (Plecko et al., 2012).

Stainless steel is much easier to handle in surgeries requiring plastic deformation compared to titanium (Kheirkhah et al., 2015). The problem with stainless steel is vulnerability to corrosion resulting in inadequate osseointegration (Omar et al., 2016). Nevertheless, stainless steel has other biocompatibility issues. For example, the nickel content in stainless steel has the ability to cause an adverse allergic reaction in the patient (Plecko et al., 2012).

2.4.4 Stainless Steel Surface Modifications

Brooks et al. (2017) investigated stainless steel corrosion and corresponding inflammation by placing samples in acidic environments. They found that stainless steel samples placed in acidic environments are susceptible to rapid and aggressive corrosion in crevice areas (Brooks et al., 2017). Crevice corrosion occurs due to low oxygen levels within the crevice. This results in metal dissolution inside the crevice. There are more positive ions in the crevice and chlorine ions are used to stabilize the charge. The metal dissolution creates electrons that are taken away by oxygen on the metal surface (cathode site) (Gibon et al., 2016). Figure 2.5 illustrates crevice corrosion mechanism of action.

2.4.4.1 Hydroxyapatite (HAp) Coating

In order to improve osseointegration, surface modifications are used (Zaporozhets et al., 2017). Surface modifications provide an additional new layer that interacts with the surrounding environment affecting the metal's behaviour. (Omar et al., 2016; Kheirkhah et al., 2015).

Chitosan is a natural cationic polysaccharide. Pang and Zhitomirsky (2005) investigated the electrodeposition of hydroxyapatite-chitosan films. The electrodeposition provided an increased coating thickness and found that composite of HAp-chitosan can reduce the corrosion rate and improve biocompatibility by avoiding harmful by-products from corrosion (Pang and Zhitomirsky, 2005; Mina et al., 2016). Another study conducted by Wang et al. found that β-Tricalcium phosphate (β-TCP) improves cell proliferation. Mina et al. (2016) investigated the effect of stainless steel 316 L coated with chitosan- β-TCP that was electrodeposited. X-ray

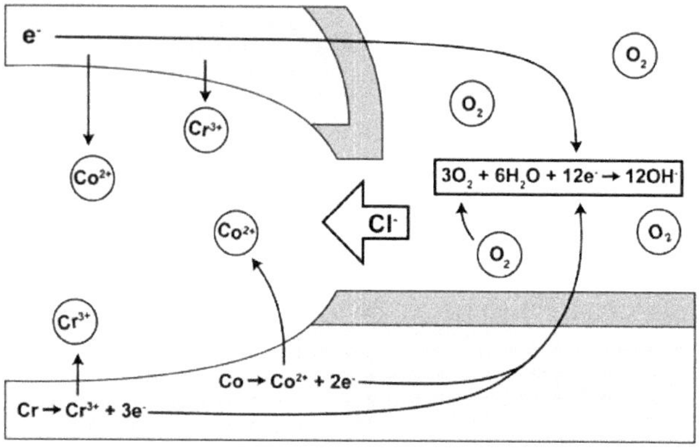

FIGURE 2.5 Displays the crevice corrosion mechanism of action. (Gibon et al., 2016).

diffraction results demonstrated a good crystalline structure that is in agreement with the Joint Committee on Powder Diffraction Standards (JCPDF). Energy dispersive X-ray spectroscopy was used to analyse the chemical composition of the stainless steel and the chitosan- β-TCP coat and found that no toxic by-products were present, suggesting that the stainless steel and the chitosan- β-TCP coat are compatible with one another. Electrochemical impedance spectroscopy was used to determine the electrochemical performance of the chitosan- β-TCP. The results demonstrated that the chitosan- β-TCP improved the stainless steel corrosion resistance when compared to the uncoated stainless steel. Figure 2.6 illustrates the relationship between the percentage of chitosan (by weight) and the critical load

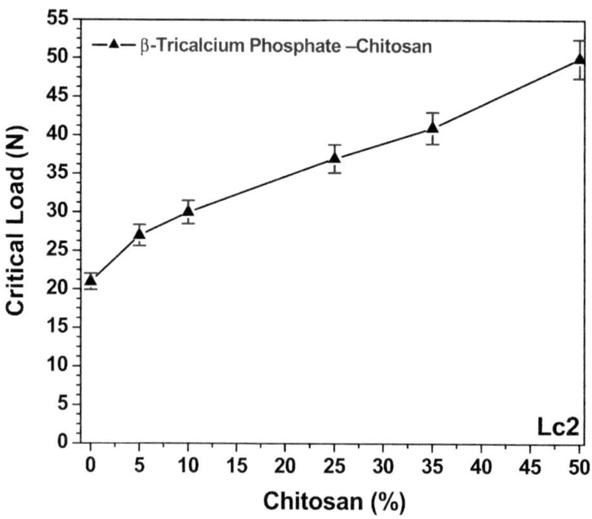

FIGURE 2.6 Displays the percentage of Chitosan by weight and compares it to the critical load of β-Tricalcium-Phosphate -Chitosan. (Mina et al., 2016).

(Mina et al., 2016). The in-vitro biocompatibility test demonstrated that an up-to-25wt.% coat can be applied to stainless steel and can increase cell viability for the cells studied (Mina et al., 2016).

Past research has shown that HAp is osteoconductive and has the ability to restore bone tissue by stimulating bone in growth. However, low strength is a disadvantage for HAp. An alternative to HAp is Forsterite. Forsterite is a novel bioactive ceramic that has the composition of Mg_2SiO_4. Forsterite is biocompatible since silicon is found in skeletal development and magnesium is involved in mineralization metabolism and bone tissue (Kheirkhah et al., 2015). Studies such as Ni, Chou and Chang (2007) determined that Forsterite's fracture toughness is higher than HAp and demonstrates a good osteoblast proliferation rate (Ni, Chou, and Chang 2007). Kheirkhah et al. (2015) investigated the use of Forsterite on the surface of stainless steel and found bioactivity and corrosion resistance improved (Kheirkhah et al., 2015).

2.4.5 Titanium

Today, titanium occupies approximately 40 % of the biomedical implant market (Jin et al., 2016). Titanium's characteristics include chemical inertness, good mechanical biocompatibility (low Young's Modulus and high fatigue resistance), and good corrosion resistance (Jin et al., 2016; Hussein et al., 2016). Most importantly, it can prevent corrosion by developing a rapid stable passive layer (Saini et al., 2015). However, other metals such as cobalt when mixed with molybdenum (CoCrMn) create a product that is mechanically stiffer than titanium. However, biocompatibility of these metals still requires more research. Plecko et al. (2012) attempted to address this issue by evaluating seven different metals (Listed in Table 2.3) in sheep. Eight weeks after implantation, the biomechanical test demonstrated that cobalt ($p=0.031$) and stainless steel had a lower torque value compared to titanium samples listed in Table 2.4. However, samples that incorporated titanium demonstrated an improved torque value resulting in an improved biocompatibility (Plecko et al., 2012).

TABLE 2.3
Describes the Different Metals That Were Analysed

Materials Used	Abbreviations
Stainless Steel	Steel
Cobalt-Chrome	CC
Cobalt-Chrome/Titanium coating	CCTC
Cobalt-Chrome/Zirconium/Titanium coating	CCZTC
Pure Titanium Standard	PTST
Titanium-Aluminium-Niobium (TAN) Standard	TANSST
Titanium-Aluminium-Niobium (TAN) new finish	TANNEW

Source: Plecko et al. (2012).

TABLE 2.4
Describes the Average Torque Value and Standard Deviation

Type of Screw	Removal Torque in Nmm	std. dev. ± in Nmm	N
PTST	270.9	87.3	9
TANNEW	259.9	100.1	9
CCZTC	246.6	96.9	9
CCTC	245.6	91.1	8
TANST	234.0	55.8	9
Steel	155.8	83.6	9
CC	130.1	45.8	7

Source: Plecko et al. (2012).

2.4.6 Titanium Surface Coating

2.4.6.1 Hydroxyapatite (HAp) Coating

Orthopedic replacement long-term success depends on the initial stable implantation of the medical device within the body. Different bone loss results in mechanical instability due to weight distribution, thus reducing long-term success of implants (Baas et al., 2011). The disadvantage of titanium is the thin passive oxide layer that is vulnerable to fibrous tissue formation. The fibrous tissue as mentioned earlier loosens implants and promotes inflammation. Hydroxyapatite (HAp) coating is sprayed onto titanium implants to prevent fibrous tissue formation (Jin et al., 2016). Baas et al. developed a titanium coat made up of Colloss E (low concentration) and HAp to stimulate integration between the bone and the titanium implant. Colloss E is an ECM lyophilized FDA approved complex that consists of BMP-2, TGF-β1, IGF-1 growth factors. Baas et al. used old American hounds' proximal tibiae to assess maximum shear strength and stiffness. The results from their study are listed in Table 2.5. The histological evaluation assesses fibrous tissue formation and inflammation around the implant. The experiment demonstrated an increase in shear strength ($p=0.21$)

TABLE 2.5
Summaries of the Mechanical Test Results for the Different Samples Used

	Ultimate Shear Strength (MPa)	Apparent Shear Stiffness (MPa/mm)	Total Energy Absorption (kJ/m^2)
Control	5.4	22.3	1.22
BMP	4.2	15.2	1.09
Zoledronate	7.6	34.5	1.26
BMP+Zoledronate	5.1	21.0	1.28
ANOVA	$P<0.001$	$P<0.001$	$P=0.794$

Source: Baas et al. (2011).

compared to control while stiffness ($p=0.117$) was significantly different to control ($p=0.117$). Baas et al. were also able to demonstrate new bone formation ($p=0.008$) with a decrease in fibrous tissue ($p=0.02$). The developed Colloss E and HAp were successful in preventing fibrous tissue formation (Baas et al., 2011).

2.4.6.2 Bisphosphonates Coating

Early osseointegration leads to long-term functionality and bisphosphonates can help. Wermelin et al. used bisphosphonate (pamidronate) covered screws, leading to mechanical enhancements, and has shown that host bone remodelling activity can continue can occur after 8 weeks with the help of bisphosphonates (Wermelin et al., 2008).

Human mesenchymal stem cells (hMSCs) have the ability to differentiate into osteoblasts (Alqhtani et al., 2017). Previous work by Alqhtani et al. demonstrated that bisphosphonates influence osteoblast activity (Alqhtani et al., 2017). Based on their previous work, Alqhtani et al. investigated the effect of bisphosphonates on hMSCs on titanium implant surfaces. Proliferation of hMSCs was assessed over a duration of 14 days. Alqhtani et al. were able to show that titanium implants coated with 100 nm pamidronate stimulated cell proliferation (compared to controlled) titanium implants (untreated pamidronate). The cell migration test was also able to demonstrate a significant improvement with provoking cell migration. The study also found that implants with pamidronate coating had increased levels of alkaline phosphatase (ALP) activity thus improving osteoinduction enhancing stability of the implant (Figure 2.7) (Alqhtani et al., 2017). The results from this study were able to demonstrate that the application of bisphosphonates (pamidronate) can, in fact, improve osseointegration and bone healing processes, therefore improving biocompatibility (Alqhtani et al., 2017).

2.4.7 Low-Cost Alternatives to Titanium

Titanium alloys are expensive to acquire and further processing is expensive. There is a need to develop β-Ti alloys like the one used by Abdel-Hady et al. which incorporated cheaper materials and cheaper processing (Gepreel and Niinomi 2013).

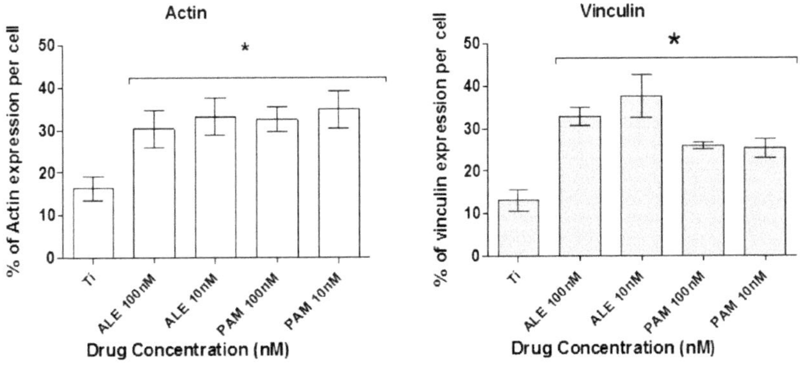

FIGURE 2.7 Displays that both vinculin and actin increase in activity one, a low dose of bisphosphonates is applied. Actin and Vinculin are markers used to identify integration between implant and host tissue (Alqhtani et al., 2017).

In 2009, Abdel-Hady et al. developed a novel β-Ti alloy consisting of $Ti_{77}Nb_{17}Ta_{6}$, which was able to enhance titanium's mechanical properties when mixed with oxygen and iron. Based on this work, Hussein et al. (2016) assessed biocompatibility, and the composition of each sample tested using X-ray diffraction are listed in Table 2.6. X-ray diffraction was able to confirm that oxygen and iron are able to increase the strength while maintaining a low Young's modulus level (Table 2.7). Figure 2.8 also compares the percentage of strain vs stress (Mpa) (Hussein et al., 2016). It was found that the titanium with iron and oxygen could handle a smaller strain (%) with a higher stress capability. The polarization test demonstrated that the novel β-Ti alloy provides ten times better corrosion resistance compared to Ti-Al6-V4. This novel β-Ti alloy demonstrated potential with its biocompatibility properties (mechanical strength, reduced cytotoxicity, and corrosion resistance) (Hussein et al., 2016)

Osteoarthritis is now affecting younger people, there is a need to develop novel long-term implants used for total knee arthroplasty (TKA). Cemented materials are the traditional method of treating TKA, however, it has a poor resistance to shear force, resulting in a loosening effect and leading to revised surgery and failed biocompatibility (Harwin et al., 2017). Newer cement-less TKA incorporates metals such as titanium. Harwin et al. compared titanium coated cement-less TKA with periapatite-coated TKA. After approximately 4.4 years, patients were followed-up and it was demonstrated that implant survivorship between the two groups was 95.5% with no significant difference in pain and other adverse reactions. More

TABLE 2.6
XRF Results and Designed Composition Values of the Designed Alloys (in Mass %)

Sample	Nb	Ta	O	FE	Ti
TNT	30.0	17.6	0	0.0701	Equal
TNTO	30.7	17.5	N/A	0.140	
TNTFO	29.9	16.7	N/A	1.08	

Source: Hussein et al. (2016).

TABLE 2.7
Displays the Mechanical Test Results

Test	TNT	TNTO	TNTFO
Young's Modulus (GPa)	71	68	60
Strain%	47.4	43.5	11.1
UTS (MPa)	530.7	741.3	851.1
Hardness	238	335	338

Source: Hussein et al. (2016).

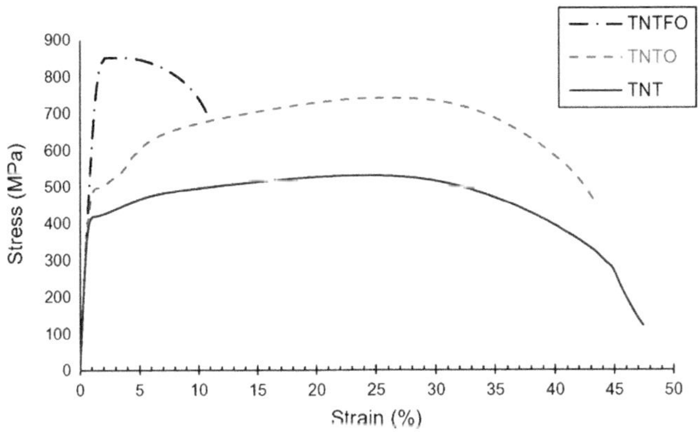

FIGURE 2.8 Compares the stress applied compared to the strain percentage for different Ti alloy compositions (Hussein et al., 2016).

research needs to be completed to determine which implants are more cost-effective (Harwin et al., 2017).

Ulta-high molecular weight polyethylene (UHMWPE) and PMMA are two commonly used polymers used in orthopedic surgeries (Cools et al., 2016) (Huang et al., 2014). UHMWPE is used for load-bearing surfaces (Huang et al., 2014; Gigante et al., 2013). UHMWPE has a good biocompatibility, fatigue, and wear resistance abilities. The problem with UHMWPE is the fabrication deficiency due to its high viscosity, so it cannot be used for screw extrusion. Current UHMWPE fabrication processes are inconvenient due to the time required and the high temperature and pressure used. The high temperatures ranging from 200 to 300°C may result in UHMWPE instability caused by molecular chain scission and oxidation. The high pressure may cause voids in the UHMWPE product that would influence the immune response in the body (Huang et al., 2014). Huang et al. suggested the use of 9.8 % low molecular weight polyethylene (LMWPE) and UHMWPE in order to decrease viscosity. They are both compatible since both have the same structure (Chemically and crystalline). Huang et al. were able to demonstrate that the combination of two polymers enhance tensile strength, fatigue and wear resistance. Their work in the future can be applied to young and active patients (Huang et al., 2014). This work can also be used as an alternative cost-saving material. However, more research is needed to prove efficacy.

3D printing has shown to be a valuable tool for both surgical planning and prosthetics (Rodriguez et al., 2017). Now research focuses on designing patient-specific implants. Advantages of 3D printing include inexpensive cost compared to traditional implants and computer software that allows the design of an endless number of shapes and sizes (Trombetta et al., 2016). Research needs to focus on the bioink used to develop the implant (Rodriguez et al., 2017). The bioinks used must be biocompatible and should be biodegradable allowing regeneration (Rodriguez et al., 2017). In Western countries, a widespread musculoskeletal disorder known as Low Back Pain (LBP) is common in adults. Lumbar fusion is a surgical procedure used to treat chronic LBP. Lumbar fusion involves the use of a bone graft inter-body cage, usually made of titanium, in order to support bone repair (Serra et al., 2016). Titanium's excessive stiffness can result in cage-vertebrate fusion failure. An increasingly used alternative to titanium cages is polyether ether ketone (PEEK). A PEEK inter-body cage demonstrates a better elastic modulus to the cortical bone and a comparable stability to titanium. Although the benefits of PEEK outweigh titanium, only a limited number of implant sizes can be manufactured (Serra et al., 2016). Increasing the number of customized PEEK cages can result in improved fusion processes by producing customized shapes that match the patient's vertebrate dimensions. Novel manufacturing procedures such as 3D printing can be used to increase the surgical availability of inter-body cage sizes/shapes by producing customized ergonomic shapes at a low cost (Serra et al., 2016). The objective of this study was to develop a polyhedral oligomeric silsesquioxane poly (carbonate-urea) urethane (POSS-PCU) lumbar cage using a 3D printer. Both structural properties and in-vitro biocompatibility capabilities were determined (Serra et al., 2016). A POSS-PCU cage was developed and modeled after the contour of the vertebrate. The printing material was made up of POSS-PCU and nHAp powder using four different filling densities (25%, 50%, 75%, and 100%). The 3D manufacturing procedure was optimized (refer to Table 2.8). After optimization, morphological-structural, mechanical, and in-vitro biological analyses were determined (Serra et al., 2016). The study demonstrated that a 25% filling density is suitable for the manufacturing of POSS-PCU cage. This provides a faster manufacturing time since less material is needed to create the cage. The 25% filling density also provided higher porosity. This was confirmed by the micro-CT evaluation, which showed a highly porous structure. The biocompatibility test showed cell area (Figure 2.9) growth and activity was higher in the POSS-PCU cage compared to other samples studied (Serra et al., 2016). In order to reduce the cost of the lumbar cage, an alternative 3D manufacturing process involving POSS-PSC had been proposed. The use of POSS-PSU had demonstrated both mechanical and biocompatibility stability. Further study is needed to determine the mechanical forces' effect on the spinal cord (Serra et al., 2016).

TABLE 2.8
3D Printer Manufacturing Procedure

Printer Used	Flash-forge Dreamer (FDM printer)
Software	Slic3r Software
Temperature of heating nozzle	265°C
Nozzle diameter	0.3 mm
Printer speed	10 mm s^{-1}
Layer height	0.2 mm
Height number of external shells	3
Filling density	25%
Temperature of heated bed	125°C

Source: Serra et al. (2016).

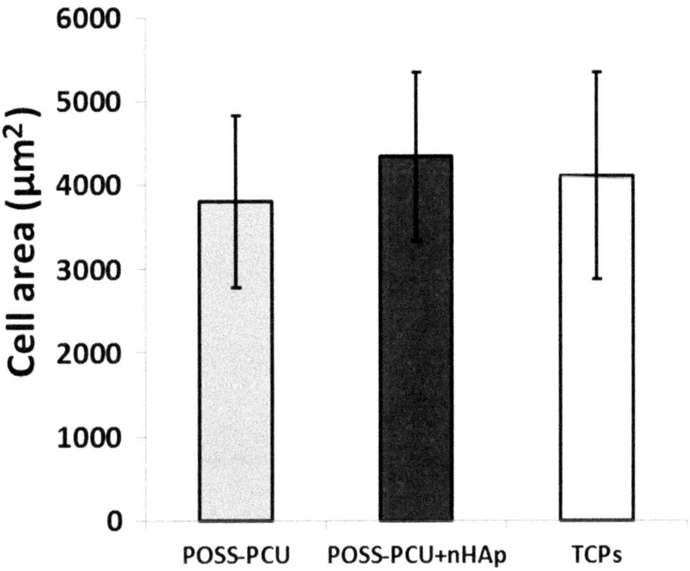

FIGURE 2.9 Compares the cell area (µm²) between POSS-PSC with POSS+nHAP and TCPs (Serra et al., 2016).

2.5 INTRAVASCULAR STENTS

Before 2003, bare metal stents were used to treat coronary artery diseases (Kim et al., 2012; Galløe et al., 2017). The main problem with bare metal stents is restenosis (Bedair et al., 2015). Restenosis is defined as a neo-intimal hyperplasia caused by smooth muscle cell migration, macrophage recruitment (inflammation), and extracellular matrix formation (thrombus reorganization) (Ma et al., 2011; Hamada et al., 2010). Restenosis can take up to several weeks to form and affects the implant's biocompatibility success (Ma et al., 2011; Heldman et al., 2001).

To reduce inflammation and prevent restenosis, bare metal stents were remodelled to drug-eluting stents that were coated with drug-in-polymer (Kim et al., 2012). Stents coated with therapeutic drugs allow local distribution of active ingredients at specific target sites reducing systemic adverse events (Heldman et al., 2001). The ideal therapeutic coating agent should be able to improve the host biocompatibility by preventing the restenosis process on targeted vascular smooth muscle cells (Scheller et al., 2005; Serruys, 2002; Heldman et al., 2001). The choice of therapeutic agent is a biocompatibility issue since the choice of drug should be biocompatible with the polymer and able to provide a sufficient integration with the host tissue (minimal adverse events) (Scheller et al., 2005). Preventing restenosis can result in better biocompatibility, reducing the risk of revised surgery along with the pain in the patient. The cost and the time spent on revised surgery would also decrease (Slee et al., 2016). Table 2.9 lists drug-eluting stents that are FDA approved.

TABLE 2.9
List Drug Eluting Stents That Are Approved by FDA

Name	Drug	Manufacturer
Cypher®	Sirolimus	Cordis
Endeavour®	Sirolimus Analogue (zotarolimus)	Medtronic Minneapolis
Xience®	Sirolimus Analogue (everolimus)	Abbott Laboratories
Taxus®	Paclitaxel	Boston Scientific, Inc.

Source: Ma et al. (2011).

2.5.1 Drug-Eluting Stents

Paclitaxel and sirolimus are two commonly used drugs incorporated into drug-eluting stents (Ma et al., 2011). Cypher is an FDA approved sirolimus coated stent used to treat neo-intimal hyperplasia and endothelial thrombosis (Ma et al., 2011, Kim et al., 2017). Sirolimus is an immunosuppressant drug that can provide a potent anti-proliferative reaction (Scheller et al., 2005). Sirolimus is used to treat organ rejection in kidney allograft (BNF 66, 2014).

Paclitaxel is a neo-intimal hyperplasia inhibitor used to treat malignant disease (Scheller et al. 2005; Heldman et al., 2001). Taxus® is the only FDA approved coated stent with Paclitaxel (Ma et al. 2011). Chemically Paclitaxel is a derivative diterpenoid that produces dysfunctional microtubules that prevent smooth cell proliferation (Scheller et al., 2005; Heldman et al., 2001). Heldman et al. investigated Paclitaxel performance in porcine and found after 4 weeks that in fact it inhibits neo-intimal hyperplasia (Heldman et al., 2001).

Restenosis has a similar growth process to tumours. Combination therapy involving multi-pharmaceutical approaches is effective in treating tumours. Stent combinations involving Paclitaxel and nitric oxide were tested but failed due to physicochemical incompatibility. This was mainly due to nitric oxide since it is hydrophilic and cannot be retained within the artery for more than a few minutes. Paclitaxel and sirolimus are hydrophobic and bind to the blood vessels for a few days after release from stents. Paclitaxel and sirolimus have a synergetic effect in anti-tumour treatment (Ma et al., 2011). Ma et al. (2011) investigated the combination of Paclitaxel and sirolimus coatings on drug-eluting stents and demonstrated different drug release proliferation, and most importantly, both were able to provide a synergic effect (Ma et al., 2011).

2.5.2 Drug-Eluting Stent Alternative

Drug-eluting stents cannot achieve full potential when the patient suffers from diseases such as diabetes, renal dysfunction, or patients with small vessel diameters (Slee et al. 2016). Stents coated with drugs also have a higher occurrence of late stent thrombosis by preventing endothelium re-growth (Slee et al., 2016). However, an alternative stent that improves biocompatibility is still required (Slee et al., 2016). Slee et al. (2016) used CD47 as an alternative coating for stents. CD47 is a transmembrane protein that, when it binds to Signal Regulatory Protein Alpha (SIRP α), provides an anti-inflammatory response. In 2011, Stachelek et al. was able to demonstrate that CD47 modified polymer surfaces reduce inflammation in-vivo and in-vitro (Stachelek et al., 2011; Slee et al., 2016). Another study completed by Finley et al. (2012) demonstrated that CD47 surfaces reduced cell adhesion and platelets and neutrophils activation. Slee et al. (2016), suggest the use of CD47 on the stent's surface to enhance biocompatibility and prevent early inflammatory events that result in restenosis (Slee et al., 2016). The purpose of this study was to demonstrate an in-vitro CD47 anti-inflammatory capacity and in-vivo CD47 inhibition of ISR in rats (Slee et al., 2016). Figure 2.10 illustrates the design of CD47 stent. The in-vitro anti-inflammatory capacity of CD47 was determined using different steel foil samples by investigating their ability to inhibit macrophage attachment. The results demonstrated that CD47-steel foil samples had a lower macrophage attachment compared to other samples. This suggests that CD47 has the potential to prevent acute inflammation (Slee et al., 2016). The in-vivo efficacies of stents with and without CD47 were investigated using rats. pepCD47 was chosen due to results from Figure 2.11. The stents were placed in the rats' carotid arteries for 30 minutes, 3 days, and 14 days. After each time period, inflammatory responses (the number of white, red, and platelet cells) and thrombosis were quantified. The SEM results demonstrated that CD47 reduced platelet attachment within 30 minutes (Figure 2.12a) and macrophage recruitment after 3 days (Figure 2.12 b), indicating that CD47 can prevent early acute inflammatory responses, thus having the ability to prevent ISR (Slee et al., 2016).

2.6 OCULAR IMPLANTS

The cornea protects the eye. Keratinocytes within the cornea are found in the stromal layer. A main component of the stromal layer's extracellular matrix is collagen. Previous work conducted by Lai and Ma have used gelatin to mimic collagen function.

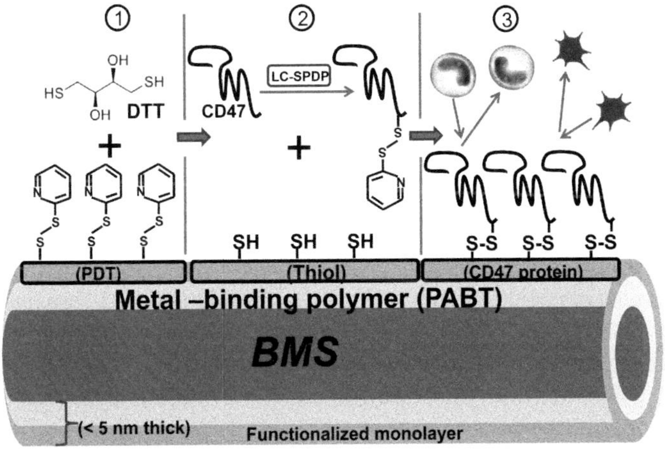

FIGURE 2.10 Design of CD47 stent. Bare metal stent (BMS). (Slee et al., 2016).

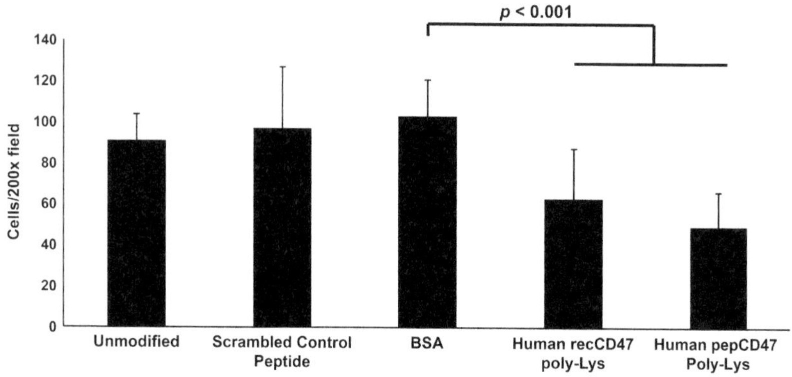

FIGURE 2.11 Displays the average number of cells that attached to macrophages. Two different CD47 were used (Recombinant (recCD47) and peptide sequence CD47 (pepCD47)) (Slee et al., 2016).

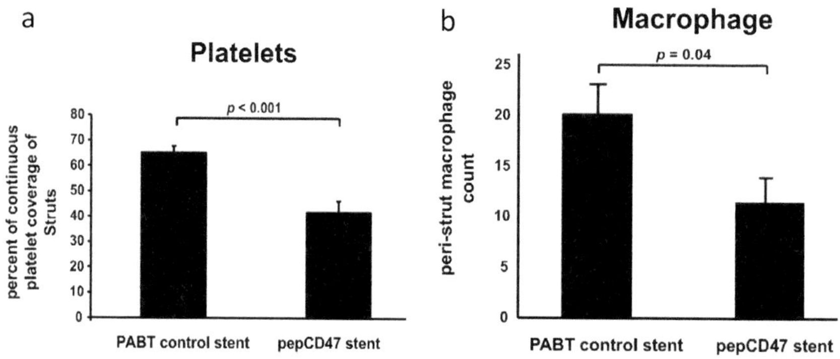

FIGURE 2.12 Compares pepCD47 and control stents. (a) Displays the percentage of platelet coverage after 30 minutes. (b) Displays the macrophage count after 3 days (Slee et al., 2016).

Gelatin incorporated chondroitin demonstrated improved light transmittance capability and nutrient permeability. However, the gelatin-chondroitin provided an unstable environment for the cornea keratinocytes (Lai and Ma, 2017).

Lai and Ma (2017) proposed gelatin with oxidized hyaluronic acid (gelatin-oHA) which helps keratinocytes proliferate and therefore help repair the cornea stromal layer. When gelatin and oHA merge, free amino acids from the gelatin react with the oHA sugar residue thus avoiding toxic aldehydes that are associated with gelatin-hyaluronic acid (gelatin-HA). The objective of this study was to determine the ocular biocompatibility of gelatin-oHA (Lai and Ma, 2017).

The in-vitro experiments required twenty white rabbits. The cornea cells were removed and prepared onto plates with a density of $5*10^4$ cells/ml. A concentration of 10 mg/ml of sample was then added to the plate. Three test groups were analysed. The control test group had no gelatin, the second test group had only gelatin and the final test group had gelatin-oHA. Cell viability (live/dead viability/cytotoxicity kit), cell growth (WST-1 assay), and degree of DNA damage (comet assay) were determined (Lai and Ma, 2017). The in-vivo experiment required eighteen white rabbits. Three test groups were analysed. The healthy eye of the rabbit was intracamerally injected with 50 μL of the sample. To determine gelatin's effect on ocular biocompatibility, slit-lamp examinations on the 7th and the 14th days were carried out. An overall ocular score was used to determine inflammation manifestation. Cornea endothelial cell count was used to determine cell loss. Cornea thickness was measured to determine hydration abnormalities from the surgical procedure (Lai and Ma, 2017). The in-vitro studies conducted by Lai and Ma (2017) demonstrated that gelatin-oHA does not cause cytotoxicity or genotoxicity. It was also shown that gelatin-oHA allows cornea cells to grow without any cellular inflammation interferences (Lai and Ma, 2017).

The in-vivo experiment conducted by Lai and Ma (2017) demonstrated the rabbits had no adverse ocular effects from intracameral injection, since no aqueous flare, cornea cloudiness, or cornea neo-vascularization was found. After calculating the overall ocular score, cornea endothelial cell count, and thickness, it was determined that no inflammation symptoms, injury, and hydration abnormalities occurred (Lai and Ma, 2017). The study demonstrated that gelatin-oHA can be tolerated by rabbits up to 14 days and has the potential to be used as a biomaterial in implants (Lai and Ma, 2017).

2.7 DENTAL IMPLANTS

Dental implants are dependable medical devices used to hold a prosthesis in place in the oral cavity for a long period of time after an oral and maxillofacial surgery (Lauer et al., 2001; Hiroshima et al., 2011). The epithelium of the human oral cavity generates immune effector cells known as human gingival keratinocytes (HGKs). HGKs guard the underlying periodontal tissue by developing a barrier that consistently defends the soft tissue from toxic environmental stimuli and bacterial penetration (Gröger et al., 2008; Grenade et al., 2017). The HGKs barrier produces calprotectin, antimicrobial peptides, and adrenomedullin. Calprotectin prevents Porphyromonas gingivitis development and resists other bacterial infiltrations (Hiroshima et al., 2011). An established HGK barrier is important for dental implants' long-lasting function (Lauer et al. 2001). Polymerization-infiltrated-ceramic-network material (PICN) is a recently developed dental composite with enhanced damage resistance (Grenade et al., 2017). The purpose of the study was to compare PICN-HGK's biocompatibility with other commonly used dental implant biomaterials. The study also determined the PICN cytotoxicity (Grenade et al., 2017). In this study, after 72 hours PICN demonstrated a higher viability (%), cell number, cell coverage (%), and cell spreading (μm^2) compared to glass-ceramic reinforced with lithium disilicate ($Li_2Si_2O_5$) (Figure 2.13) (Grenade et al., 2017). The study also demonstrated that after 48 hours Titanium (Ti), Zirconia (Zi) and PICN all have similar cell number produced and cell coverage (%) and after 72 hours demonstrate a similar viability (%) and cell spreading (μm^2), cell morphology relating to HGK (Grenade et al., 2017). The study confirmed that there was no cytotoxicity nor monomer release (Grenade et al., 2017).

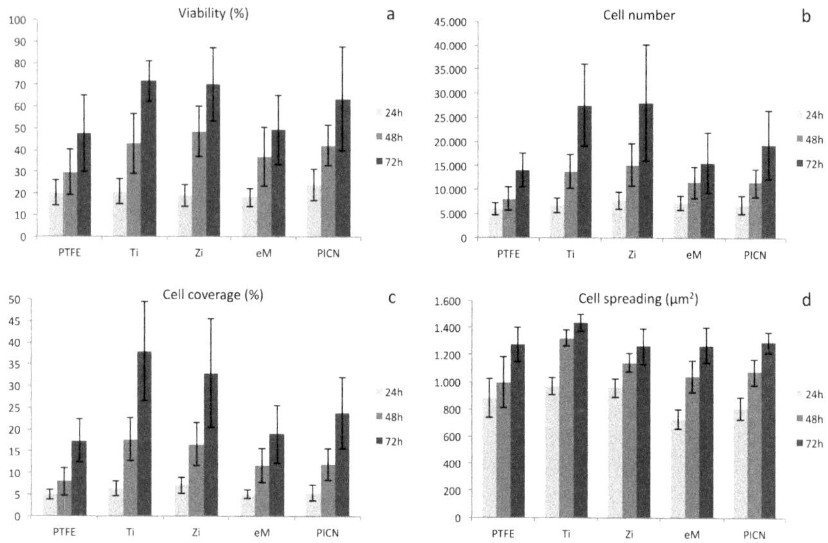

FIGURE 2.13 Displays cell viability (%), cell number, cell coverage (%), and cell spreading (μm^2) (Grenade et al., 2017).

Research and development from rival dental implant manufacturers have revolutionized the dental implant industry by innovating exciting new materials along with surface tissue treatments (Villar et al., 2012). Dental implants are used to replace natural teeth in patients with poor oral hygiene or dental health (Poskevicius et al., 2015). To enhance dental implant functionality and patient safety, it is important to understand the wound healing process and the material's biocompatibility (Villar et al., 2012; Chaturvedi, 2016). The mouth is a wet, hostile, electrolytic environment which is consistently attacked by different chemical and physical stimuli (Chaturvedi, 2016). Corrosion is one of the main concerns for dental implants since it results in adverse health reactions, implant disintegration, or in some cases overall implant failure (Chaturvedi, 2016). Corrosion in implants can be influenced by several different factors including saliva properties (pH, quantity, etc.), food, and oral health (Chaturvedi, 2016). Dental implants mainly use titanium due to its biocompatibility. Lin et al. (2012) applied titanium nitride (TiN) over the implant to increase the solubility layer of TiO2 and successfully reduced bacterial adherence to the implant. However, the TiN coating required continuous administration and the research needs to be completed in a saliva-simulated environment (Chaturvedi, 2016). Another study completed by Ye et al. (2014) used high-energy electro-pulsing treatment on the surface of the TiO_2 implant. Ye et al. (2014) developed the treatment under artificial saliva conditions and had also demonstrated an increase of TiO2 solubility layer over the implant, thus reducing corrosion over time (Chaturvedi, 2016). More research needs to be completed on preventing corrosion by implant coating techniques in order to suggest a universal coating process that can be placed safely into the patient for a long period of time (Poskevicius et al., 2015; Chaturvedi, 2016).

REFERENCES

Abdel-Hady, M. and Morinaga, M. (2009). Modification of phase stability and mechanical properties by the addition of O and Fe into β-Ti alloys. *International Journal of Modern Physics B*, 23(6), 1559.

Alqhtani, N., Logan, N., Meghji, S., Leeson, R., and Brett, P. (2017). Low dose effect of bisphosphonates on hMSCs osteogenic response to titanium surface in-vitro. *Bone Reports*, 6, 64–69.

Baas, J., Jakobsen, T., Elmengaard, B., Bechtold, J., and Soballe, K. (2011). The effect of adding an equine bone matrix protein lyophilisate on fixation and osseointegration of HA-coated Ti implants. *Journal of Biomedical Materials Research Part A*, 100A(1), 188–194.

Bedair, T., Yu, S., Im, S., Park, B., Joung, Y., and Han, D. (2015). Effects of interfacial layer wettability and thickness on the coating morphology and sirolimus release for drug-eluting stent. *Journal of Colloid and Interface Science*, 460, 189–199.

Brooks, E., Brooks, R., and Ehrensberger, M. (2017). Effects of simulated inflammation on the corrosion of 316L stainless steel. *Materials Science and Engineering: C*, 71, 200–205.

Chaturvedi, T. (2016). Corrosive behaviour of implant biomaterials in oral environment. *Materials Technology*, 31(12), 689–695.

Chen, Q., and Thouas, G. (2015). Metallic implant biomaterials. *Materials Science and Engineering: R: Reports*, 87, 1–57.

Cools, P., De Geyter, N., Vanderleyden, E., Barberis, F., Dubruel, P., and Morent, R. (2016). Adhesion improvement at the PMMA bone cement-titanium implant interface using methyl methacrylate atmospheric pressure plasma polymerization. *Surface and Coatings Technology*, 294, 201–209.

Dapunt, U., Giese, T., Lasitschka, F., Reinders, J., Lehner, B., Kretzer, J. and Hänsch, G. (2014). On the inflammatory response in metal-on-metal implants. *Journal of Translational Medicine*, 12(74), 1–9.

Dyer, O. (2016). Johnson & Johnson is ordered to pay $1bn over faulty hip implants. *British Medical Journal*, 355, i6551.

Finley, M., Rauova, L., Alferiev, I., Weisel, J., Levy, R., and Stachelek, S. (2012). Diminished adhesion and activation of platelets and neutrophils with CD47 functionalized blood contacting surfaces. *Biomaterials*, 33(24), 5803–5811.

Gepreel, M., and Niinomi, M. (2013). Biocompatibility of Ti-alloys for long-term implantation. *Journal of the Mechanical Behavior of Biomedical Materials*, 20(1), 407–415.

Gibon, E., Amanatullah, D., Loi, F., Pajarinen, J., Nabeshima, A., Yao, Z., and Goodman, S. (2016). The biological response to orthopaedic implants for joint replacement: Part I: Metals. *Journal of Biomedical Materials Research Part B: Applied Biomaterials*, 1(1), 1–12.

Gigante, A., Cigna, V., Manzotti, S., Villa, T., Salvolini, E., and Mattioli-Belmonte, M. (2013). Total knee prosthesis polyethylene wear reduction by a new metal part finishing method. *Journal of Applied Biomaterials & Functional Materials*, 11(2), 99–105.

Galløe, A. M., Kelbæk, H., Thuesen, L., Hansen, H. S., Ravkilde, J., Hansen, P. R., and Bligaard, N., (2017). 10-year clinical outcome after randomization to treatment by sirolimus- or paclitaxel-eluting coronary stents. *Journal of the American College of Cardiology)*, 69(6), 616–624.

Grenade, C., De Pauw-Gillet, M., Pirard, C., Bertrand, V., Charlier, C., Vanheusden, A., and Mainjot, A. (2017). Biocompatibility of polymer-infiltrated-ceramic-network (PICN) materials with Human Gingival Keratinocytes (HGKs). *Dental Materials*, 33(3), 333–343.

Gröger, S., Michel, J., and Meyle, J. (2008). Establishment and characterization of immortalized human gingival keratinocyte cell lines. *Journal of Periodontal Research*, 43(6), 604–614.

Hamada, N., Miyata, M., Eto, H., Shirasawa, T., Akasaki, Y., Nagaki, A. and Tei, C. (2010). Tacrolimus-eluting stent inhibits neointimal hyperplasia via calcineurin/NFAT signaling in porcine coronary artery model. *Atherosclerosis*, 208(1), 97–103.

Harwin, S., Patel, N., Chughtai, M., Khlopas, A., Ramkumar, P., Roche, M., and Mont, M. (2017). Outcomes of newer generation cementless total knee arthroplasty: Beaded periapatite-coated vs highly porous titanium-coated implants. *The Journal of Arthroplasty*, 32(7), 2156–2160.

Heldman, A., Cheng, L., Jenkins, M., Heller, P., Kim, D., Ware, M., and Froehlich, J. (2001). Paclitaxel stent coating inhibits neointimal hyperplasia at 4 weeks in a Porcine model of coronary restenosis. *Circulation*, 103(18), 2289–2295.

Hiroshima, Y., Bando, M., Kataoka, M., Inagaki, Y., Herzberg, M., Ross, K., and Kido, J. (2011). Regulation of antimicrobial peptide expression in human gingival keratinocytes by interleukin-1α. *Archives of Oral Biology*, 56(8), 761–767.

Hussein, A., Gepreel, M., Gouda, M., Hefnawy, A., and Kandil, S. (2016). Biocompatibility of new Ti–Nb–Ta base alloys. *Materials Science and Engineering: C*, 61, 574–578.

Huang, Y., Xu, J., Li, J., He, B., Xu, L., and Li, Z. (2014). Mechanical properties and biocompatibility of melt processed self-reinforced ultrahigh molecular weight polyethylene. *Biomaterials*, 35(25), 6687–6697.

Jin, M., Yao, S., Wang, L., Qiao, Y., and Volinsky, A. (2016). Enhanced bond strength and bioactivity of interconnected 3D TiO2 nanoporous layer on titanium implants. *Surface and Coatings Technology*, 304, 459–467.

Kheirkhah, M., Fathi, M., Salimijazi, H., and Razavi, M. (2015). Surface modification of stainless steel implants using nanostructured forsterite (Mg2SiO4) coating for biomaterial applications. *Surface and Coatings Technology*, 276, 580–586.

Kim, S., Kim, T., Choi, J., and Kwon, I. (2012). Current perspectives of biodegradable drug-eluting stents for improved safety. *Biotechnology and Bioprocess Engineering*, 17(5), 912–924.

Kim, S., Park, S., Bedair, T., Kim, M., Park, B., Joung, Y., and Han, D. (2017). The effect of solvents and hydrophilic additive on stable coating and controllable sirolimus release system for drug-eluting stent. *Materials Science and Engineering: C*, 78, 39–46.

Lai, J., and Ma, D. (2017). Ocular biocompatibility of gelatin microcarriers functionalized with oxidized hyaluronic acid. *Materials Science and Engineering: C*, 72, 150–159.

Lauer, G., Wiedmann-Al-Ahmad, M., Otten, J., Hübner, U., Schmelzeisen, R., and Schilli, W. (2001). The titanium surface texture effects adherence and growth of human gingival keratinocytes and human maxillar osteoblast-like cells in-vitro. *Biomaterials*, 22(20), 2799–2809.

Lin, N., Huang, X., Zhang, X., Fan, A., Qin, L., and Tang, B. (2012). In-vitro assessments on bacterial adhesion and corrosion performance of TiN coating on Ti6Al4V titanium alloy synthesized by multi-arc ion plating. *Applied Surface Science*, 258(18), 7047–7051.

Ma, X., Oyamada, S., Gao, F., Wu, T., Robich, M., Wu, H., and Laham, R. (2011). Paclitaxel/sirolimus combination coated drug-eluting stent: In-vitro and in vivo drug release studies. *Journal of Pharmaceutical and Biomedical Analysis*, 54(4), 807–811.

Mahmood, B., and Hammert, W. (2015). Metal implant allergy. *The Journal of Hand Surgery*, 40(4), 831–833.

Mina, A., Caicedo, H., Uquillas, J., Aperador, W., Gutiérrez, O., and Caicedo, J. (2016). Biocompatibility behavior of β-tricalcium phosphate-chitosan coatings obtained on 316L stainless steel. *Materials Chemistry and Physics*, 175, 68–80.

Ni, S., Chou, L., and Chang, J. (2007). Preparation and characterization of forsterite (Mg2SiO4) bioceramics. *Ceramics International*, 33(1), 83-88.

Pang, X., and Zhitomirsky, I. (2005). Electrodeposition of composite hydroxyapatite–chitosan films. *Materials Chemistry and Physics*, 94(2-3), 245–251.

Plecko, M., Sievert, C., Andermatt, D., Frigg, R., Kronen, P., Klein, K., and von Rechenberg, B. (2012). Osseointegration and biocompatibility of different metal implants - a comparative experimental investigation in sheep. *BMC Musculoskeletal Disorders*, 13(32), 1–12.

Omar, S., Ballarre, J., and Ceré, S. (2016). Protection and functionalization of AISI 316L stainless steel for orthopedic implants: hybrid coating and sol gel glasses by spray to promote bioactivity. *Electrochimica Acta*, 203, 309–315.

Poskevicius, L., Sidlauskas, A., Galindo-Moreno, P., and Juodzbalys, G. (2015). Dimensional soft tissue changes following soft tissue grafting in conjunction with implant placement or around present dental implants: a systematic review. *Clinical Oral Implants Research*, 28(1), 1–8.

Ratner, B., and Bryant, S. (2004). Biomaterials: Where we have been and where we are going. *Annual Review of Biomedical Engineering*, 6(1), 41–75.

Rodriguez, M., Brown, J., Giordano, J., Lin, S., Omenetto, F., and Kaplan, D. (2017). Silk based bioinks for soft tissue reconstruction using 3-dimensional (3D) printing with in-vitro and in vivo assessments. *Biomaterials*, 117, 105–115.

Saini, M., Singh, Y., Arora, P., Arora, V., and Jain, K. (2015). Implant biomaterials: A comprehensive review. *World Journal of Clinical Cases*, 3(1), 52–57.

Scheller, B., Grandt, A., Wnendt, S., Lorenz, G., Böhm, M., and Nickenig, G. (2005). Comparative study of tacrolimus and paclitaxel stent coating in the porcine coronary model. *Zeitschrift für Kardiologie*, 94(7), 445–452.

Serra, T., Capelli, C., Toumpaniari, R., Orriss, I., Leong, J., Dalgarno, K., and Kalaskar, D. (2016). Design and fabrication of 3D-printed anatomically shaped lumbar cage for intervertebral disc (IVD) degeneration treatment. *Biofabrication*, 8(3), 035001.

Serruys, P. (2002). Rapamycin eluting stent: the onset of a new era in interventional cardiology. *Heart*, 87(4), 305–307.

Slee, J., Alferiev, I., Nagaswami, C., Weisel, J., Levy, R., Fishbein, I., and Stachelek, S. (2016). Enhanced biocompatibility of CD47-functionalized vascular stents. *Biomaterials*, 87, 82–92.

Sridharan, R., Cameron, A. R., Kelly, D. J., Kearney, C. J., and O'Brien, F. J. (2015). Biomaterial based modulation of macrophage polarization: A review and suggested design principles. *Materials Today*, 18(6), 313–325.

Stachelek, S., Finley, M., Alferiev, I., Wang, F., Tsai, R., Eckells, E., and Levy, R. (2011). The effect of CD47 modified polymer surfaces on inflammatory cell attachment and activation. *Biomaterials*, 32(19), 4317–4326.

Trombetta, R., Inzana, J., Schwarz, E., Kates, S., and Awad, H. (2016). 3D Printing of calcium phosphate ceramics for bone tissue Engineering and drug delivery. *Annals of Biomedical Engineering*, 45(1), 23–44.

Trombetta, R., Inzana, J., Schwarz, E., Kates, S., and Awad, H. (2016). 3D Printing of calcium phosphate ceramics for bone tissue Engineering and drug delivery. *Annals of Biomedical Engineering*, 45(1), 23–44.

Villar, C., Huynh-Ba, G., Mills, M., and Cochran, D. (2012). Wound healing around dental implants. *Endodontic Topics*, 25(1), 44–62.

Wang, J., de Boer, J., and de Groot, K. (2004). Preparation and characterization of electrodeposited calcium phosphate/chitosan coating on Ti6Al4V Plates. *Journal of Dental Research*, 83(4), 296–301.

Wermelin, K., Suska, F., Tengvall, P., Thomsen, P., and Aspenberg, P. (2008). Stainless steel screws coated with bisphosphonates gave stronger fixation and more surrounding bone Histomorphometry in rats. *Bone*, 42(2), 365–371.

Williams, D. (2008). On the mechanisms of biocompatibility. *Biomaterials*, 29(20), 2941–2953.

Williams, D. (2014). There is no such thing as a biocompatible material. *Biomaterials*, 35(38), 10009–10014.

Ye, X., Yang, Y., and Tang, G. (2014). Microhardness and corrosion behavior of surface gradient oxide coating on the titanium alloy strips under high energy electro-pulsing treatment. *Surface and Coatings Technology*, 258, 467–484.

Zaporozhets, T., Puz', A., Sinebryukhov, S., Gnedenkov, S., Smolina, T., and Besednova, N. (2017). Biocompatibility of modified osteoinductive calcium-phosphate coatings of metal implants. *Bulletin of Experimental Biology and Medicine*, 162(3), 366–369.

Zinar, R., and Schmalzried, T. (2015). Why hip implants fail: Patient, surgeon, or device? *Seminars in Arthroplasty*, 26(3), 118–120.

3 Polymeric-Based Biomaterials

3.1 INTRODUCTION

Biomaterials have been used for hundreds of years, but it was only within the last century that polymeric-based materials started to become commonly used within the biomaterial field. Before the use of polymers, the biomaterial field was limited to using materials sourced from plants, ore, and animals (Moukwa, 1997). It was only in the 1920s that synthetic polymers started to become feasible and became commonly used for medical purposes (Moukwa, 1997). In the 1930s–1950s the most commonly used polymeric-based biomaterial was poly(methyl methacrylate) due to its biocompatibility and versatility. This material is still used today for prosthetic eyeballs, bone cements, and intraocular lenses (Jia et al., 1999). Since the inception of polymeric biomaterials, there have been ample advancements that have revolutionized the medical field and have laid much of the groundwork for the creation of the biomaterial field in general (Moukwa, 1997).

3.2 STRUCTURE AND POLYMERIZATION

Polymeric biomaterials are created from long chains of carbon molecules strung together in sequence. This is usually done by condensation and addition reactions that force the small molecules to lose electrons and bond together (Griffith, 2000). This reaction is controlled through the manipulation of temperature, pressure, time, and the presence of a catalyst. The altering of these factors provides control over the extent of polymerization and the length of chains being created (Griffith, 2000). Figure 3.1 below shows the formation of polyethylene, a polymeric biopolymer.

The carbon backbone chain that makes up a polymer is held together through covalent bonds or secondary forces; such as van der Waals or hydrogen bonds. The length of the carbon backbone and its side chains will affect the overall molecular interactions of the polymer and will dictate its structure and functions (Moukwa, 1997). For instance, a polymer such as paraffin wax, that consists of repeating units of $(CH_2CH_2)_n$, does not have any large, branched, or highly charged side groups. This allows the chains in this polymer to fit tightly together and crystallize almost completely. Slightly altering the carbon chains in paraffin wax will affect its overall properties. For example, the lengthening of its chains (>50) will decrease the level of crystallinity achievable (80–90% crystallization) (Jia et al., 1999). The addition of branches to the carbon backbone will act in a similar manner to that of lengthening the chains and will decrease the level of crystallization attainable because the molecules cannot fit together as tightly as before. Side chains added to the carbon

by Yaser Dahman and Luqmaan Moolla

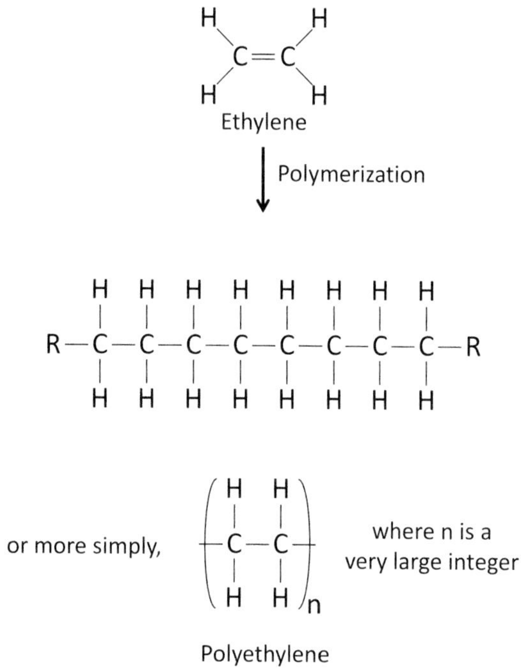

FIGURE 3.1 Representation of the creation of the common polymeric biomaterial polyethylene (PE) (MPCFaculty).

backbone decrease the level of crystallization even further due to steric hindrance and because of other intermolecular interactions that occur within the polymer structure (Jia et al., 1999). Through manipulating the length of a polymer chain, its branching, and the side chains present, the polymeric properties can be altered to fit a wide variety of applications. This is what makes polymers so versatile and an important player in the field of biomaterials (Jia et al., 1999).

3.3 CLASSIFICATION OF POLYMERIC BIOMATERIALS

The structure and function of polymeric biomaterials range immensely, but most polymers share some general characteristics. The first is that they are resistant to chemicals (Griffith, 2000). Most polymers are thermal or electrical insulators and are also comparatively light with varying amounts of strengths. Most polymers can be processed easily and in several different ways allowing them to be versatile and processed to fit their intended application (Griffith, 2000).

There are many ways that polymers can be classified, but for this paper they will be classified based on certain important properties and will be grouped into 7 different subgroups. The 7 classifications that will be used are thermosets, thermoplastics, elastomers, hydrogels, polyelectrolytes, natural polymers, and biodegradable polymers. Each of these subgroups will be overviewed below and will be explored for current advancements in the field.

3.4 THERMOSETTING POLYMERS

Thermosetting prepolymers or thermosetting resins are viscous soft fluids or solids that cure into insoluble polymer networks called thermosets (Jolly et al., 1996). This process is generally irreversible and is accomplished through exposing the prepolymer or resin to heat, pressure, or some form of radiation (PCC, 2006). The change from prepolymer to thermoset can also be accomplished through the addition of a crosslinking agent or a catalyst (PCC, 2006). Changing the composition of the prepolymer or thermosetting resin or adding fibers into the curing process to strengthen or support the thermoset can vary the properties of the thermoset. An overview of the thermoset process is visualized in Figure 3.2.

Thermosetting polymers form three-dimensional bond networks through crosslinking and strong covalent bonds that allow these polymers to maintain their structural integrity when exposed to high temperatures and other means of stress (Dodiuk and Goodman, 2013). Thermosets that have a higher density of crosslinking, aromatic groups are generally stronger, can withstand higher temperatures, and are resistant to a wider range of chemicals (Jolly et al., 1996). What is important to note is that crosslinking is not all beneficial, and that it comes as a price. The greater the extent of crosslinking the more brittle the thermoset becomes, which might mitigate some of the beneficial side effects associated with a higher crosslinking density (Dodiuk and Goodman, 2013).

There are several different types of thermosetting polymers. High weight thermosets have a rigid three-dimensional structure that tends to deform under stress and continues to remain deformed, even after the stressor is removed (Lonescu, 2005). This type of thermoset is also irreversible and cannot be melted and reformed once cured (Lonescu, 2005). There are some special thermosets that are made of softer material and can return to their original shape once deformed. Some of these thermosets are also special in the sense that they can be melted and reformed even after they have been cured (Fink, 2013). Most conventional thermosets, once cured, are

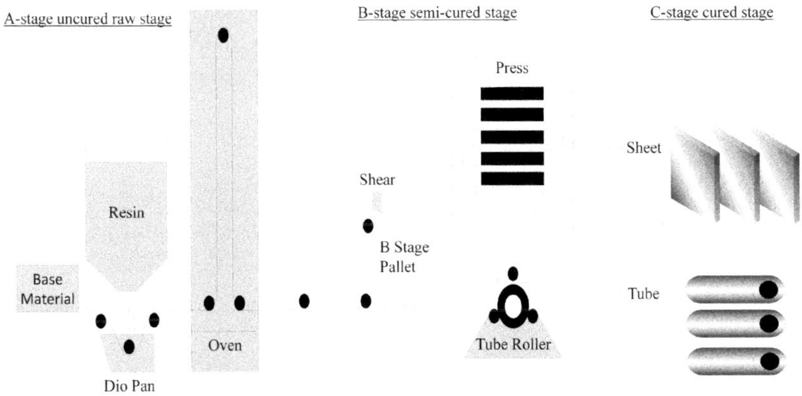

FIGURE 3.2 Overview of the thermoset polymer activation process (Acculam).

stuck in that shape and cannot be melted and moulded. This is sometimes an issue, because thermosets that are irreversibly cured cannot be recycled for future use (Fink, 2013).

Current research in the field is looking at creating the next generation of thermosets that have the ability to be remoulded and reused after being cured. An excellent example of this is the work of Montarnal et al. (2011) and his group of researchers, creating silica like thermoset polymer from organic networks. Materials that are permanently crosslinked have excellent mechanical properties and are resistant to solvents. The problem with these materials is that they are hard to process and cannot be remoulded once made. On the other hand, materials that can be reprocessed and remoulded are generally not as mechanically strong as their permanently crosslinked counterparts. These researchers found a way to use organic networks to create silica like structures that can be reshaped and changed after curing. Their networks were synthesized through a classical epoxy reaction between diglycidyl ether of bisphenol A, and a mixture of fatty dicarboxylic and tricarboxylic acids. This was done in a 1:1 ratio to ensure hydroxyl and ester groups in the final material. The created material when left at room temperature maintains its shape and structural properties, but when exposed to heat for a prolonged period of time becomes mouldable and can be reshaped to a new desired structure. This works through topological rearrangements via exchange reactions that occur to preserve network integrity.

Another recent advancement in the field of thermosets was the creation of a citric acid and glycerol polymer that does not need a catalyst to cure (Halpern et al., 2014). This process combined glycerol and citric acid, which are both non-toxic and renewable agents, and crosslinked them through a polymerization procedure that exposed the polymer to temperatures between 90–150°C. Condensation reactions were the driving force behind the curing of this polymer and the main by-product was water. This polymer is highly biocompatible and can be altered to have a controlled degradation rate within the body (Halpern et al., 2014). The rate of degradation of this polymer is inversely related to the amount of crosslinking present. This thermoset was tested as a drug delivery system, that administering an antibiotic in-vitro, and was found to be significantly more effective than the control (Halpern et al., 2014). Many other advancements like this are being made across the field of thermosetting polymers, making this type of material a key player in the biomaterial field.

3.5 THERMOPLASTIC POLYMERS

Thermoplastic polymers are generally materials that become mouldable and pliable above a certain temperature and solidify when cooled (Baeurle et al., 2006). The polymer chains in thermoplastics are held together through intermolecular forces that are rapidly weakened when heated, creating a viscous polymer fluid. This property of thermoplastic polymers allows them to be reshaped and used in processing techniques such as compression moulding, injection moulding, and calendaring (Baeurle et al., 2006). What sets thermoplastics apart from thermosets are the bonds: in thermoplastics, the bonds are generally reversible, while those in thermosets are generally irreversible (Shenoy and Saini, 1996). Table 3.1 represents some of the

TABLE 3.1
Summary of the Advantages and Disadvantages of Thermoplastics and Thermosetting Polymers

Material		Advantages	Disadvantages
Physically crosslinked (Thermoplastic)	Plastic	• Easy fabrication (by melt or solvent processing) • Tunable mechanical properties and degradation kinetics	• Rigid • Lack of flexibility • Release of acidic degradation products • Possibility of foreign body response
	Elastomer	• Easily fabrication • Flexible • High elongation • Tuneable mechanical properties • Degradation kinetics	• Heterogeneous degradation profile • Mechanical failure much faster than material degradation • Release of acidic degradation products • Possibility of foreign body response
Chemically crosslinked (Thermoset)	Synthetic elastomer	• Flexible • Tightly controlled purity, structure mechanical properties, and degradation kinetics • Good maintenance of form stability during degradation • No immunogenicity issues	• Relatively poor processability • Possibility of foreign body response • Release of acidic degradation products
	Biologically occurring elastomer	• Flexible • Similar structure to natural extracellular matrix • Bioactive • Able to present receptor-binding ligands to cells • Natural remoulding	• Costly purification • Strong immunogenic response • Unpredictable degradation • Possible calcification • Complexities in processing • Possibilities of disease transmission

Source: Chen et al. (2013).

differences and advantages/disadvantages of thermoplastics and thermosets (Chen et al., 2013).

The properties of thermoplastics vary greatly depending on composition and temperature. Extreme property changes generally occur in thermoplastics when they are above their glass transition temperature, and below their melting temperature. Within this temperature range they are in their amorphous state (Gupta, 2000). Some thermoplastics do not fully crystallize under their glass transition temperature and retain certain amorphous properties. Amorphous or semi-amorphous thermoplastics generally maintain a high degree of clarity, are less resistant to chemical attack and less likely to crack from environmental stress (MacDermott and Shenoy, 1997). Plasticizers can be added to reduce the brittleness of a thermoplastic by making the chain segments that constitute the polymer more mobile, essentially increasing its amorphous nature. The addition of plasticizers, or other means of increasing the amorphic nature, works by effectively reducing the glass transition temperature (MacDermott and Shenoy, 1997). The reduction of the glass transition temperature decreases the brittleness of the polymer and brings about many of the amorphous characteristics of the polymer. The addition of plasticizers is usually done to make the polymer more pliable and reduce its chances of creaking once stressed (MacDermott and MacDermott, 1997). This is how many thermoplastics are altered to tailor the polymer to the specific needs of the application at hand.

A lot of recent research in the biomaterial field is focusing on the creation of scaffolds that can be used for the regeneration of soft tissue. Thermoplastics are one of the polymers that are being employed to achieve this goal. Researchers from the University of Pittsburgh were able to synthesize a family of polyurethane thermoplastics that included partial substitutions of polyester segments into the polymer make-up to create a more robust scaffold. Their scaffold was able to handle higher levels of mechanical stress without jeopardizing the soft, pliable properties of the thermoplastic scaffold, making this scaffold compatible for several soft tissue-engineering applications (Hong et al., 2010).

Thermoplastics are also finding their way into other avenues of the biomaterial field. Other researchers from the Cape Western Reserve University's Department of Biomedical Engineering were looking into the efficacy of the fracture toughness and flexural properties of short-fiber thermoplastic composites. These scientists wanted to use these composites as enhanced fracture fixation devices. The issue with the current fixation devices is that they are highly rigid. The problem with rigid fixation devices is that it takes longer to heal due to the forces that the rigid fixator places on the bones (Brown et al., 1990). Other research has shown that less rigid and more flexible bone fixator devices, such as a plastic, increases the rate of healing and external callus formation, while simultaneously increasing the rate of bone remodelling (Brown et al., 1990). These researchers tested three different types of thermoplastics and evaluated their efficacy as bone fixators (Uhthoff et al., 1981). The testing processes that they used in the study included sterilization, heat contouring, and saline soaking. The three materials tested in this study were polyether ether ketone, polybutylene terephthalate, and polysulfone. These thermoplastics were also reinforced with 30% short carbon fibers. The results of this experiment showed that the polysulfone and polybutylene terephthalate composites lost mechanical properties once exposed to the saline soaking test. The polyether ether ketone thermoplastic composite exhibited no degradation in mechanical properties under all tests, which means that this composite should be viable under biological conditions (Uhthoff et al., 1981). The results of this study show that thermoplastics could potentially be used as a new means of bone fixators that better facilitates bone regeneration. These are just a few examples of how thermoplastics have advanced, and there is a lot of other research similar to what is explored above, that is working on creating the next generation of thermoplastic polymers with a wider variety of uses.

3.6 ELASTOMERIC POLYMERS

Elastomers are a special class of polymers that exhibit the special property of viscoelasticity. A polymer that exhibits viscoelasticity is both elastic and viscous (Abdullateef et al., 2012). Elastomers have very weak intermolecular bonds with a high failure strain. Elastomers or elastic polymers are also generally known as rubber (Alemán et al., 2007). These polymers are composed of monomer units, usually carbon, hydrogen, oxygen, or silicon, that are weakly bonded together (Alger, 1996). Most elastomers exist above their glass transition temperature, so they are usually found in their amorphous form at ambient temperatures (Alemán et al., 2007). The weak binding of the monomers is what gives these polymers their amorphic properties and allows them to be stretched and deformed easily (Alger, 1996). When elastomers are unstressed their monomers are clumped together randomly. Their monomer arrangement changes when they are stressed, and the monomers stretch out and reorganize to accommodate the stress on the polymer, as shown in Figure 3.3.

Elastomers play an important role in the field of biomaterials and in the engineering of tissue constructs. A common issue with engineering biological tissue is mimicking the mechanical stretchability of the native tissue (Chen et al., 2013). Elastomers, because of their elastic properties are important in the development of tissue replacements and the creation of constructs that can mimic the complex elasticity of native tissue (Chen et al., 2013). Some key elastomeric biomaterials used today are silicones, polyurethanes, styrene-based rubbers, polyesters, and copolyesters (Lanza et al., 2011). These elastomers are used across the biomaterial field from catheters and blood bags to heart valves and wound dressings (Langer and Vacanti, 1993). Table 3.2 is a chart that contains many of the uses of these elastomer materials and some of the properties, advantages, and disadvantages of each type (Chen et al., 2013).

Since elastomers hold the promise of being adapted for tissue augmentation and support, there has been much emphasis placed on advancing this type of biomaterial. Examples of recent developments include the highly tuneable elastomeric silk biomaterial created by researchers from the Department of Biomedical Engineering in Tuffs University (Partlow et al., 2014). These researchers found it difficult to functionalize currently available elastomeric polymers and decided to create their own. They used a new method to covalently crosslink tyrosine residues in silk proteins via the use of hydrogen peroxide, to generate highly tuneable polymers with excellent elastomeric properties (Partlow et al., 2014). The polymers created through this process had gelatinous characteristics and could be tuned to control the amount

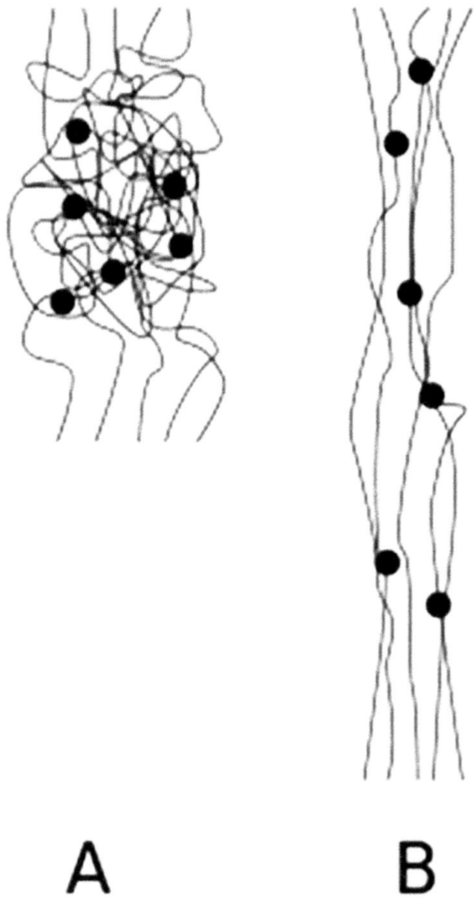

FIGURE 3.3 (a) An elastomer that is unstressed and the monomers can arrange randomly. (b) An elastomer that is stressed and the monomers rearrange to allow the for the elastomer to stretch (Wikipedia).

of swelling that the polymer undergoes once in aqueous solution. These polymers could also withstand shear forces in the magnitude of 100% and compressive strains greater than 70%, all while displaying a stiffness between 500–10,000 Pa (Partlow et al., 2014). With these characteristics, the created polymer covers the mechanical properties of a significant portion of native tissue types. This means that this polymer could be adapted for a wide variety of biological applications (Partlow et al., 2014). These researchers also tested the biocompatibility of this polymer through encapsulating mesenchymal stem cells (MSC) and introducing them into a rat model. The results from their in-vivo study show that the polymer was highly biocompatible and was able to protect the encapsulated cells in-vivo during the term of the experiment (Partlow et al., 2014). This is a significant advancement in this field, because not only are these polymers tuneable, but they are also highly biocompatible and viable in a biological setting.

TABLE 3.2
Summary of Properties and Biomedical Application of Key Elastomeric Biomaterials

Elastomer Class	Medical Device Usage (Current and Potential)	Major Properties (Advantages and Disadvantages)
Silicones (chemically crosslinked)	USFDA class I or II: • Catheters, drains, tubes, and cannulas • In extracorporeal equipment such as cardiovascular by-pass machines USFDA class III: • Programmable value shunts • Finger joint implants • Cosmetic implants (e.g. breasts, chin, nose, cheek, calf, and buttock)	• Stable at high temperature (suitable for autoclaving) • Excellent biostability (resistance to inflammatory response or metabolism with minimal deterioration of properties over time) • Excellent resistance to ageing (the self-life at ambient conditions being virtually indefinite) • Moderate biocompatibility (minimal tissue reaction as implants) • Excellent dielectric behaviour over a wider range of temperatures (suitable for usage as heart value shunts) • Good elasticity • Hydrophobicity (causing dryness when used as a contact lens) • Good oxygen-permeability (suitability as a membrane) • Excellent resistance to oxidation and ultraviolet light (suitable for various sterilization methods)
Polyurethanes (physically crosslinked)	USFDA class I or II: • Catheters, drains, medical tubing • Blood bags • Transdermal drug delivery patches • Surgical drapes and gowns • Transient cardiovascular devices • Intra-aortic balloon pumps • Temporary left ventricular assist devices (LVADs) and biventricular assist devices USFDA class III: • Interim cardiovascular devices • Ventricular assist devices • Total artificial hearts used as a temporary measure prior to organ transplantation (generally intended for less than a month) • Permanent cardiovascular devices (e.g. totally implantable ventricles, artificial heart valves) • Blood conduits and access devices • Breast implants with polyurethane overlays (involved in a scientific controversy) • Wound dressing and barrier scaffolds • Contraceptives	• Enormous diversity of chemical compositions and mechanical properties • Tissue-specific biocompatibility (e.g. potentially not suitable breast augmentation • Widely tuneable biodegradability • Moderate blood compatibility • Tuneable rate of moisture transmission • Mechanical flexibility

(Continued)

TABLE 3.2 (CONTINUED)

Elastomer Class	Medical Device Usage (Current and Potential)	Major Properties (Advantages and Disadvantages)
Styrene-based rubbers (physically crosslinked)	USFDA class I or II: • Catheters, tubing, intravenous solution containers, syringe tips, stoppers, dropper blubs, baby bottle nipples, flexible caps, medical seals, and eye dropper bulbs • Films: blood and platelet bags, surgical drapes • Replacement of latex rubber in condoms and surgical gloves	• Soft and elastic without vulcanization • Good sterilizability • Poor tissue compatibility
Polyesters and copolyesters (physically crosslinked)	USFDA class I or II: • Transdermal patches for drug delivery • Intravenous site dressings • Surgical drapes • Ostomy site dressings • Operating room garments • Packing usages range from bottles to bags USFDA class III: • Wound care dressings	• Natural softness, good contact to surfaces • Tactile comparable to that of dry human skin • No contamination to wounds • No plasticizers used • Excellent resistance to a wide range of solvents and chemicals • High tear strengths • Low-friction surfaces (suitable for adherence to wound areas, clothing, or bedding) • Excellent breathability • Good processability • Good sterilizability • Low toxicity

Source: Chen et al. (2013).

Another notable advancement in the field of elastomers is their alteration to contain properties or factors that are not normally present. Scientists from the Department of Surgery from the Feinberg School of Medicine in Chicago identified that the major reason for the failure of vascular grafts is that they do not have the same antithrombogenic properties as native blood vessels (Hoshi et al., 2013). The lack of these antithrombogenic properties leads to the development of thrombosis, restenosis, and eventually the failure of the implant (Hoshi et al., 2013). These researchers tackle this issue by integrating bioactive heparin, a strong anticoagulant, to the lumen of expanded polytetrafluoroethylene (ePTFE) vascular grafts. This newly designed elastomer was then tested under platelet adhesion and anti-clotting assays to determine its effectiveness. Their results showed that the heparin graft remained viable under in-vitro conditions for at least one month, and were also able to show reduced platelet adhesion, and the inhibition of blood clotting kinetics under physiologically relevant conditions. These results can be visually seen in Figure 3.4 (Hoshi et al., 2013).

Their elastomer was also tested with endothelial progenitor cells and was able to support adhesion and viability, and to show an increase in proliferation, NO production, and an increased expression of the endothelial cell-specific markers von

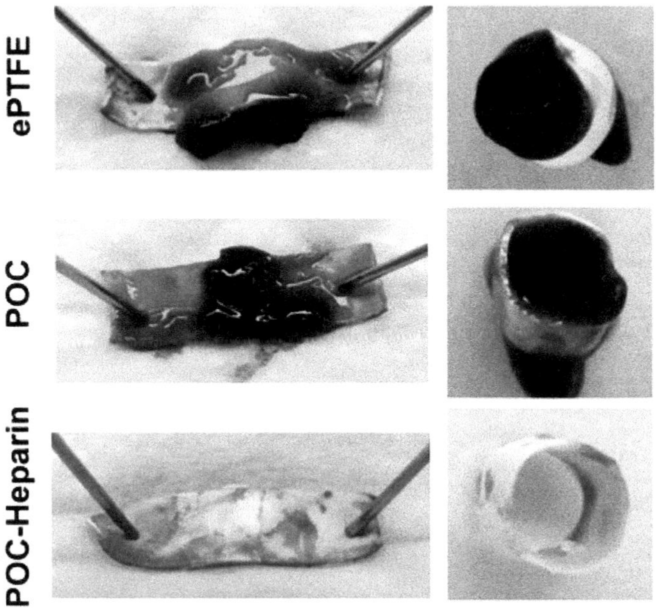

FIGURE 3.4 A visual representation of the clot formation on the basic untreated vascular graft segments of (ePTFE and POC) vs the heparin integrated vascular graft (Heparin–POC), after 28 days of incubation in-vitro in PBS (Hoshi et al., 2013).

Willebrand factor (vWF) and vascular endothelial-cadherin (VE-cadherin) (Hoshi et al., 2013). The importance of this work is that this advancement can be adapted to fit many other medical devices that have similar issues, such as stents and catheters. Many other advancements are also being made across the elastomer field and are seeing similarly promising results to the studies mentioned above.

3.7 HYDROGELS

Hydrogels are a class of polymers that consists of polymers made up of hydrophilic chains. Hydrogels are highly absorbent and can consist of natural or synthetic polymer chains (Policarpo et al., 2011). Some hydrogels are excellent in absorbing water and have the ability to contain over 90% water (Policarpo et al., 2011). Hydrogels are important for the field of tissue engineering because they can be applied to a wide variety of biological uses due to their natural flexibility and their high-water content (Mellati et al., 2014). Currently, hydrogels are used for cell culture applications, as smart gels to sense pH or temperature changes, as drug release systems, biosensors, contact lenses, and even breast implants (Discher et al., 2005).

Hydrogels are also being experimented with for tissue engineering applications and are being used as scaffolds to direct and enhance cell growth of tissue systems (Drury and Mooney, 2003). Hydrogels show promise as a scaffold material because they can be delivered in a minimally invasive way, be processed under relatively mild conditions, and are structurally similar to the extracellular matrix of many native tissue types (Drury and Mooney, 2003). There have been many recent advancements in this field, and it is leading to the creation of more tailored and functional scaffolds. Researchers from the Department of Material Science from the University of Illinois were able to create functional 3D scaffolds that could direct and enhance cell growth (Barry et al., 2009). These scientists used a direct ink writing process, which is a layer-by-layer technique that allowed them to create a scaffold that is patterned in both planar and 3D form (Barry et al., 2009). Mixing acrylamide, glycerol, and water created the ink that they used. The polymer was then left to age for several hours to yield a gel composed of 30 w/o polyacrylamide chains. Using a crosslinking agent, N, N methylene bisacrylamide, and diethoxyacetonephenone as a photo-indicator further optimized the ink. The optimized gel was then printed to create a 3D structure that is shown in Figure 3.5 (Barry et al., 2009).

The created 3D structures were tested for cytocompatibility by being seeded with fibroblasts. This test was also used to determine if the created scaffolds might be viable for tissue engineering applications. What they noticed was that their scaffolds were able to uptake the fibroblast well and direct them to fill the complete structure of the scaffold (Barry et al., 2009). The fibroblasts were also able to compartmentalize themselves within the scaffold and take on a similar orientation to what is found in native tissue. Figure 3.6 is a visual representation of fibroblasts filling the created hydrogel scaffold (Barry et al., 2009).

The importance of this work is that it is helping progress current hydrogel technologies so that they can be used to create more advanced scaffolds for several different tissue-engineering applications. Hydrogels are not only being used for scaffolds; there is also a lot of emphasis being placed on them as drug delivery devices. The reasons why hydrogels are excellent polymers for drug delivery devices are because they can have a sustained protein release and are capable of targeted drug delivery (Ashley et al., 2013). Hydrogels also have excellent biocompatibility and can be integrated into biological systems easily (Jiang et al., 2014). Hydrogels are

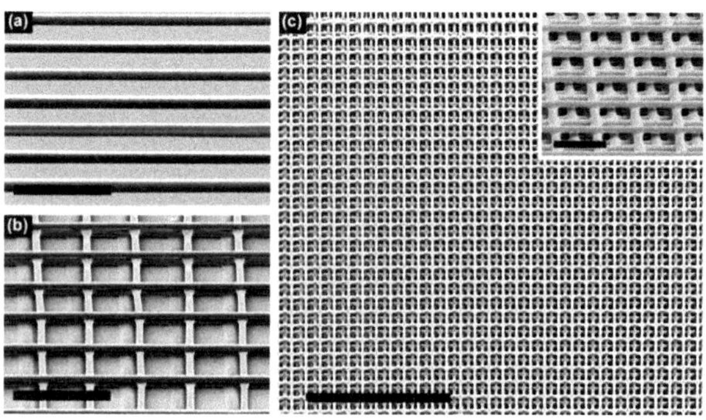

FIGURE 3.5 (a) SEM image of the 1D scaffold composed of 5 μm filaments that are spaced 20 μm apart. (b) 3D hydrogel scaffold that is composed of four of the 1D layers seen in (a). (c) a 3D hydrogel scaffold that is created by six of the 1D layers that are seen in (a). Scale bars are 50 μm and 6 μm respectively (Barry et al., 2009).

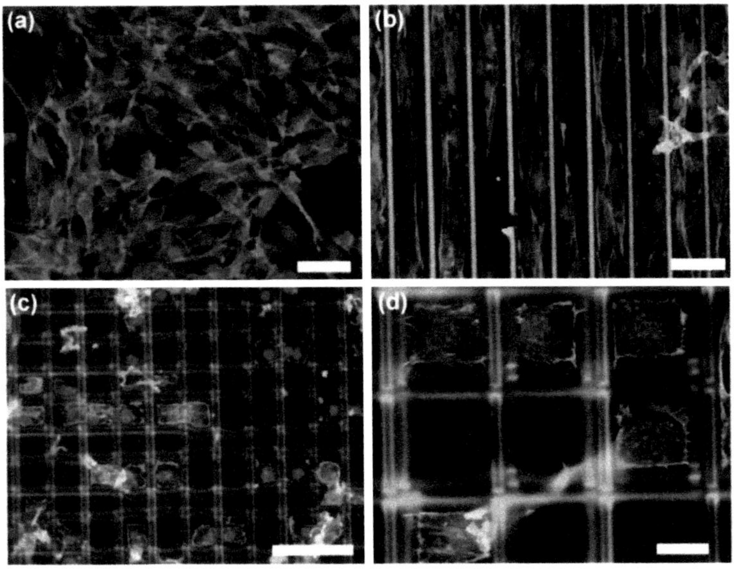

FIGURE 3.6 Optical fluorescence microscopy images of fibroblasts plated on (a) flat glass control. (b) A 1D hydrogel scaffold. (c,d) 3D hydrogel scaffold consisting of four layers. (d) A closer image of the scaffold found in (c) to exhibit the compartmentalization of the fibroblasts in the 3D hydrogel scaffold. The scale bars are 100 μm (a–c) and 2 μm (d) respectively (Barry et al., 2009).

appealing because their platform is versatile and can be tuned in many ways to tailor the hydrogel to the intended application (He et al., 2008). Some tuneable aspects of a hydrogel can include pore size, absorbability, and their rate of drug release (Jiang et al., 2014). Current advancements are working on tailoring hydrogels to have a constant controlled drug delivery under biological conditions in order to reduce the chances of misdosing (Rosenthal et al., 2003). Other advancements also include the creation of hydrogels that can respond to certain stimuli and release proteins or drugs when triggered (Hoare and Kohane, 2008). These are just a few examples of progressions happening across the hydrogel field. In total, the hydrogel field is a large part of the polymeric biomaterial family and is progressing due to all the applications that this polymer can be used for (Nochi et al., 2010).

3.8 POLYELECTROLYTES

Polyelectrolytes are a special group of polymers that consist of repeating monomer units that have an electrolyte group (Jenkins et al., 1996). Examples of electrolyte groups are seen in Figure 3.7 (Kebes, 2005).

When polyelectrolytes are in aqueous solution their electrolyte groups dissociate and leave the polymers charged (Podgornik and Ličer, 2006). Polyelectrolytes are sometimes called polysalts. The reason why they are called polysalts is that they have properties of both polymers and salts (Podgornik and Ličer, 2006). Since they have

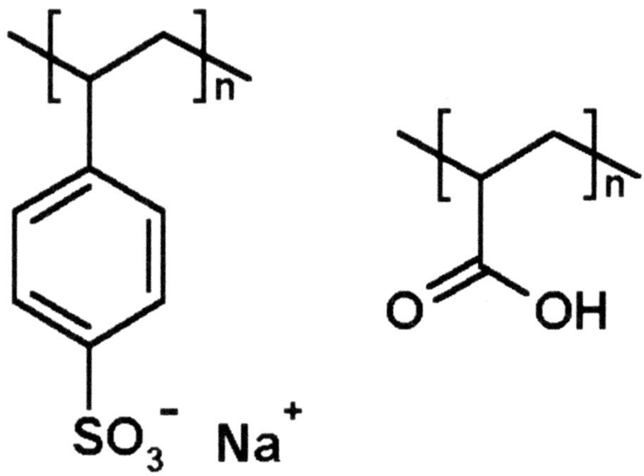

FIGURE 3.7 Chemical structure of two polyelectrolytes. The right structure is polyacrylic acid (PAA) and the structure on the left is poly(sodium styrene sulfonate) (PSS). Both polyelectrolytes are negatively charged (Kebes, 2005).

salt like properties, they are generally electroconductive in aqueous solutions, while their polymeric properties make them viscous and give them a high molecular weight. Many biological molecules can be considered polyelectrolytes such as polypeptides and DNA (Tatykhanova et al., 2012). The conformation of polyelectrolytes is affected by many factors, such as solvent affinity, polymer architecture, and charge. Generally, when in solution, polyelectrolytes adopt an expanded, rod-like configuration due to the interactions of the charges on the monomer units (Tatykhanova et al., 2012). This is much different from the random conformation that uncharged linear polymers take on once in solution. The conformation that a polyelectrolyte takes can greatly affect its bulk properties, such as turbidity and viscosity (Tatykhanova et al., 2012).

There are many applications for polyelectrolytes. They can be used to modify the flow and stability of solutions and gels (Tatykhanova et al., 2012). They can also be used as a means of imparting a charge on naturally uncharged surfaces. Some polyelectrolytes are also highly biocompatible and allow for cellular uptake (Berillo et al., 2012). A lot of current research is focusing on using polyelectrolytes to functionalize the surface of implantable medical devices. Functionalizing the surface of implants helps with their integration in the body and reduces their chances of being rejected (Berillo et al., 2012). Polyelectrolytes are also being explored for other purposes such as protecting implants from microbial infections. Bacteria adhering to the surface of implantable materials are the first step of infection (Boulmedais et al., 2004). In order to reduce this, some researchers developed a multilayer film of PEGylated polypeptides to cover the implantable device. This reduces the chance of bacterial infection by blocking the adhesion of bacteria and the adsorption of proteins to the implant (Boulmedais et al., 2004). The functionalization of medical devices with polyelectrolytes is currently a major factor driving the development of this segment of polymeric materials, and many new advancements are being made and integrated into medical practices such as the examples mentioned above.

3.9 NATURAL POLYMERS

Natural polymers are just a subgroup of polymers that originate from natural sources. Natural polymers can also fit into many of the groups above but fall into this category if they are naturally sourced (Kim et al., 2010). Examples of natural polymers include starch, cotton, and natural rubber. Many polymers used today are natural polymers that are chemically modified to enhance or change certain characteristics of the polymer (Kim et al., 2010). An example of this is the reaction of nitric acid with cellulose to form nitrocellulose and vulcanized rubber shown in Figure 3.8 (Firey Deity, 2017).

A lot of recent research in the field of tissue engineering is focusing on the creation of 3D scaffolds that can be used as a template for cellular repair, and for the regeneration of damaged and lost tissue (Ko et al., 2010). There are many biomaterials that can be used to create these scaffolds, but it is generally tough to create scaffolds that are biocompatible, have excellent cell surface chemistry, and have suitable mechanical and physical properties. These issues are bringing more attention to naturally derived polymers that can be used for 3D scaffolds since many already exhibit those properties (Ko et al., 2010). Natural polymers are also generally biocompatible, easy to process, and are highly cost-effective. These favourable attributes are driving researchers to experiment with natural polymers and integrate them for newer

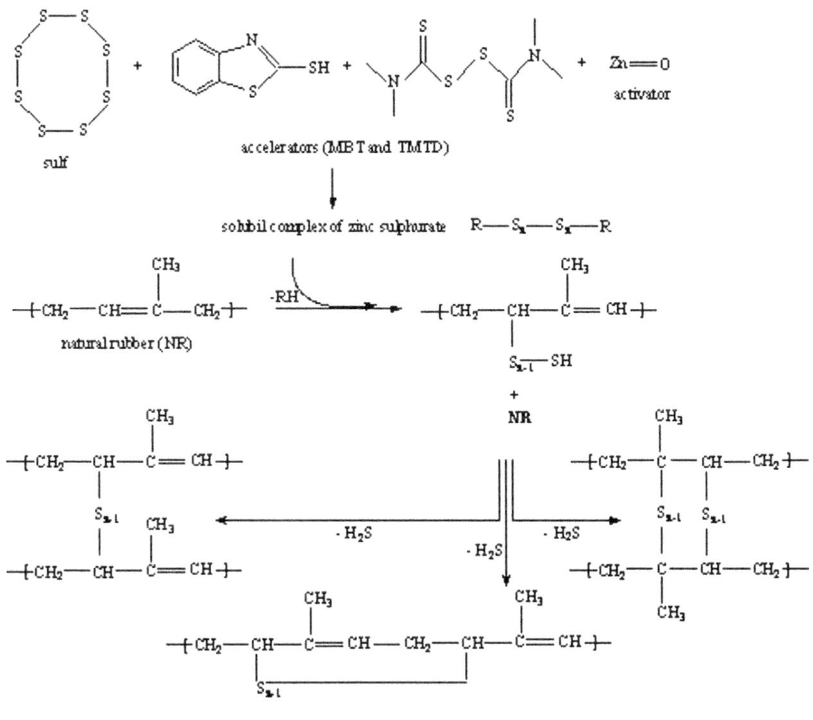

FIGURE 3.8 General reaction of natural rubber being changed into vulcanized rubber (Firey Deity, 2017).

uses. Natural polymers that are currently being used are cellulose, chitosan, alginate, and agarose (Ko et al., 2010). Chitin is finding many uses due to its ability to be processed into many different forms. Chitin can be processed to also have various properties ranging from being water soluble to being soluble in organic solvents, as shown below in Figure 3.9 (Wan and Tai, 2013).

Chitin is the second most abundant natural polymer after cellulose; it is cheap and easy to obtain (Wan and Tai, 2013). Chitin is being researched for use as a polymer for tissue engineering scaffolds. Chitin is biocompatible and has a low immunogenicity, which makes it an excellent material to integrate into biological systems (Wan and Tai, 2013). There has been a lot of success around using chitin as a scaffold for cartilage tissue regeneration, and many other studies are being conducted on modified chitin scaffolds for other forms of tissue (Suzuki et al., 2008). Cellulose is also being experimented with as a natural biomaterial for tissue engineering. Like chitin, cellulose is biocompatible and easy to obtain. Cellulose also has many desirable properties such as a high water-holding capacity, a high tensile strength, a fine fiber network, and a high level of crystallinity (Dugan et al., 2013). An excellent example of cellulose being used as a biomaterial is when it is milled into cellulose nanowhiskers to be used in scaffolds for tissue engineering purposes. The porosity, strength, and biocompatibility allow for the easy uptake of cells and integration into biological systems (Dugan et al., 2013). These are just a few examples of natural polymers being used today as biomaterials, and interest will remain in this field as long as they remain easily accessible and cheap to use (Dugan et al., 2013).

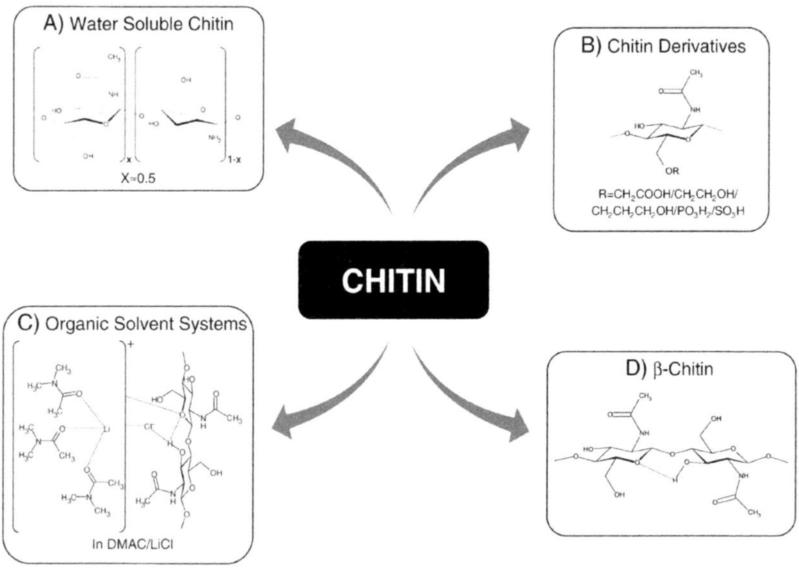

FIGURE 3.9 Examples of how chitin can be processed into polymers with different properties or structural configurations (Wan and Tai, 2013).

3.10 BIODEGRADABLE POLYMERS

Biodegradable polymers are a large class of polymers that encompasses any polymer that degrades over time after an intended purpose (Bastioli, 2005). Biodegradable polymers can originate from natural sources or can be synthetically made (Bastioli, 2005). Generally, biodegradable polymers consist of ester, amide, and ether functional groups (Vroman and Tighzert, 2009). Most biodegradable polymers are created via condensation reactions, metal catalysts, and via ring-opening polymerizations. Their rate of degradation and properties can vary greatly and are generally dictated by their structure and the type of environment that they are in (Vroman and Tighzert, 2009).

Currently, much emphasis is being placed on degradable biomaterials because of the promise they hold in advancing several different medical processes. The two main medical applications for which biodegradable polymers are being sought after are the development of biodegradable scaffolds for tissue engineering and for biodegradable drug delivery devices (Kumar et al., 2011). Much emphasis is being placed on creating biodegradable scaffolds because they can be used to help regenerate tissue and structures without having to be removed after the process (Ratner et al., 2004). The ideal properties of good biodegradable scaffolds are that they need to be non-toxic and biocompatible, during implantation and after they are broken down (Lendlein and Sisson, 2011). A good scaffold also needs to be able to maintain its structural integrity as it degrades so that it will be able to support the tissue even after it is being broken down (Amass et al., 1998). Another important property of an ideal biodegradable scaffold is that it needs to have a degradation rate that matches the regeneration rate of the tissue around it (Lendlein and Sisson, 2011). This ensures that it is properly achieving its task of supporting the surrounding tissue and promoting growth as it is being broken down. There is a lot of work that has been done to create biodegradable scaffolds with these properties, and several researchers have come very close to creating efficient models (Amass et al., 1998). Excellent examples are the biodegradable nanofiber scaffold created by researchers from Harvard Medical School (Yoshimoto et al., 2003) and the 3D printed polymer scaffolds created by researchers from the University of Texas (Trachtenberg et al., 2014). The scaffold created by the researchers from Harvard used a special electrospinning process that created microporous, non-woven poly(e-caprolactone) scaffolds (Yoshimoto et al., 2003). Their process allowed them to form fibers that had diameters in the nanometer range by subjecting a fluid jet to an electric field. Their scaffold was then seeded with MSC and was left to culture in media with MSC. What they noticed was that their scaffold was effective in housing the MSC and mineralization was apparent after 4 weeks of culture (Yoshimoto et al., 2003). This advancement is important because these results show that the created scaffold could be a good candidate to use for bone tissue engineering. The researchers from the University of Texas took a different approach and created a system that could be used to develop 3D scaffolds via a cost-effective printing method (Trachtenberg et al., 2014). These researchers also used a type of poly(e-caprolactone) scaffold and paired it with an open-source 3D printing system. The results of this study showed that the pore size, fiber diameter, fiber spacing, and porosity could all be controlled via the basic 3D printing process and that this process could be used to create viable 3D scaffolds easily (Trachtenberg et al., 2014). This work is important because it represents the progressions in the field of biodegradable polymers and exemplifies how they are becoming more effective and easier than ever to use.

Another major interest of biodegradable polymers is for their use in drug delivery systems. An ideal polymer drug delivery system is one that can be incorporated into the body easily, has a controlled rate of drug release, and can be broken down without being toxic or eliciting negative effects on the body (Ratner et al., 2004). A lot of work is being done to create these degradable drug delivery systems and many researchers have created promising systems. An excellent example of this is the nanocomposite hydrogel created by researchers from the University of Manchester in the United Kingdom (Ratner et al., 2004). The drug delivery system created by these researchers incorporated a polymer scaffold, the target drug, and nanoparticles to create a biodegradable drug delivery system that could deliver the drug on demand. This multidimensional approach has been tested for cancer therapy and other injectables and has shown promising results and better means of drug delivery than the conventional approaches (Ratner et al., 2004). Advancements like this are the future of drug delivery since they provide better control of the drug delivery and can be used to integrate other properties that might have other therapeutic effects in the body, such as what was done above with the incorporated nanoparticles in the scaffold.

Not all biodegradable polymers are used for the creation of scaffolds or for drug delivery systems. They are a highly diverse group and are commonly used for many other purposes. These uses involve degradable stitches that melt over time, or even certain types of bone cements that support bone as it is healing (Lendlein and Sisson, 2011). In the end, the field of biodegradable polymers is one of the largest and fastest developing subgroups of the polymeric biomaterial field and will only grow in future due to all the interest being placed on the advancement of these polymers.

3.11 CONCLUSION

Polymeric biomaterials are a massive family of materials that exhibit a wide range of properties. This family encompasses any type of polymer ranging from hydrogels to thermoplastics. The immense variety of polymeric materials makes this family very versatile and allows them to be used for many different applications in the biomaterial field, ranging from implants and scaffolds to drug delivery devices. The field of polymeric biomaterials is also a major player in the biomaterial field and is currently the fastest growing segment (Rohan, 2017). A lot of this growth is stemming from interest in biodegradable polymers, and from trying to replace conventional biomaterials with newer polymers that are generally cheaper and more efficient than their conventional counterparts (Griffith, 2000). The recent interest in the polymeric biomaterial field is causing this field to advance quickly, creating more sophisticated polymer materials than ever before. This can be seen in many aspects of the field but is especially apparent with the new types of scaffolds and smart drug delivery systems that are currently being created. In the end, polymeric biomaterials are a highly diverse field that is expanding quickly due to its versatility and excellent properties as a biomaterial. These materials are a major player in the biomaterial field and are only set to grow bigger over the next few years.

REFERENCES

Abdullateef, A. A., Thomas, S. P., Al-Harthi, M. A., De, S. K., Bandyopadhyay, S., Basfar, A. A., and Atieh, M. A. (2012). Natural rubber nanocomposites with functionalized carbon nanotubes: mechanical, dynamic mechanical, and morphology studies. *Journal of Applied Polymer Science*, 125(S1), E76–E84.

Alemán, J. V., Chadwick, A. V., He, J., Hess, M., Horie, K., Jones, R. G., and Penczek, S. (2007). Definitions of terms relating to the structure and processing of sols, gels, networks, and inorganic-organic hybrid materials (IUPAC Recommendations 2007). *Pure and Applied Chemistry*, 79(10), 1801–1829.

Alger, M. (1996). *Polymer Science Dictionary*. Springer Science and Business Media.

Amass, W., Amass, A., and Tighe, B. (1998). A review of biodegradable polymers: uses current developments in the synthesis and characterization of biodegradable polyesters, blends of biodegradable polymers and recent advances in biodegradation studies. *Polymer International*, 47(2), 89–144.

Ashley, G. W., Henise, J., Reid, R., and Santi, D. V. (2013). Hydrogel drug delivery system with predictable and tunable drug release and degradation rates. *Proceedings of the National Academy of Sciences*, 110(6), 2318–2323.

Baeurle, S. A., Hotta, A., and Gusev, A. A. (2006). On the glassy state of multiphase and pure polymer materials. *Polymer*, 47(17), 6243–6253.

Barry, R. A., Shepherd, R. F., Hanson, J. N., Nuzzo, R. G., Wiltzius, P., and Lewis, J. A. (2009). Direct-write assembly of 3D hydrogel scaffolds for guided cell growth. *Advanced Materials*, 21(23), 2407–2410.

Bastioli, C. (Ed.). (2005). *Handbook of Biodegradable Polymers*. iSmithers Rapra Publishing.

Berillo, D., Elowsson, L., and Kirsebom, H. (2012). Oxidized dextran as crosslinker for chitosan cryogel scaffolds and formation of polyelectrolyte complexes between chitosan and gelatin. *Macromolecular Bioscience*, 12(8), 1090–1099.

Boulmedais, F., Frisch, B., Etienne, O., Lavalle, P., Picart, C., Ogier, J., and Egles, C. (2004). Polyelectrolyte multilayer films with pegylated polypeptides as a new type of antimicrobial protection for biomaterials. *Biomaterials*, 25(11), 2003–2011.

Brown, S. A., Hastings, R. S., Mason, J. J., and Moet, A. (1990). Characterization of short-fiber reinforced thermoplastics for fracture fixation devices. *Biomaterials*, 11(8), 541–547.

Cameron, S. S. (2010). *Polymer picture*. [Online Image]. Retreived from https://commons.wikimedia.org/wiki/File:Polymer_picture.svg.

Chen, Q., Liang, S., and Thouas, G. A. (2013). Elastomeric biomaterials for tissue engineering. *Progress in Polymer Science*, 38(3), 584–671.

Discher, D. E., Janmey, P., and Wang, Y. L. (2005). Tissue cells feel and respond to the stiffness of their substrate. *Science*, 310(5751), 1139–1143.

Dodiuk, H., and Goodman, S. H. (Eds.). (2013). *Handbook of Thermoset Plastics*. William Andrew.

Drury, J. L., and Mooney, D. J. (2003). Hydrogels for tissue engineering: scaffold design variables and applications. *Biomaterials*, 24(24), 4337–4351.

Dugan, J. M., Gough, J. E., and Eichhorn, S. J. (2013). Bacterial cellulose scaffolds and cellulose nanowhiskers for tissue engineering. *Nanomedicine*, 8(2), 287–298.

Fink, J. K. (2013). *Reactive Polymers Fundamentals and Applications: A Concise Guide to Industrial Polymers*. William Andrew.

Firey Deity. (2017). *Vulcanisation skeletal*. [Online Image]. Retrieved from https://commons.wikimedia.org/wiki/File:Vulcanisation_skeletal.svg.

Griffith, L. G. (2000). Polymeric biomaterials. *Acta Materialia*, 48(1), 263–277.

Gupta, R. K. (2000). *Polymer and Composite Rheology*. CRC Press.

Halpern, J. M., Urbanski, R., Weinstock, A. K., Iwig, D. F., Mathers, R. T., and Recum, H. A. (2014). A biodegradable thermoset polymer made by esterification of citric acid and glycerol. *Journal of Biomedical Materials Research Part A*, 102(5), 1467–1477.

He, C., Kim, S. W., and Lee, D. S. (2008). In situ gelling stimuli-sensitive block copolymer hydrogels for drug delivery. *Journal of Controlled Release*, 127(3), 189–207.

Hoare, T. R., and Kohane, D. S. (2008). Hydrogels in drug delivery: progress and challenges. *Polymer*, 49(8), 1993–2007.

Hong, Y., Guan, J., Fujimoto, K. L., Hashizume, R., Pelinescu, A. L., and Wagner, W. R. (2010). Tailoring the degradation kinetics of poly(ester carbonate urethane) urea thermoplastic elastomers for tissue engineering scaffolds. *Biomaterials*, 31(15), 4249–4258.

Hoshi, R. A., Van Lith, R., Jen, M. C., Allen, J. B., Lapidos, K. A., and Ameer, G. (2013). The blood and vascular cell compatibility of heparin-modified ePTFE vascular grafts. *Biomaterials*, 34(1), 30–41.

Jenkins, A. D., Kratochvíl, P., Stepto, R. F. T., and Suter, U. W. (1996). Glossary of basic terms in polymer science (IUPAC Recommendations 1996). *Pure and Applied Chemistry*, 68(12), 2287–2311.

Jia, Z., Wang, Z., Xu, C., Liang, J., Wei, B., Wu, D., and Zhu, S. (1999). Study on poly(methyl methacrylate)/carbon nanotube composites. *Materials Science and Engineering: A*, 271(1), 395–400.

Jiang, Y., Chen, J., Deng, C., Suuronen, E. J., and Zhong, Z. (2014). Click hydrogels microgels and nanogels: emerging platforms for drug delivery and tissue engineering. *Biomaterials*, 35(18), 4969–4985.

Jolly, M. R., Carlson, J. D., and Munoz, B. C. (1996). A model of the behaviour of magnetorheological materials. *Smart Materials and Structures*, 5(5), 607.

Kebes. (2005). *Polyelectrolyte examples*. [Online Image]. Retrieved from https://commons.wikimedia.org/wiki/File:Polyelectrolyte_examples.png.

Kim, J. Y., Xin, X., Moioli, E. K., Chung, J., Lee, C. H., Chen, M., and Mao, J. J. (2010). Regeneration of dental-pulp-like tissue by chemotaxis-induced cell homing. *Tissue Engineering Part A*, 16(10), 3023–3031.

Ko, H. F., Sfeir, C., and Kumta, P. N. (2010). Novel synthesis strategies for natural polymer and composite biomaterials as potential scaffolds for tissue engineering. *Philosophical Transactions of the Royal Society of London A: Mathematical Physical and Engineering Sciences*, 368(1917), 1981–1997.

Kumar, A. A., Karthick, K., and Arumugam, K. P. (2011). Properties of biodegradable polymers and degradation for sustainable development. *International Journal of Chemical Engineering and Applications*, 2(3), 164.

Langer, R., and Vacanti, J. P. (1993)Tissue engineering. *Science*, 260(5110), 920–926.

Lanza, R., Langer, R., and Vacanti, J. P. (Eds.). (2011). *Principles of Tissue Engineering*. Academic Press.

Lendlein, A., and Sisson, A. (2011). *Handbook of Biodegradable Polymers: Isolation, Synthesis, Characterization and Applications*. John Wiley and Sons.

Lonescu, M. (2005). *Chemistry and Technology of Polyols for Polyurethanes*. iSmithers Rapra Publishing.

MacDermott, C. P., and Shenoy, A. V. (1997). *Selecting Thermoplastics for Engineering Applications*. CRC Press.

Mellati, A., Dai, S., Bi, J., Jin, B., and Zhang, H. (2014). A biodegradable thermosensitive hydrogel with tuneable properties for mimicking three-dimensional microenvironments of stem cells. *RSC Advances*, 4(109), 63951–63961.

Montarnal, D., Capelot, M., Tournilhac, F., and Leibler, L. (2011). Silica-like malleable materials from permanent organic networks. *Science*, 334(6058), 965–968.
Moukwa, M. (1997). The development of polymer-based biomaterials since the 1920s. *JOM Journal of the Minerals, Metals and Materials Society*, 49(2), 46–50.
Nochi, T., Yuki, Y., Takahashi, H., Sawada, S. I., Mejima, M., Kohda, T., and Tokuhara, D. (2010). Nanogel antigenic protein-delivery system for adjuvant-free intranasal vaccines. *Nature Materials*, 9(7), 572–578.
Partlow, B. P., Hanna, C. W., Rnjak-Kovacina, J., Moreau, J. E., Applegate, M. B., Burke, K. A., and Kaplan, D. L. (2014). Highly tunable elastomeric silk biomaterials. *Advanced Functional Materials*, 24(29), 4615–4624.
Podgornik, R., and Ličer, M. (2006). Polyelectrolyte bridging interactions between charged macromolecules. *Current Opinion in Colloid and Interface Science*, 11(5), 273–279.
Policarpo, H., Neves, M. M., and Maia, N. M. M. (2011). On the determination of storage and loss Moduli for cork composition materials. In *Proceedings of ICEDyn2011–International Conference on Structural Engineering Dynamics*, Tavira, Portugal.
Ratner, B. D., Hoffman, A. S., Schoen, F. J., and Lemons, J. E. (2004). *Biomaterials Science: An Introduction to Materials in Medicine*. Academic Press.
Ratner, B. D., Hoffman, A. S., Schoen, F. J., and Lemons, J. E. (2004). *Biomaterials Science: An Introduction to Materials in Medicine*. Academic Press.
Rohan (2017). Biomaterials Market worth USD 149.17 Billion by 2021. Retrieved from http://www.marketsandmarkets.com/PressReleases/global-biomaterials.asp
Rosenthal, A., Barry, J. J., and Sahatjian, R. (2003). U.S. Patent No. 6,524,274. U.S. Patent and Trademark Office.
Shenoy, A. V., and Saini, D. R. (1996). *Thermoplastic Melt Rheology and Processing*. CRC Press.
Suzuki, D., Takahashi, M., Abe, M., Sarukawa, J., Tamura, H., Tokura, S., and Nagano, A. (2008). Comparison of various mixtures of β-chitin and chitosan as a scaffold for three-dimensional culture of rabbit chondrocytes. *Journal of Materials Science: Materials in Medicine*, 19(3), 1307–1315.
Tatykhanova, G. S., Sadakbayeva, Z. K., Berillo, D., Galaev, I., Abdullin, K. A., Adilov, Z., and Kudaibergenov, S. E. (2012). Metal complexes of amphoteric cryogels based on allylamine and methacrylic acid. In *Macromolecular Symposia*, 317(1), 18–27. Wiley-VCH Verlag.
Trachtenberg, J. E., Mountziaris, P. M., Miller, J. S., Wettergreen, M., Kasper, F. K., and Mikos, A. G. (2014). Open-source three-dimensional printing of biodegradable polymer scaffolds for tissue engineering. *Journal of Biomedical Materials Research Part A*, 102(12), 4326–4335.
Uhthoff, H. K., Bardos, D. I., and Liskova-Kiar, M. (1981). The advantages of titanium alloy over stainless steel plates for the internal fixation of fractures. An experimental study in dogs. *Bone & Joint Journal*, 63(3), 427–484.
Vroman, I., and Tighzert, L. (2009). Biodegradable polymers. *Materials*, 2(2), 307–344.
Wan, A. C., and Tai, B. C. (2013). CHITIN—A promising biomaterial for tissue engineering and stem cell technologies. *Biotechnology Advances*, 31(8), 1776–1785.
Yoshimoto, H., Shin, Y. M., Terai, H., and Vacanti, J. P. (2003). A biodegradable nanofiber scaffold by electrospinning and its potential for bone tissue engineering. *Biomaterials*, 24(12), 2077–2082.

4 Ceramic Based Biomaterials

4.1 INTRODUCTION OF BIOCERAMICS

Ceramic biomaterials, also called bioceramics, are a class of materials used for the application of repairing or replacing damaged parts of the musculoskeletal system (Best et al., 2008). Ceramics are refractory polycrystalline compounds that are inorganic, hard, brittle, and have high compressive strengths. They also have many advantages that make them great biomaterials, such as having a high resistance against temperature, hard environments, and wear, as well as being able to be highly polished on surfaces (Oh et al., 2006). Some disadvantages of bioceramics are their difficult fabrication and reshaping process. Some bioceramics also have the disadvantageous characteristics of low tensile strength, brittleness, and low fracture toughness (Dubok, 2000).

Bioceramics can be classified into three groups: bioinert ceramics, bioactive ceramics, and biodegradable ceramics (Best et al., 2008). The first group of ceramics are materials that are stable and inert in the body, these are ceramics such as alumina and zirconia. Bioinert materials do not react or elicit a response in the body after being introduced, making them excellent biocompatible materials to use for short and long-term applications (Kokubo, 2008). Bioceramics such as bioglass have the capability of being bioactive; this means that the bioceramics can release ions such as Ca^{++}, K^{+} and, Na^{+} into the environment resulting in a direct physiological bonding of tissue to the material. This bonding allows for the fixation of implants and for better osteointegration (Tilocca, 2010). Lastly, some bioceramics are biodegradable, this means the material will slowly dissolve in the body. This will allow for endogenous tissue to proliferate and slowly replace the degradable material, it is beneficial since a revision surgery can be avoided (Lin et al. 2005). Table 4.1 is a summary of different bioceramics and their characteristics.

Bioceramics have many different applications in the biomedical field. These ceramics have been used as bone plates, bone screws, and femoral head implants (Dubok, 2000). They are also commonly used for dental applications such as for tooth implants and maxillofacial reconstruction. Another use of bioceramics is surface coatings for joint replacements; this helps increase wear resistance and decrease the chance of inflammation in the body (Best et al., 2008). Outside-the-body applications of bioceramics are devices like pacemakers, respirators, and kidney dialysis machines (Oh et al., 2006).

by Yaser Dahman and Daniela Herrera

TABLE 4.1
Bioceramics and Some Characteristics

Bioceramics	Characteristics
Alumina	Bioinert
Zirconia	Bioinert – The steel of bioceramics
Pyrolytic Carbon	Bioinert – Thromboresistance: The ability to resist blood clotting. This type of bioceramic is most commonly used in artificial heart valves
Bioglass	Bioactive
Hydroxyapatite (sintered at high temperature)	Bioactive
Tricalcium Phosphate	Biodegradable with a rapid degradation rate
Hydroxyapatite (sintered at low temperature)	Biodegradable with a slow degradation rate

4.2 ALUMINA

4.2.1 History of Alumina

Alumina (Al_2O_3) is an oxide that can be naturally found as a mineral, diaspore, gibbsite, and bauxite. It was first synthetically created in the 1700s from bauxite, a naturally found ore that contains variable amounts of hydrous aluminum oxides (Davis, 2010). Marggraf first isolated alumina in 1754, extracting it from natural clay using sulphuric acid. The substance was later coined "alumina" by Guyton de Morveau in 1761. The first application of alumina was for the use of dental and jaw implants (Rock, 1933). In 1972, Boutin wrote a paper about the unique mechanical and thermal properties of alumina, this led to an increase in popularity of alumina and to the creation of femoral heads using alumina biomaterials for hip implants (Davis, 2010).

4.2.2 Production of Alumina

One popular method of creating alumina is through the Bayer process, which was developed in 1888 by an Austrian in the aluminum industry named Bayer. In this process, contact surface is increased by grinding the bauxite to form smaller grains with a diameter of <30 mm. Lime (CaO) and soda (NaOH) are then mixed into the bauxite grains and pumped into high-pressure containers where they are heated for several hours. This step allows the alumina in the bauxite to get dissolved into the caustic soda and become separated from the mix through precipitation, leaving behind wet alumina hydrate ($Al_2O_3.3H_2O$). Calcination of the wet hydrate allows for alumina formation; this is done in inclined rotary kilns. The high vapour release of 250–400°C fluidizes the alumina, which is then heated at a temperature of 1000–1250°C for one hour for the gibbsite to be completely transformed into alumina. Mineralizers are used in this stage to catalyse the transformation. Soda is the main impurity of this mix; it is removed by hydrate washing using silica (Year, 1995).

The result of this process is calcined alumina, which is found in the form of alpha alumina crystallite agglomerates, which have a varying size, from 0.5 to 10 μm. The conditions such as the amount of soda added, temperature, technique, and other factors of the process determine the particle size, impurities, and calcine alumina produced (Year, 1995). Overall, this process can create three types of alumina; activated alumina, which is used as substrate catalysts, smelter-grade alumina, to create aluminum metal, and lastly, calcined alumina, which is used to create bioceramic products and is the one of interest for biomaterial applications such as bone and dental implants (Davis, 2010). Ever since the discovery of alumina, engineers have worked to improve its compressive strength and fracture toughness by decreasing the grain size all the way down to nanoscale level or by mixing alumina with strong materials such as zirconia particles to increase toughness (Veljović et al., 2010).

4.2.3 Characteristics of Alumina

Alumina has advantageous characteristics that make it a great biomaterial for in-vivo and in-vitro applications. Alumina has low electrical conductivity, high strength, corrosion, and temperature resistance in the body, with a high melting point of approximately 2050°C (Rieger, 2001). Alumina's surface energy and smoothness allow for high wear and friction resistance. This bioceramic can also be polished to high surface finish, making it a great candidate for sprayed bioceramic coatings (Landers et al., 2004). The structure of alumina has two sites, a hexagonal, and an octahedral where it holds its atoms. The hexagonal sites are the corner atoms, while the octahedral site is found in the middle. The octahedral site is made up of 1/3 oxygen anions and 2/3 aluminum cations. The oxygen anions allow for strong bonding and give rise to the strong properties of alumina (Levin and Brandon, 2009). These characteristics, as well as alumina's inertness and insolubility in the presence of chemical reagents, make it a great candidate for the biomedical field (Rieger, 2001).

Alumina can be combined with other bioceramics such as zirconia to improve properties of the biomaterial. Al_2O_3/ZrO_2 microcomposites with monolithic alumina of the same grain size have been used to improve the mechanical and tribological properties of bioceramics for future biomedical applications (Bartolome et al., 2007). This combination allows the wear rate to increase as contact loads increase; it also displays a clear transition in friction and wear behaviour. Overall, combining zirconia and alumina results in a higher sliding wear resistance in terms of fracture resistance properties and residual stress compared to just using monolithic alumina alone (Bartolome et al., 2007).

4.2.4 Current Applications of Alumina

Alumina bioceramics have been used in many different medical applications. One of the applications of alumina that has been looked at for many years is the creation of prosthetic implants. Alumina–alumina combination bioceramics have been used to create hip prostheses since the 1970s, and these materials have shown to be a highly successful treatment for the application of total hip replacements (Hatton et al., 2002). Boutin et al. found that to have a successful implant geometry, the sphericity deviation should be ~ ±1 μm with a radius tolerance between components of 7–10 μm. They also suggest that to have a high-performance alumina microstructure, the implant should have a small grain size, uniform distribution, and minimal porosity. The long-term behaviour of the prosthesis depends on stability and the initial positioning of the implant. Alumina–alumina hip prostheses have been shown to have 88% survival probability after 8 years for patients over the age of 50, and a 100% survival probability after 8 years for patients under the age of 50 (Boutin et al., 2000). Not only is alumina used as a total hip prosthesis, but it has also been used for total knee, shoulder, elbow, and wrist implants (Davis, 2010).

Porous alumina scaffolds can be used for the reconstruction/replacement of damaged body parts. Indirect fused depositing modelling (FDM), which is a rapid prototyping technique, can be used to create a bioinert porous alumina material on a scaffold (Bose et al., 2002). Hydroxyapatite (Hap) can be used to coat the alumina to increase biocompatibility since it has many characteristics that mimic bone. Bose et al. used two models, a porous alumina scaffold and a porous alumina scaffold coated with HAp. They then tested the models in-vitro with rat pituitary tumour cells (PR1) to compare effectiveness between the scaffolds. The results showed that both scaffolds provided a non-toxic surface that allowed for cell attachment and proliferation of PR1 cells. When the efficacy of both scaffolds was compared (Figure 4.1), the researchers found that the alumina-HAp scaffold allowed for higher proliferation of the cells compared to the alumina scaffold by itself. This was due to an advantageous combination between the bioceramics that provided a more nurtured environment and better mimicked bone. Overall, this study shows that alumina-HAp scaffolds can be used as a potential future treatment of tissue reconstruction (Bose et al., 2002).

Alumina is also widely used in dentistry due to its strength, biocompatibility, durability, and esthetic benefits. Alumina is one of the only true crystalline ceramics that

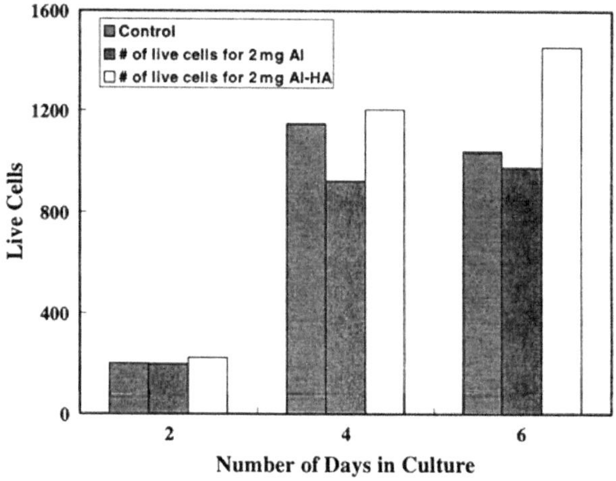

FIGURE 4.1 Cell proliferation data for 2 mg alumina and HAp coated alumina samples for 2, 4, and 6 days in culture (Bose et al., 2002).

is currently being used in restorative dentistry (Denry and Kelly, 2014). Aluminous porcelain is a type of bioceramic that contains a glass matrix phase, and a volume of 35% alumina. It is currently used to create dental cores for dental implants and conventional crowns (Raghavan, 2012). A mix of glass silica, alumina, and fluoride powders can be mixed with an aqueous solution of phosphoric acid and buffer salts to make dental cement. This type of cement can be used for direct restorative dentistry since they can be directly placed on teeth. The advantages of this include that it's quick to set, it has anti-carcinogenic actions due to fluoride release, can chemically bond to tooth materials, and has good esthetics (Raghavan, 2012). Unfortunately, silicates have been discontinued since they have low pH when setting, which affects the dental pulp. Future research on how to stabilize the pH can benefit the characteristics of silicate cement and may allow for the cement to be returned into the market (Raghavan, 2012).

4.3 CALCIUM PHOSPHATE

4.3.1 History of Calcium Phosphates

Calcium phosphates are bioceramics that contain calcium minerals with inorganic phosphate anions. Two of the most important bioceramics currently being used are HAps $Ca_{10}(PO_4)_6(OH)_2$ and tricalcium phosphates $Ca_3(PO_4)_2$ (Samavedi, Whittington, and Goldstein, 2013). In 1769, Johan Gottlieb Gahn and Carl Wilhem Scheele isolated tricalcium phosphate by burning bone in a search to isolate phosphorus. Later, X-ray diffraction technology was used in 1930 to study the structural chemistry of β-tricalcium phosphate (β-TCP) and HAp (Samavedi, Whittington, and Goldstein, 2013). In 1976, synthetic HAp was considered as a potential compatible material that could be implemented in the body. The first application of bioceramics was the plasma-sprayed coating with HAp for dental implants and hip-ball implants, which was noted to improve the biocompatibility of the material (Ducheyne et al., 1980). Other early applications of calcium phosphate ceramics were for root replacement implants, bone cavity fillers, and skeletal prostheses moulding (Samavedi, Whittington, and Goldstein, 2013). Today, calcium phosphate bioceramics are mainly used for the replacement and augmentation of bone tissue due to its biocompatibility and degradation characteristics.

4.3.2 Production of Calcium Phosphates

There are various methods used to create calcium phosphate bioceramics. One of them is the wet method reaction that creates HAp powders. The wet method consists of an acid-base titration (co-precipitation) of an aqueous solution that contains calcium nitrate and diammonium hydrogen phosphate. The product is then sintered at a temperature of 900°C to produce crystalline HAp. The rate at which pH is introduced and the sintering temperature of this method influences the overall nature of the HAp being made (Mostafa, 2005). Other methods that are also used are solid-state reactions, electrocrystallization, freeze-drying, mechanochemical methods, and emulsion processing (Thamaraiselvi and Rajeswari, 2004).

β-TCP can be mixed with HAp to form a more stable bioceramic using the co-precipitation reaction discussed above. Studies have shown that the dissolution of HAp in the human body after implantation is too low to achieve optimum results, and the dissolution rate of β-TCP ceramics is too fast for bone bonding (Bartolome et al., 2007). In a study by Kwon et al. (2002), HAp, TCP, and HAp/TCP composite powders were synthesized by a co-precipitation method to compare their dissolution rate and compressive strength. The compressive strength was measured using a universal testing machine at a crosshead speed of 0.5 mm/min by immersing the bioceramics in a commercially available ringer's solution-tested dissolution. The results (Figure 4.2) showed a negative correlation between porosity and compression strength, meaning that as porosity increased, the strength of the composite decreased. HAp had a greater compressive strength than TCP, and HAp/TCP showed an intermediate value when it came to compressive strength. Overall, the compressive strength of the bioceramics was dependent on the porosity more than the type of bioceramic material used (Kwon et al., 2002). The results of the dissolution test showed HAp to be more stable in the ringer solution and TCP to have a much higher dissolution rate. HAp/TCP again showed an intermediate dissolution rate. This study demonstrates that adjusting the amount of HAp can control the dissolution rate of calcium phosphates and TCP added into the mixture. For a material designed to have a slow dissolution rate, more HAp should be added, and for one with fast dissolution, TCP should be increased in the mixture. Having the right porosity and disintegration rate can optimize the effectiveness of the material and overall treatment (Kwon et al., 2002).

4.3.3 Characteristics of Calcium Phosphate

HAp and β-TCP contain a calcium/phosphate ratio of 1.64 and 1.5, respectively. They are both made up of a pure hexagonal crystal structure (Champion, 2013). HAp is the most stable phase of different calcium phosphates, and it becomes very stable

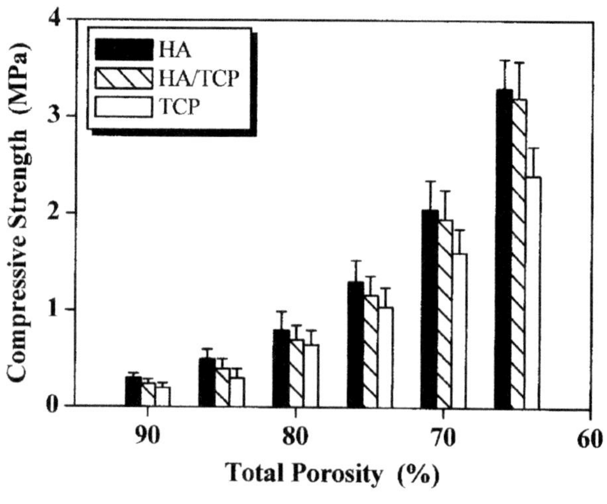

FIGURE 4.2 Compressive strength of porous HAp, TCP, and HAp/TCP composite ceramics as a function of total porosity (Kwon et al., 2002).

in body fluid up to 1200°C. It has similar properties to that of the mineral phase of bone and teeth, making it a great material for hard tissue replacement (Dorozhkin and Epple, 2002). When it is sintered at a high temperature, HAp takes on the characteristic of being bioactive due to its resorbable behaviour, and when HAp is sintered at low temperature it becomes a stable biodegradable material in the body (Champion, 2013). Calcium phosphate bioceramics are also osteoconductive, meaning the biomaterial interacts with the bone, causing guided bone formation on the material surfaces, which results in better bone bonding (LeGeros, 2002).

One important characteristic of calcium phosphate bioceramics is the porosity of the material. Finding the optimum porosity can increase healing time in bone tissue engineering. This is because pore size determines bone formation, provides surface, space for cell adhesion, and allows bone ingrowth. Pore interconnection allows for cell distribution, migration, and permits blood vessel formation (Mastrogiacomo et al., 2006). The pore size, pore shape, and pore interconnection pathway of bioceramics are important to improve biomaterial efficiency for in-vivo bone tissue ingrowth.

In a study by Mastrogiacomo et al. (2006), vascularization and extent of bone matrix deposition were measured on two different HAp porous devices. The first bioceramic (A) was synthesized by sponge matrix embedding, a process in which cellulose matrix with controlled pore morphology and distribution is embedded into a HAp slurry. This structure is then dried, and a decomposition reaction of the organic phase is performed by exposing it to a 600°C at atmospheric pressures for 2 hours. The second bioceramic (B) was synthesized using the foaming method. In this method, surfactants and high concentration of HAp powder were mixed in slurry and expanded using a known volume (40–60% of total) to control morphology and porosity. Bone marrow was then taken from female sheep, the bone marrow stem cells (BMSC) were isolated and loaded into the two scaffolds, which were later implanted in mice. Morphology of the structures of both scaffolds was analysed using scanning electron microscopy (SEM) and is shown in Figure 4.3 below (Mastrogiacomo et al., 2006).

Calcium phosphate bioceramic A had a density of 1.26 g/cm^3 with a porosity of 65%. Bioceramic B had a density of 0.63 g/cm^3 with a porosity of 80%, this scaffold had smaller pores and smaller interconnection size than scaffold A. Different interconnection pathways were seen in both scaffolds. Bioceramic A had a tunnel pore interconnection that was about 200 μm; it had a low percentage of smaller interconnections between adjacent tunnel walls. Bioceramic B has an interconnection pathway of 100 μm and had converging adjacent pore walls.

2 weeks after implantation, bone formation was seen in scaffold B, but no bone formation was detected in bioceramic A. 4 weeks after implantation, bioceramic A showed low amounts of bone formation (22%), scaffold B had much more bone with evenly distributed pores (35%). Scaffold A and B both displayed a significant amount of vascularization, but larger size blood vessels were noted in scaffold A compared to B (Mastrogiacomo et al., 2006). Overall, porosity allowed for vascularization, osteoblast, mesenchymal cell migration, cell proliferation, and bone deposition in the empty spaces of the implant. It was found that less dense materials and higher amount of porosity allowed for a more rapid and efficient increase in

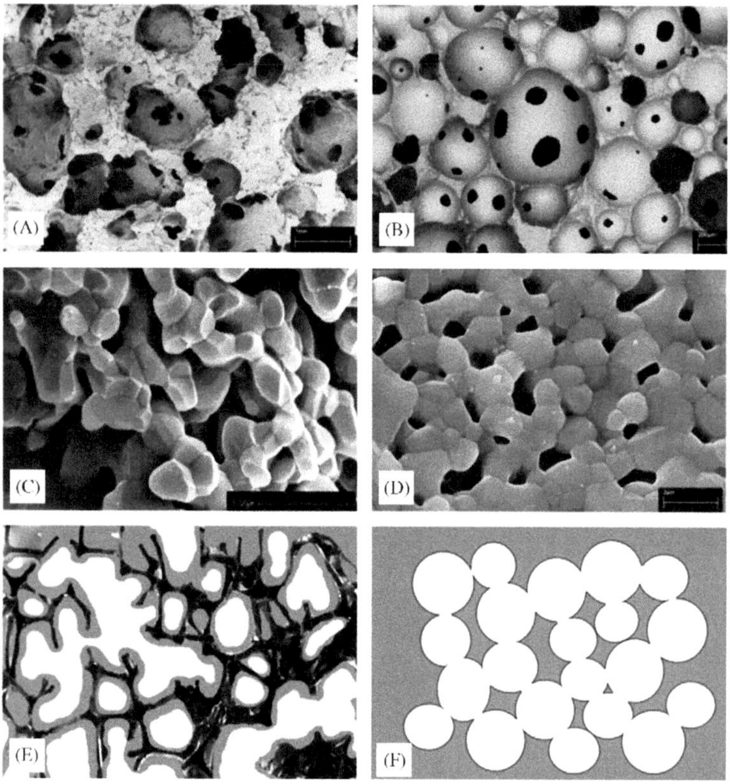

FIGURE 4.3 Left column is scaffold A, right column is scaffold B. Differences between Scaffold A and Scaffold B in porosity, pore interconnection, and size (Mastrogiacomo et al., 2006).

bone formation. Larger pores allowed for the penetration of larger blood vessels but decreased compressive strength of the scaffold. One thing that can improve vascularization and bone deposition is to use more resorbable bioceramics, such as a mix of HAp and tricalcium phosphate scaffolds. This would allow bone to grow proportionally to the biomaterial resorption (Mastrogiacomo et al., 2006).

4.3.4 Applications of Calcium Phosphate

Currently, there are many different applications of calcium phosphates bioceramics in the medical field. One of the main applications of calcium phosphates is in tissue engineering due to their potential to become biocompatible and form degradable scaffolds (Dorozhkin and Epple, 2002). Bone loss is a big problem seen today in the orthopedic field due to a lack of predictability when it comes to the reconstruction of functional bone. There are different methods of repairing long bones and repair can be divided into two methods. The first method used is an excluded graft transplant, it consists of separating the bone or removing damaged parts to allow bone regeneration. This method is inconvenient for the patients and there is a risk that the bone will grow in the wrong direction. The second method used is an included graft transplant, in which the patient is implanted with a vascularized or nonvascularized bone graft. This method has a high success rate, but also a risk of infections, immune response, and nonunions of the bone (Marcacci et al., 2007).

Bone marrow derived pluripotent mesenchymal stem cells can be combined with HAp and other calcium phosphate scaffolds to create a large bone repair method. This material allows for conductive properties, unlimited availability, and absence of an immune response. The long-term durability study of four patients with large bone diaphysis defects that have been treated with the bone-engineered method mentioned above will be analysed. (Marcacci et al., 2007). In the study four patients between the ages of 19–45 with no neoplastic pathologies participated. These patients have had other alternative surgical methods that failed. The materials used to create the scaffold consisted of porous HAp bioceramic material with an average density of 1.26 g/cm^3 and a total porosity of 60%. The material was shaped and sized to the bone gap of the patients. Human BMSC were cultured and placed in the scaffold by capillarity, it was then incubated at 37°C for 60 minutes. Thrombonin was then added to undergo fibrinogen polymerization. The scaffold with the cells was placed in a nutrient medium. Surgery was performed, and the scaffold was placed on the problematic bone (Marcacci et al., 2007). After surgery, the patients underwent short and long-term analysis of their bone. During the short-term analysis, the researchers found no major complications. There were no complaints from the patients, signs of pain, or swelling, and infections were not seen in the implantation site. After 1–2 months, a callus was seen, and during the following months peri-implant bone formation was detected. 5–7 months after the surgery, full integration of the implant and the bone was seen, all patients recovered well and those with surgery on the legs could walk without support. The implants showed some bioceramic disintegration while formation of the bone continued. A 7 year follow up was conducted; the implant showed good integration with the patients and vascularization on the grafted bone was seen. The patients were all pain free and satisfied with the results.

This long-term study was a success and showed that HAp and other calcium phosphate ceramics could create a bioceramic scaffold that was able to fully integrate into the bone and allow vascularization. The importance of this study is that this technology may be used in the creation of future tissue engineered implantations since it showed promising results. (Marcacci et al., 2007).

Another current technology that is being researched is the potential use of porous calcium phosphates scaffolds, as a drug delivery system. One study evaluated microporous β-TCP scaffolds as a potential carrier and delivery system for bone morphogenetic protein-2 (BMP-2). The scaffold had interconnected porosity of 64% and a microporous surface. 15–30 μg of BMP-2 was loaded into the scaffold and implanted into femoral defects of rabbits. The calcium phosphate scaffold successfully released the BMP-2 drug into the affected bone. This resulted in a 1.3-times increase of newly formed bone compared to the unloaded scaffold treatments. The successful results of the study indicate that the β-TCP scaffolds as a drug delivery system can be used for a future approach of bone regeneration treatments (Sohier et al., 2010).

4.4 ZIRCONIA

4.4.1 History of Zirconia

Zirconia (ZrO_2) is a strong ceramic biomaterial that was synthesized and researched in 1969. It was first introduced to target ceramic brittleness and mechanical failure of ceramic implants (Helmer and Driskell, 1969). This material was initially introduced in the mid-eighties as a femoral head implant for the treatment of total hip replacements, and zirconia was later introduced into dentistry, being used in the making of dental parts such as dental roots (Cales and Stefani, 1995). Today, zirconia has been used for more than 600 000 femoral heads worldwide and is mainly seen in the US and Europe (Kokubo, 2008). It is still a major biomaterial used in the field of dentistry. The most attractive quality of Zirconia is that it can be alloyed with oxides like magnesia and calcia to increase toughness that surpasses crack resistance levels of regular ceramics. (Nielsen, 2000) This improves the quality of the bioceramics and allows for biomedical applications that require strong materials.

4.4.2 Production of Zirconia

Zirconia is made through the decomposition reaction of zircon ($ZrSiO_4$) (Ayala et al., 2003). The first step of the process is called zircon disintegration; it is composed of mixing zircon with soda ash or with lime. Thermal treatment of the product is then performed, where the temperature can range from 1100 to 1600°C. It takes from 5 minutes to 3 hours to dry depending on the amount of soda ash used. The result of this step is a mixture of monoclinic zirconia, Na-silicozirconate, and amorphous silica. To remove the amorphous silica, a wet chemistry process is performed, called alkali leaching. This process consists of the mixture being treated with NaOH solution in an acid digestion bomb, which is later heated at a temperature of 200°C. Finally, the last step is acid leaching using hydrochloric acid. This is done to remove the Na-silicozirconate from the mixture leaving behind the desired zirconia product. Overall, the zirconia product created is dependent on the concentration of the soda ash, the temperature of the reaction, and the overall reaction time (Ayala et al., 2003).

4.4.3 Characteristics of Zirconia

Zirconia can be considered the ceramic steel of bioceramics since it has a small grain size, high purity, density, elicits strength, and fracture toughness (8–10 MPa). Its fracture toughness occurs due to a stress-induced phase transformation from a metastable tetragonal phase to a stable monoclinic phase at ambient temperatures. (Chevalier et al., 2004). Zirconia also has a flexural strength that ranges from 900 to 1100 MPa. Another characteristic of zirconia is that it has a high thermal expansion similar to steel and has excellent thermal insulation with low thermal conductivity (Luthardt et al., 2002). Zirconia's characteristics can be improved by creating composites. Zirconia can be partially stabilized with other metal oxides to create a material that can resist even higher stress than zirconia. For example, usually, 3% by weight of yttria is added to zirconia for dental applications since it results in greater strength and resistance to fatigue (Ahn et al., 2001). One disadvantage of zirconia bioceramics over metals is their brittleness and breakdown when exposed to high temperatures (Chevalier et al., 2004).

4.4.4 Applications of Zirconia

Zirconia is a biomaterial that is mostly used for the applications of endoprosthesis and dentistry. Ceramic dental implants have many advantages over metal implants such as minimal ion release, better aesthetics, elimination of potential allergy to metals, and most importantly it does not undergo corrosion (Zarone et al., 2011). Clinical studies have also been able to show that there is no difference in the performance of dental implants between zirconia and titanium (Apratim et al., 2015). Zirconia has gained popularity as a ceramic crown over metals since zirconia crowns are equally effective, improve the appearance of teeth due to their translucent properties, are durable, easy to wear, and long lasting. Zirconia crowns also last longer than other types of crown and help to preserve more of the original tooth since minimal preparation for zirconia crowns is required (Zarone et al., 2011). One of its downfalls is that zirconia can cause friction against other teeth, and the root of the tooth due to its strength and the nature of its surface (Apratim et al., 2015).

Zirconia can also be combined with alumina to create a bioceramic that exhibits higher hydrothermal stability and superior wear resistance. Zirconia-toughened alumina (ZTA) is composed of 80% tetragonal zirconia polycrystal and 20% alumina ceramics. This component is made of an alumina rich composition with evenly disperses zirconia in the alumina matrix. ZTA has superior tribological and mechanical properties compared to zirconia and alumina (Kurtz et al., 2014). ZTA has been used for implants for load-bearing applications such as total hip replacements. Since ZTA materials have higher fracture toughness, it is possible to create thinner liners and larger femoral heads, which allows for a greater range of motion in the joint. In a study by Lombardi et al. (2010) the performance of 65 ZTA femoral head implants was monitored for 6.1 years. They recorded the ZTA implants to have 95% survivorship with no cases of osteolysis recorded. The implants also had no squeaking noise reported, which is a common problem for metal implants (Lombardi et al., 2010). This study helps to demonstrate the advantageous properties found in ZTA composites and how they can be used as an alternative to alumina or metals to treat load-bearing complications such as a total hip replacement.

4.5 BIOGLASS BIOCERAMICS

4.5.1 History of Bioglass

Bioglass is a type of glass ceramics composed of different proportions of SiO_2, Na_2O, CaO, and P_2O_5. There are many types of bioglass that have been created such as bioglass 45S5, 8625, 5S4.3, etc. (Goller, 2004). The first bioceramic bioglass to be created was bioglass 45S5, which was synthesized by Professor Larry Hench. He was inspired to create a biocompatible glass material that will not be rejected by the body, after hearing about many soldiers undergoing amputations and their bodies rejecting the prosthetics being used at that time. In 1969, Hench created small rectangles of 45S5 glass, which he implanted into mice to test for biocompatibility. In this study, Hench realized the bioactive effects that this glass had on the organisms. He noted that the surface of the bioglass transformed from a silicate-rich composition into a phosphate rich structure (Hench, 2006). Bioglass 45S5 was first successfully used in the replacement of ossicles in the middle ear to treat conductive hearing loss. Today, bioglasses are used for small bone repair and encapsulation of implanted devices.

4.5.2 Production of Bioglass

Bioactive glass can be synthesized through sol-gel processing methods, melt quench synthesis, and microwave synthesis. The first method used to create bioglass was the melt quench synthesis, in which related oxide precursors were melted at high temperatures. The compositions of the precursors used are 46.1% of SiO_2, 24.4% of Na_2O, 26.9% of CaO, and 2.6% of P_2O_5 (Goller, 2004). The sol-gel processing synthesis is another method used to create bioglasses, composed of a solution made up of metal-organics and metal salt precursors. These precursors undergo chemical reactions or aggregation resulting in a gel formation. This gel mixture is thermally treated to dry, remove organic waste, and crystallize. This method allows for the creation of various types of bioglasses but is time-consuming and limited when it comes to engineering the composition of the bioglass produced (Li et al., 1991). Finally, microwave synthesis can be performed to obtain bioglasses. In this method, precursors are dissolved in deionized water and placed in an ultrasonic bath to undergo irradiation. Microwave operations are then performed on the resulting powder and then placed in an oven to dry. Calcination of the material occurs at a temperature of 700°C. This method has gained popularity since it has a short reaction time and can create nanophase powders (Hong et al., 2010).

4.5.3 Characteristics of Bioglass

Glasses are made from silica, calcium oxide, and sodium oxide. To be classified as a bioglass the material must contain at least 65% of silica. Bioglasses are inert and have excellent thermal properties, but their main advantage and most important characteristic is the ability to be bioactive in the body, meaning that bioglasses actively take part in physiological processes (Rahaman et al., 2011).

There are many compositions of bioglasses that act differently when introduced in the body. There are some bioglasses such as bioglass 45S5 that can bind to both bone and soft tissue (Xynos et al., 2000). Other bioglasses such as bioglass 5S4.3 does not bond, but instead becomes encapsulated in fibrous tissue. Some bioglass can even interact with the body's environment and become resorbable (Silver et al., 2001). Bioglasses can also be combined with other biomaterials to enhance their properties, for example, a thin layer of apatite can be added to the surface of bioglass to allow for a stronger bond to bone (Goller, 2004). One disadvantage that limits the use of bioglasses in biomedical applications is that they display mechanical weakness and low fracture resistance. This results in bioglass bioceramics being used for nonload-bearing applications, encapsulating devices, and as an additive material for composites that result in an increase of bioactivity in the body (Thompson and Hench, 2001). Currently, the FDA has approved for melt-derived compositions of 45S5, S53P4, and bioglass 8625 to be used for biomedical purposes.

4.5.4 Applications of Bioglass

Bioglass is currently used for various applications, such as bone plating, dental implants, spinal fusions, and bone graft studies. One of the main uses of Bioglass 45S5 is to produce bioactive bone graft products. This type of bioglass functions by undergoing a dissolution process after being implanted into the body. The dissolution process of the bioglass consists of the material releasing ions (silica, phosphate, calcium, and sodium ions) that are resorbed and replaced by bone. The process is dependent on the size and shape of the particles; materials that are smaller have a higher surface area while materials with a spherical shape will tend to be resorbed faster and release more ions (Rahaman et al., 2011). In-vitro studies have shown that bioglass 45S5 dissolution ions positively affect bone formation involved cells (Figure 4.4). The ability of the bioglass to create a bone-like mineral layer on its surface is what gives the bioactive properties to this material. In the body, the particles that begin to dissolve combine with ions from the environment to form a hydroxy-carbano-apatite (HCA) coating on the surface. The similarity between HCA and bone mineral characteristics creates the optimal surface coating for new bone formation, resulting in a chemical bond between the bioactive glass ions and the surrounding bone. Bone replacement of the bioglass will take approximately 9–12 months (Bosetti, 2005).

Bioglass can also be a potential treatment for spinal fusion; this material was compared against iliac crest autograft as a treatment for thoracic adolescent idiopathic scoliosis (AIS). In this study, 40 patients underwent the iliac crest autograft

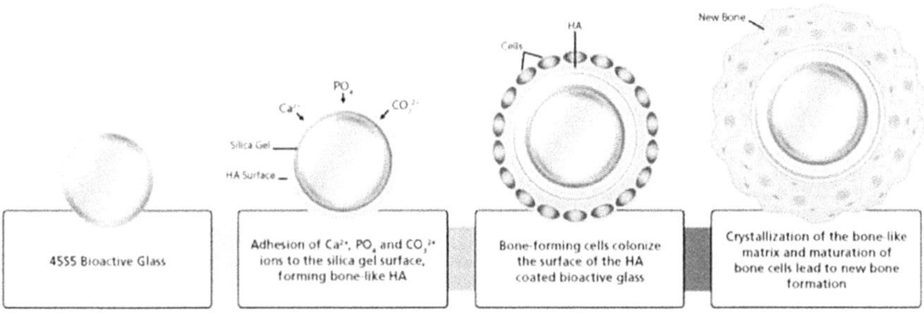

FIGURE 4.4 The bioactive process of introduced bioglass 45S5 (Velez, 2016).

and 48 underwent the bioglass spinal fusion. The patients were monitored for 2 years, and statistical data was compared using the Mann-Whitney U test and an unpaired t test. The results recorded displayed a 5% infection rate and 7.5% mechanical failure of the implant in the autograft group. The bioglass group was more successful with only 2% of the population developing an infection and 2% of the implants undergoing a mechanical failure. Although bioglass had more effective results, the overall data did not reach statistical significance, meaning the material is not more successful but just equally as effective as an autograft treatment. Overall, this study was able to show that bioglass is an effective biomaterial that can be used to maintain fusion and achieve correction in AIS patients. (Ilharreborde et al., 2008).

4.6 CHALLENGES OF BIOCERAMICS

Every bioceramic used has some type of challenge that needs to overcome to create a successful biomedical application. For example, bioglass has mechanical weakness and low fracture resistance, while zirconia displays brittleness and is vulnerable to degradation. To overcome this problem, we need to fully understand what characteristics and properties are needed for the application at hand, and then select the appropriate bioceramic that will fit its needs. This problem can also be overcome by experimenting with composite biomaterials that are made up of different bioceramics to create a specialized bioceramic for its intended application.

One characteristic that needs to be examined to avoid mechanical failure is the critical load-bearing value of the ceramic biomaterials. In a study by Lawn et al. (2004) zirconia, alumina, glass ceramics, and porcelain were tested and placed under stress to obtain the critical load. The critical load is the highest load the ceramic could support before damage occurred. The thickness of the ceramics was also evaluated against the critical load the material could bear. The result of the study showed a positive correlation between critical load and thickness of the biomaterial, meaning that as the thickness of the ceramics increased, there were higher critical loads that could be supported. This experiment also allowed the scientists to see where the materials ranked in strength. The results (Figure 4.5) showed high-strength ceramics such as zirconia, intermediate strength such as alumina and glass ceramics, and low strength such as porcelain. (Lawn et al., 2004).

Another characteristic that needs to be investigated to avoid failure is the fatigue and fracture resistance of bioceramics. Ceramics are most likely to undergo a brittle fracture and need absolute quality control to overcome this issue (Theo, 2000). If failure occurs, it can lead to rejection of the bioceramic implants/devices in the body. When rejection occurs, the body will try to digest the material, excrete it, or engulf it in a cellular fibrous lining. This is especially problematic if the device is critical for the person's well being, such as a pacemaker or a mechanical heart valve. If devices like this fail, it can result in a fatal catastrophe and the death of the patient. To avoid this problem, an increase in bioengineering studies can be done to improve tools that can accurately predict how the bioceramic implants will work to avoid failure of the devices (Theo, 2000).

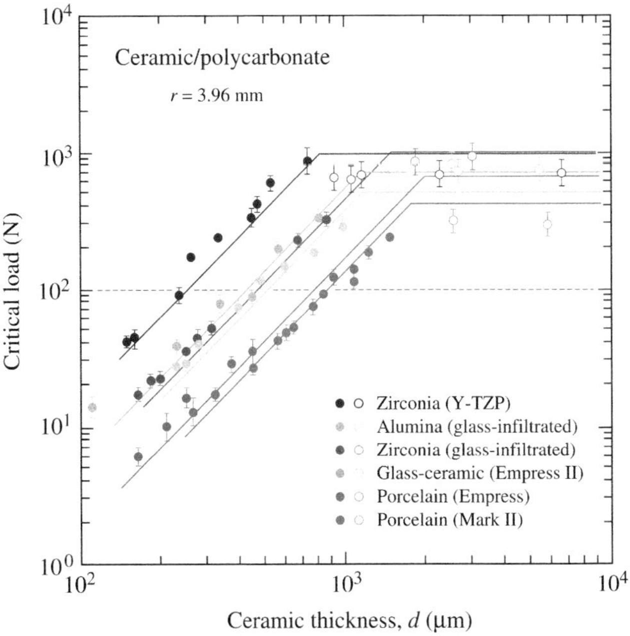

FIGURE 4.5 Change in critical load of different bioceramics with the change in their thickness (Deng, Lawn, and Lloyd, 2002).

4.7 FUTURE APPLICATIONS OF BIOCERMICS

Ceramics have become a popular biomaterial in dentistry and for other biological applications, but newer technologies are bound to supersede their use. Improving the mechanical performance, brittleness, degradation, and strength of bioceramic materials can be achieved by experimenting with a combination of bioceramics that have different characteristics. Bioceramics can also be synthesized to create the right composition, microstructure, and molecular surface chemistry of various types of ceramics to tailor the implant to the specific biological and metabolic requirements of the targeted tissue (Dubok, 2000).

Enhanced bioactivity is also something that will be aimed for in the future. Bioglass can directly bond and connect well with the tissue around it. In the future, bioglass can be engineered to not only improve osteointegration, but to also allow for gene activation of the tissue around it. This could be used to increase overall cell proliferation and improve healing of the area (Best et al., 2008). Since some bioceramics make great scaffolds for tissue engineering, in the future they can be used to create 'smart scaffolds'. These smart scaffolds will not only directly repair injured tissue but will be implanted with cells and osteoinductive agents that will trigger regeneration of the tissue. This can improve and speed up the healing process of the tissue being supported (Daculsi, 2015). Another way to improve bioactivity is by creating plasma treated bioceramics. This creates thick ceramic coatings on orthopedic applications that enhance surface bioactivity to improve the biomedical treatment at hand (Dinarvand et al., 2011). The future is bright for bioceramics and it will only begin to improve with newer bioceramics being created.

REFERENCES

Ahn, E. S., Gleason, N. J., Nakahira, A., and Ying, J. Y. (2001). Nanostructure processing of hydroxyapatite-based bioceramics. *Nano Letters*, 1(3), 149–153.

Apratim, A., Eachempati, P., Salian, K. K. K., Singh, V., Chhabra, S., and Shah, S. (2015). Zirconia in dental implantology: a review. *Journal of International Society of Preventive & Community Dentistry*, 5(3), 147.

Ayala, J. M., Verdeja, L. F., Garcia, M. P., Llavona, M. A., and Sancho, J. P. (2003). Production of zirconia powders from the basic disintegration of zircon and their characterization. *Journal of Materials Science*, 27(2), 458–464.

Bartolome, J. F., De Aza, A. H., Martin, A., Pastor, J. Y., Llorca, J., Torrecillas, R., and Bruno, G. (2007). Alumina/zirconia micro/nanocomposites: a new material for biomedical applications with superior sliding wear resistance. *Journal of the American Ceramic Society*, 90(10), 3177–3184.

Best, S. M., Porter, A. E., Thian, E. S., and Huang, J. (2008). Bioceramics: past, present and for the future. *Journal of the European Ceramic Society*, 28(7), 1319–1327.

Bose, S., Darsell, J., Hosick, H. L., Yang, L., Sarkar, D. K., and Bandyopadhyay, A. (2002). Processing and characterization of porous alumina scaffolds. *Journal of Materials Science: Materials in Medicine*, 13(1), 23–28.

Bosetti, M. (2005). The effect of bioactive glasses on bone marrow stromal cells differentiation. *Biomaterials*, 26(18), 3873–3879.

Boutin, P., Christel, P., Dorlot, J. M., Meunier, A., De Roquancourt, A., Blanquaert, D., and Witvoet, J. (2000). The use of dense alumina–alumina ceramic combination in total hip replacement. *Journal of Biomedical Materials Research*, 22(12), 1203–1232.

Cales, B., and Stefani, Y. (1995). *Yttria-Stabilized Zirconia for Improved Orthopedic Prostheses*. In Wise, D. L. (Ed.). *Encyclopedic Handbook of Biomaterials and Bioengineering*. Marcel Dekker.

Champion, E. (2013). Sintering of calcium phosphate bioceramics. *Acta Biomaterialia*, 9(4), 5855–5875.

Chevalier, J., Deville, S., Münch, E., Jullian, R., and Lair, F. (2004). Critical effect of cubic phase on aging in 3mol% yttria-stabilized zirconia ceramics for hip replacement prosthesis. *Biomaterials*, 25(24), 5539–5545.

Daculsi, G. (2015). Smart scaffolds: the future of bioceramic. *Journal of Materials Science: Materials in Medicine*, 26(4), 1–4.

Davis, K. (2010). Material Review: Alumina (Al_2 O_3). *School of Doctoral Studies European Union Journal*, 2, 109–114.

Deng, Y., Lawn, B. R., and Lloyd, I. K. (2002). Characterization of damage modes in dental ceramic bilayer structures. *Journal of Biomedical Materials Research*, 63(2), 137–145.

Denry, I., and Kelly, J. R. (2014). Emerging ceramic-based materials for dentistry. *Journal of Dental Research*, 93(12), 1235–1242.

Dinarvand, P., Seyedjafari, E., Shafiee, A., Babaei Jandaghi, A., Doostmohammadi, A., Fathi, M. H., and Soleimani, M. (2011). New approach to bone tissue engineering: simultaneous application of hydroxyapatite and bioactive glass coated on a poly(L-lactic acid) scaffold. *ACS Applied Materials & Interfaces*, 3(11), 4518–4524.

Dorozhkin, S. V., and Epple, M. (2002). Biological and medical significance of calcium phosphates. *Angewandte Chemie International Edition*, 41(17), 3130–3146.

Dubok, V. A. (2000). Bioceramics——Yesterday, today, tomorrow. *Powder Metallurgy and Metal Ceramics*, 39(7), 381–394.
Ducheyne, P., Hench, L. L., Kagan, A. 2nd, Martens, M., Bursens, A., and Mulier, J. C. (1980). Effect of hydroxyapatite impregnation on skeletal bonding of porous coated implants. *Journal of Biomedical Materials Research Part A*, 14(3), 225–237.
Goller, G. (2004). The effect of bond coat on mechanical properties of plasma sprayed bioglass-titanium coatings. *Ceramics International*, 30(3), 351–355.
Hatton, A., Nevelos, J. E., Nevelos, A. A., Banks, R. E., Fisher, J., and Ingham, E. (2002). Alumina–alumina artificial hip joints. Part I: a histological analysis and characterisation of wear debris by laser capture microdissection of tissues retrieved at revision. *Biomaterials*, 23(16), 3429–3440.
Helmer, J. D., and Driskell, T. D. (1969). *Research on Bioceramics*. In *Symposium on Use of Ceramics as Surgical Implants*. Clemson University.
Hench, L. L. (2006). The story of Bioglass®. *Journal of Materials Science: Materials in Medicine*, 17(11), 967–978.
Hong, Y., Chen, X., Jing, X., Fan, H., Guo, B., Gu, Z., and Zhang, X. (2010). Preparation, bioactivity, and drug release of hierarchical nanoporous bioactive glass ultrathin fibers. *Advanced Materials*, 22(6), 754–758.
Ilharreborde, B., Morel, E., Fitoussi, F., Presedo, A., Souchet, P., Penneçot, G. F., and Mazda, K. (2008). Bioactive glass as a bone substitute for spinal fusion in adolescent idiopathic scoliosis: a comparative study with iliac crest autograft. *Journal of Pediatric Orthopaedics*, 28(3), 347–351.
Kokubo, T. (Ed.). (2008). *Bioceramics and their clinical applications*. Elsevier.
Kurtz, S. M., Kocagöz, S., Arnholt, C., Huet, R., Ueno, M., and Walter, W. L. (2014). Advances in zirconia toughened alumina biomaterials for total joint replacement. *Journal of the Mechanical Behavior of Biomedical Materials*, 31, 107–116.
Kwon, S. H., Jun, Y. K., Hong, S. H., Lee, I. S., Kim, H. E., and Won, Y. Y. (2002). Calcium phosphate bioceramics with various porosities and dissolution rates. *Journal of the American Ceramic Society*, 85(12), 3129–3131.
Landers, W. F., Rutten, M. J., Fisher Jr, T. R., and Schaffer, D. A. (2004). U.S. Patent No. 5,676,587. Washington, DC: U.S. Patent and Trademark Office.
Lawn, B. R., Pajares, A., Zhang Y., Deng, Y., Polack, M. A., and Lloyd, I. K. (2004). Materials design in the performance of all-ceramic crowns. *Biomaterials*, 25(14), 2885–2892. doi:10.1016/j.biomaterials.2003.09.050
LeGeros, R. Z. (2002). Properties of osteoconductive biomaterials: calcium phosphates. *Clinical Orthopaedics and Related Research*, 395, 81–98.
Levin, I., and Brandon, D. (2009). Metastable alumina polymorphs: crystal structures and transition sequences. *Journal of the American Ceramic Society*, 81(8), 1995–2012.
Li, R., Clark, A. E., and Hench, L. L. (1991). An investigation of bioactive glass powders by sol-gel processing. *Journal of Applied Biomaterials*, 2(4), 231–239.
Lin, K. L., Chang, J., and Wang, Z. (2005). Fabrication and the characterization of the bioactivity and degradability of macroporous calcium silicate bioceramics in vitro. *Journal of Inorganic Materials*, 20, 692–698.
Lombardi, Jr, A. V., Berend, K. R., Seng, B. E., Clarke, I. C., and Adams, J. B. (2010). Delta ceramic-on-alumina ceramic articulation in primary THA: Prospective, randomized FDA-IDE study and retrieval analysis. *Clinical Orthopaedics and Related Research*, 468(2), 367–374. doi:10.1007/s11999-009-1143-9
Luthardt, R. G., Holzhüter, M., Sandkuhl, O., Herold, V., Schnapp, J. D., Kuhlisch, E., and Walter, M. (2002). Reliability and properties of ground Y-TZP-zirconia ceramics. *Journal of Dental Research*, 81(7), 487–491.
Marcacci, M., Kon, E., Moukhachev, V., Lavroukov, A., Kutepov, S., Quarto, R., and Cancedda, R. (2007). Stem cells associated with macroporous bioceramics for long

bone repair: 6- to 7-year outcome of a pilot clinical study. *Tissue Engineering*, 13(5), 947–955.
Mastrogiacomo, M., Scaglione, S., Martinetti, R., Dolcini, L., Beltrame, F., Cancedda, R., and Quarto, R. (2006). Role of scaffold internal structure on in vivo bone formation in macroporous calcium phosphate bioceramics. *Biomaterials*, 27(17), 3230–3237.
Mostafa, N. Y. (2005). Characterization, thermal stability and sintering of hydroxyapatite powders prepared by different routes. *Materials Chemistry and Physics*, 94(2), 333–341.
Nielsen, R. (2000). Zirconium and zirconium compounds. In *Ullmann's Encyclopedia of Industrial Chemistry*. J. Wiley & Sons.
Oh, S., Oh, N., Appleford, M., and Ong, J. L. (2006). Bioceramics for tissue engineering applications—a review. *American Journal of Biochemistry and Biotechnology*, 2(2), 49–56.
Raghavan, R. N. (2012). *Ceramics in Dentistry*. INTECH Open Access Publisher.
Rahaman, M. N., Day, D. E., Bal, B. S., Fu, Q., Jung, S. B., Bonewald, L. F., and Tomsia, A. P. (2011). Bioactive glass in tissue engineering. *Acta Biomaterialia*, 7(6), 2355–2373.
Rieger, W. (2001). Ceramics in orthopedics–30 years of evolution and experience. In Rieke, C., Oberholzer, S., and Wyss, U., (Eds.). *World Tribology Forum in Arthroplasty*, 283–294. Hans Huber.
Rock, M. (1933) Artificial spare parts for the interior and exterior of the human and animal body. *German Reichspatent DRP 583* 589, 24 August 1933.
Samavedi, S., Whittington, A. R., and Goldstein, A. S. (2013). Calcium phosphate ceramics in bone tissue engineering: a review of properties and their influence on cell behavior. *Acta Biomaterialia*, 9(9), 8037–8045.
Silver, I. A., Deas, J., and Erecińska, M. (2001). Interactions of bioactive glasses with osteoblasts in vitro: effects of 45S5 Bioglass®, and 58S and 77S bioactive glasses on metabolism, intracellular ion concentrations and cell viability. *Biomaterials*, 22(2), 175–185.
Sohier, J., Daculsi, G., Sourice, S., De Groot, K., and Layrolle, P. (2010). Porous beta tricalcium phosphate scaffolds used as a BMP-2 delivery system for bone tissue engineering. *Journal of Biomedical Materials Research Part A*, 92(3), 1105–1114.
Thamaraiselvi, T., and Rajeswari, S. (2004). Biological evaluation of bioceramic materials–a review. *Carbon*, 24(31), 172.
Thompson, I. D., and Hench, L. L. (2001). Mechanical properties of bioactive glasses glass–ceramics and composites. *Proceedings of the Institution of Mechanical Engineers, Part H: Journal of Engineering in Medicine*, 212(2), 127–136.
Tilocca, A. (2010). Models of structure dynamics and reactivity of bioglasses: a review. *Journal of Materials Chemistry*, 20(33), 6848–6858.
Veljović, D., Zalite, I., Palcevskis, E., Smiciklas, I., Petrović, R., and Janaćković, D. (2010). Microwave sintering of fine grained HAP and HAP/TCP bioceramics. *Ceramics International*, 36(2), 595–603.
Xynos, I. D., Hukkanen, M. V. J., Batten, J. J., Buttery, L. D., Hench, L. L., and Polak, J. M. (2000). Bioglass® 45S5 stimulates osteoblast turnover and enhances bone formation in vitro: implications and applications for bone tissue engineering. *Calcified Tissue International*, 67(4), 321–329.
Year, C. G. N. (1995). Bayer's process for alumina production: a historical perspective. *Bulletin for the History of Chemistry*, 17(18), 15.
Zarone, F., Russo, S., and Sorrentino, R. (2011). From porcelain-fused-to-metal to zirconia: clinical and experimental considerations. *Dental Materials*, 27(1), 83–96.

5 Biocomposites as Implantable Biomaterials

5.1 BIOMATERIALS

Biomaterials adopted for medical purposes interact with biological systems for therapeutic purposes such as treatment, augmentation, repair or replacement of tissue, or diagnostic functions. They can be natural or synthesized in laboratory. Common medical devices made by biomaterials may be dental implants, artificial heart, heart valves, kidney dialysers, intraocular lenses, artificial hip and knee joints, fracture fixators, etc.

5.2 DEFINITION OF BIOCOMPOSITES

Biocomposites comprise two or more physically and chemically dissimilar constituent materials, appropriately arranged and distributed, having separating interfaces characterized by combined material rather than individual material characteristics. Composite materials are stronger, lighter, and often less expensive, and are thus preferred over individual or traditional materials. The term "biocomposites" specifies that those composites are used for biomedical engineering applications. Composites' constituents do not merge. Properties of composites are improved over the constituents' properties. Composites are of an anisotropic structure possessing different mechanical properties in diverse directions. Living tissues like skin, bone, or collagen cartilage are natural biocomposites. Synthetic composites consist of two component phases, a reinforcing phase (fiber) and a continuous phase (matrix). The primary objective of developing a biocomposite is to obtain improved mechanical, biological, and thermal properties. Optimization of structure and performance of a composite implant can be done. The most important characteristic of biocomposites for biomedical application is biocompatibility.

Engineered composite materials include:

Fiber-reinforced polymer
Ceramic–matrix composite
Metal–matrix composite
Polymer–matrix composite

In the way that a thin coating on a material will not make a composite, similarly, a structure being constructed of accumulated components will not make it a

by Yaser Dahman and Umme SharmeenHyder

composite. A pacemaker having metal as a central part in a casing made of polymer will not be considered as a composite, while a catheter tube made of reinforced polymer with implanted metal wire will be.

Composites consist of a matrix and reinforcement. The matrix is the bulk phase, which is continuous, and the reinforcement is the dispersed phase, which is non-continuous. Reinforcement incorporates improved mechanical properties into the matrix while the bulk phase carries load over a large surface area. The bulk phase transmits a load to the reinforcement phase and helps improve the mechanical properties of the composite such as strength, toughness, fatigue resistance, etc., as seen in Figure 5.1 above. The matrix and reinforcement phases are separated by an interface phase. In the case of a structural polymer composite, reinforcement is stiff which makes the composite stiff. Numerous types of matrix and reinforcement are combined in innumerable methods to produce biocomposite materials with desired properties. Advanced uses of biocomposites are observed mainly in structural applications like designing dental or orthopedic implants. Improved strength and stiffness are not always a concern while designing a biomedical composite; other properties like biocompatibility, mimicking of natural structures, or precise property matching are also significant.

Biocomposites composed of hydroxyapatite (HAp) and natural polymers are used for hard tissue engineering, such as bone (Swethaa, 2010). Chitin, chitosan (CS), and collagen biopolymers are based on the same structure and composition of human bone tissue with biocompatibility and biodegradability properties, but they cannot be used as scaffolds due to their poor mechanical strength. Again, tissue engineering demands that biodegradability and biocompatibility feature in biocomposites. Therefore, nanoparticles like HAp (ceramic) when combined with a biopolymer matrix in nanoscale result in improved mechanical strength and better thermal stability, incorporating nanotopographic features in the composite that mimic the nanostructure of the bone, and can be used for bone defects and bone regeneration. The polymer acts as matrix in the composite preparation and has features like porosity, mechanical strength, hydrophobicity, biocompatibility and biodegradability. These properties can be altered by changing monomer constituents' ratios or by incorporating different functional groups. Nano-HAp (nHAp) is popular as bone

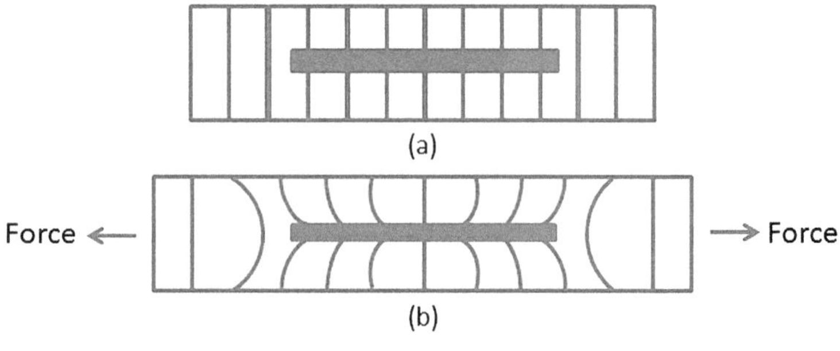

FIGURE 5.1 High-modulus fiber opposes strain around it in a low-modulus matrix. (a) Before deformation. (b) After deformation. Arrows indicate force direction.

repair material as it is bioactive and biocompatible. Collagen fiber, which is tough and flexible, has similar fracture toughness and tensile strength as that of bone when reinforced with Hap; therefore, nHAp composites improve the mechanical properties and biocompatibility of bone-fixing material. (Swethaa, 2010).

Scaffolds that contain nHAp and polymers have improved mechanical strength as well as biodegradation and biocompatibility. The biocompatibility property of a composite characterizes how well the composite implant would perform in a human body in terms of structural and surface compatibility. Surface morphology and biological, physical, and chemical suitability of an implant refer to its surface compatibility, while the structural compatibility is the extent to which the composite adapts to the mechanical behaviour of the surrounding host tissue or organ. (Swethaa, 2010)

A scaffold was prepared by mixing CS with ceramic. CS was crosslinked to the scaffold to expand the artificial bone matrix property of the composite. Hydroxyapatite/chitosan–gelatin (HAp/CS-Gel) has a similar structure to bone because collagen and HAp are the main constituents of bone. Chitosan–gelatin (CS-Gel) and HAp/CS-Gel composites were fabricated and tested by in-vitro experiments. The tests proved the scaffold to be non-toxic with good biocompatibility. Test results (Figure 5.2) show that HAp in CS-Gel polymer has increased protein and calcium ion absorption properties, thereby improving osteoblastic cell attachment, spreading, and proliferation (Swethaa et al., 2010).

5.3 POTENTIAL OF BIOCOMPOSITES FOR MEDICAL APPLICATIONS

Tissues are of two types: hard tissue and soft tissue. Teeth and bones are hard tissue, while blood vessels, skin, and ligaments are soft tissue. Hard tissues are stronger and stiffer than soft ones. Human hard tissues and soft tissues are composite

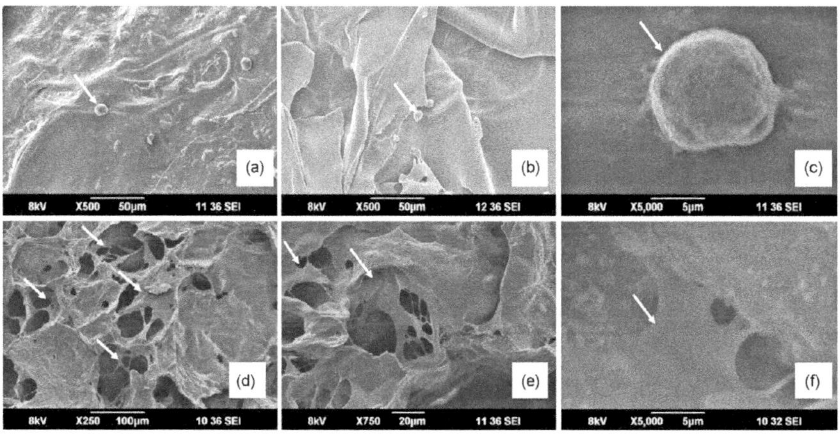

FIGURE 5.2 Scanning electron microscopic images showing osteoblastic cell attachment on the (a–c) chitosan/gelatin and (d–f) chitosan/gelatin/nHAp composite scaffolds (Swethaa et al., 2010).

materials with properties that are anisotropic. Again, elastic and anisotropic properties of tissues are major design criteria for composite implants. From a mechanical point of view, polymers are considered more suitable for soft tissue application, while metals and ceramics more suitable for hard tissue. ceramic–polymer biocomposites with porous morphology and mechanical strength mimic the natural bone structure. ceramic–polymer biocomposites show high strength and low elastic modulus, therefore are suitable for orthopedic application. ceramic–polymer composites are stiffer than adhered bone tissue since their elastic moduli are at least 20 times higher than bone.

The disparity of stiffness between bone and metal /ceramic composite transplants impacts load-bearing and distribution capacity of implants. Composites show high resistance and fracture toughness against fatigue failures. These biocomposite implants are highly compatible to diagnostic approaches like computed tomography (CT) and magnetic resonance imaging (MRI) and expose low X-ray scattering with similar magnetic exposure to human tissue. For instance, dental implants offer better aesthetic features. Examples of biocomposites used in hard tissue applications are orthodontic arch wires, orthodontic brackets, dental posts, prosthetic sockets, fixators, bone plates, hip replacements, screws, etc. Examples of biocomposite uses in clinical application are faster bone healing without any risk of pathogen transfer, and cheaper and quicker surgery without pain. The use of biocomposites in dentistry and dental cements is considered the most successful use of biocomposite materials. Carbon-reinforced polymer composites with low elastic moduli are mostly used for bone restoration and joint replacement. Composites are extensively used in prosthetic limbs due to their higher strength, lower density, and lower weight. Figure 5.3 depicts the use of biocomposites in human hard tissue reconstruction, repair, and replacement (Ramakrishna et al., 2001).

Some key factors for biocomposite implants are

- Composition
- Bulk matrix
- Fiber
- Surface topography
- Shape, geometry, distribution of constituents
- Porosity
- Crystal structure
- Chemistry
- Bonding in atomic scale
- Protein interaction
- Conductivity
- Surface energy
- Corrosion properties
- Mechanical properties
- Young's modulus
- Strength properties
- Wear resistance

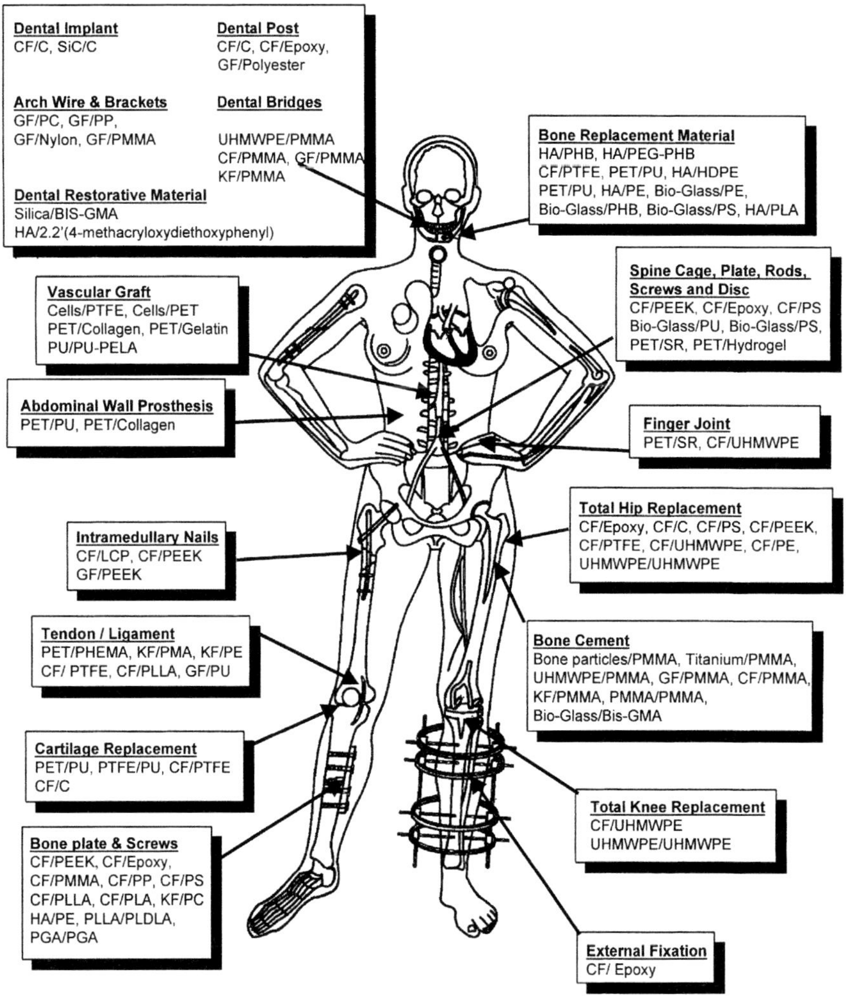

FIGURE 5.3 Applications of composite biomaterials (Ramakrishna et al., 2001).

5.4 CLASSIFICATION OF COMPOSITE MATERIALS

The factors contributing to the performance of the composite include

- Individual component materials
- Arrangement and quantity of constituents
- Collaboration among components

Reinforcement materials mainly determine the properties of composites. Composites are classified based on the features of the reinforcement materials, including size, shape, composition, distribution, orientation, and incorporation manner. Biomedical composites are classified as

- Fiber-reinforced composites
- Particle-reinforced composites

A classification table is shown in Figure 5.4. They are also classified depending upon the matrix material such as ceramic-matrix composites (CMC), polymer-matrix composites (PMC), and metal-matrix composites (MMC).

5.4.1 Particle-Reinforced Composite

Particles are harder and stiffer than the matrix and can be micro-, macro- or nano-scale. Improvements of mechanical strength depend on interface bonding (Figure 5.5). For effective reinforcement, particles should be small and evenly distributed in the

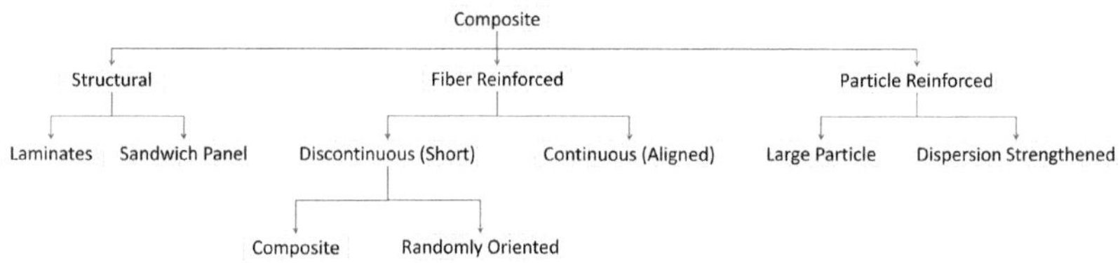

FIGURE 5.4 Classification of biocomposites depending on their reinforcement form (Lee, 2014).

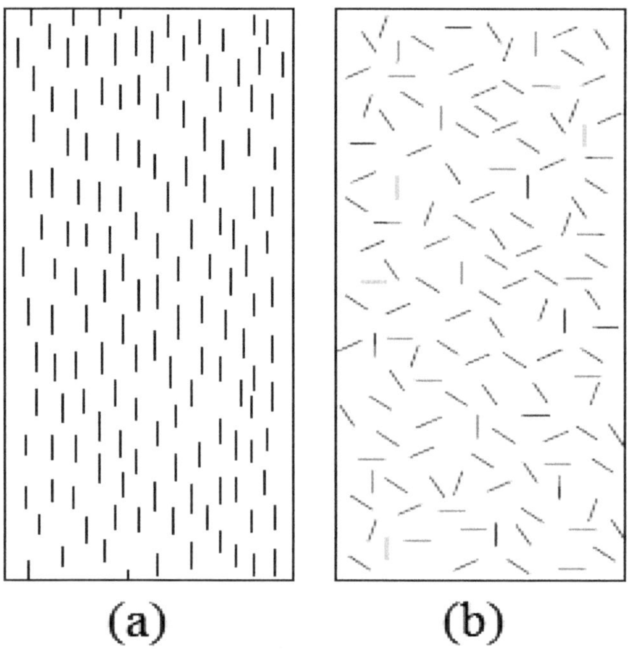

FIGURE 5.5 Particle reinforcement in matrix (Rao, Kiran, and Prasad, 2017).

matrix. Dispersed particles are known to help with the strengthening of materials. This is due to very small (<0.1 mm), inert, hard particles being homogeneously spread in the matrix. These particles can be made of metallic or non-metallic materials.

5.4.2 Fiber-Reinforced Composite

Fiber-reinforced composites (FRC) are of three types, as shown in Figure 5.6 (Pugliesi, 2010). Continuous and aligned fibers are the most effective in the strengthening of fiber composites. Fibers should be at least 15 times longer than the critical length to transmit the entire load from the matrix to the fibers. If fibers are shorter than the critical length, only some of the load will be transmitted. A mechanical property of an FRC depends on the fiber properties and the degree of transmission of applied load. Good interfacial bonding between the fibers and the matrix is necessary.

Particles shorter than critical length are less effective in strengthening, but the resultant composite's modulus and tensile strength can be 50–90% of their continuous counterparts and are easier to fabricate.

Discontinuous and random composites have chopped and separate fibers. Strength is not as high as with aligned fibers, but the composite will be isotropic.

5.4.3 Structural Composite

Structural composites can be made of laminar, sandwich panels as shown in Figure 5.7 (Mosallam et al., 2015). Structural composites consist of 2D panels or sheets in a preferred direction to attain more strength. Successive layers are stacked one on top of another in a favoured direction and tiled together. These preparations or orientations confer a high strength.

This composite is kind of layered, consisting of faces and a core. The faces are two strong outer sheets and the core is a layer of dense material with a honeycomb combination. Table 5.1 lists examples of different polymer composites.

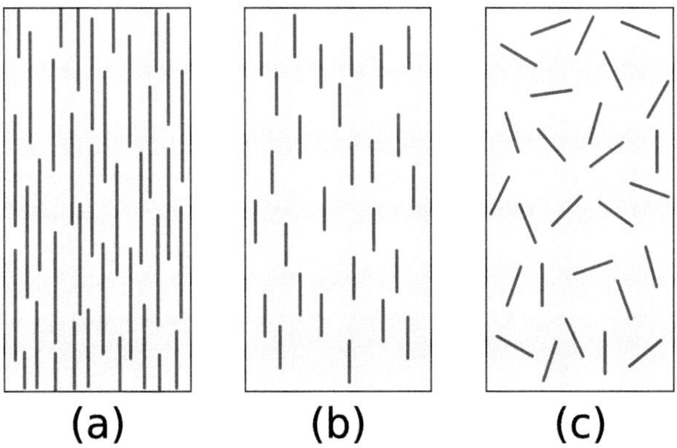

FIGURE 5.6 Different types of fiber reinforced composites (FRC) (Pugliesi, 2010).

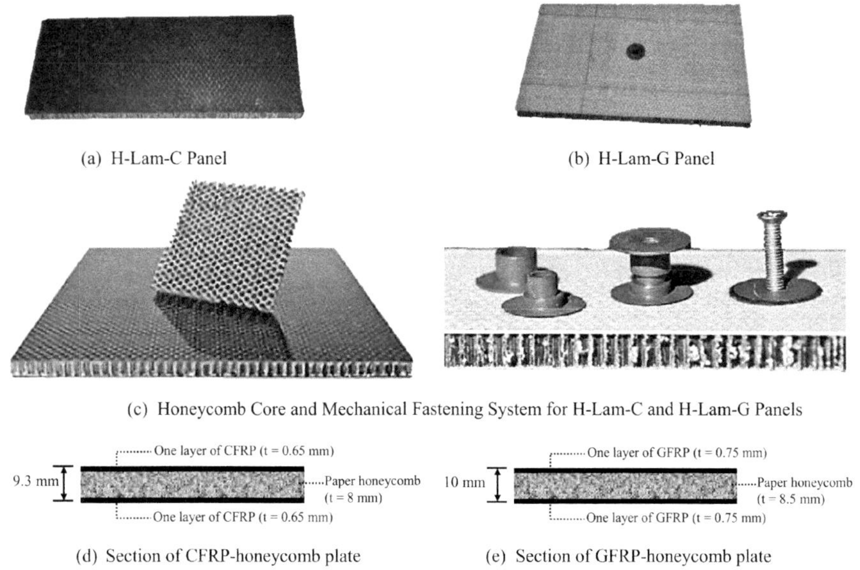

FIGURE 5.7 Different types of structural composites (Mosallam et al., 2015).

5.4.4 Hybrid Composites

Hybrid composites are made for improved material and structural performance. Studies show that higher structural performances can be achieved in hybrid composites in optimized configurations. Although FRCs are becoming popular in their usage, they are outclassed by conventional glass-reinforced plastics with biocomposites in terms of exhibiting structural and functional stability and degradability, and therefore, cannot fully replace them. Biofiber-synthetic hybrid composites are blends of synthetic and natural biofiber matrices with the presence of one or more biofibers in the matrix. Hybridization with glass fiber delivers improved mechanical properties in thermoset biocomposites. Epoxy resin reinforced with glass fiber shows improved tensile strength. If more glass fibers are incorporated, they increase the Young's modulus and impact strength. A study shows that if only a small amount of glass fiber is added, the impact performance of glass fiber and oil palm empty fruit bunch (OPEFD) fiber hybrid phenol formaldehyde (PF) composites increased by more than 100% (John, Anandjiwala, and Thomas, 2009). If only 10% glass fiber is added, the hybrid composite shows an optimum tensile and impact strength for hybrid composites for both treated and untreated types. A coupling agent is added to increase tensile strength. These properties were revealed in a study of thermoplastic natural rubber-reinforced short glass fiber hybrid composites tested for tensile and impact properties (John, Anandjiwala, and Thomas, 2009). Figure 5.8 shows the tricorner approach used to design high performance composites (John and Thomas, 2008).

Biocomposite implants can be bioinert or bioactive, listed below with examples.

TABLE 5.1
Polymer Composite Biomaterials

Non-resorbable Composites	Partially Resorbable Composites	Fully Resorbable Composites
Alumina/PMMA	CF/PGA	PGA/PGA
Bioglass/PS	CF/PLA	PLA-PGA/PLA
Bioglass/PU	CF/PLLA	PLA/PLDLLA
Bone/PMMA	HAp/Alginate	
CF/C	HAp/PBT	
CF/Epoxy	HAp/PEG-PHB	
CF/Nylon	HAp/PLA	
CF/PBT	PET/Collagen	
CF/PEEK	PET/PHEMA	
CF/PMMA	PU/PU-PELA	
CF/PP	Alumina/PLLA	
CF/PS		
CF/PTFE		
CF/UHMWPE		
GF/bis-GMA		
GF/PMMA		
GF/PP		
GF/PU		
HAp/HDPE		
HAp/UHMWPE		
KF/PC		
KF/PMA		
PE/PMMA		
PET/PU		
PMMA/PMMA		
PTFE/PU		
Silica/bis-GMA		
Silica/SR		
UHMWPE/bis-GMA		
UHMWPE/UHMWPE		

Source: Ramakrishna et al. (2001).

5.5 CONSTITUENTS OF BIOCOMPOSITES

5.5.1 Matrices

The matrix is the bulk phase for a composite and is continuous. The matrix covers the reinforcement phase and influences the processing. It holds the reinforcement, that is, the fibers. The matrix holds the fiber direction. The matrix transmits loads and distributes stress to the reinforcement. If brittle fibers are used, the matrix increases fracture toughness because the matrix contains low stiffness and can tolerate greater shear force and elongation. The matrix is resistant to thermal stress and chemical reaction. It also protects the fibers from stress. So, the matrix determines the durability of the composite within the body environment.

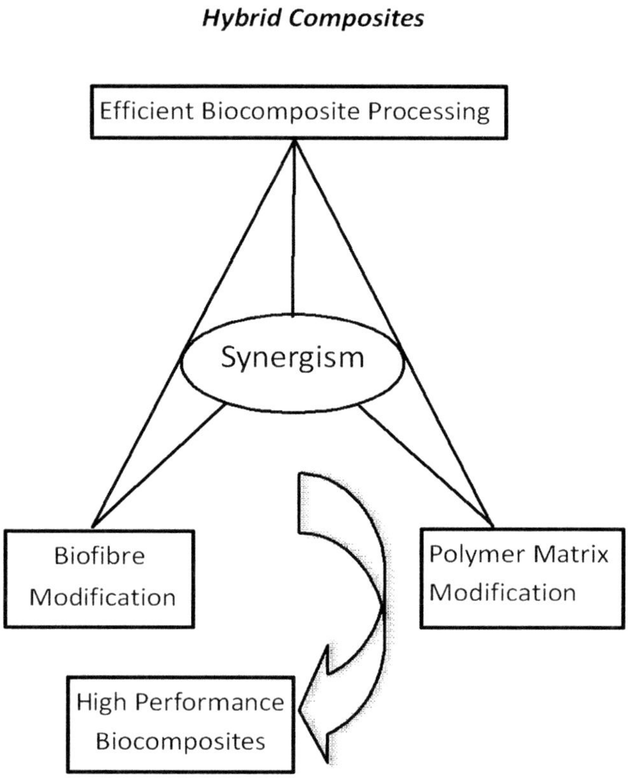

FIGURE 5.8 Tricorner approach to designing high performance hybrid biocomposites (John and Thomas, 2008).

5.5.2 Fibers

Some composites are stronger and stiffer in fibrous form. Fibers are the reinforcement phase and are non-continuous. They are the load-carrying element. The length to diameter aspect ratio of fibers is very high. Strength of the fiber increases as diameter decreases. As diameter decreases, surface flow decreases and strength increases. Fiber type, length, orientation, and volume fraction determine the important properties of a composite. The diameter of the fibers is 5 to 50 μm, which are used to make continuous fiber composites. Short fibers with length 3–50 mm are used to make short fibers or discontinuous composites. Continuous fiber composites are preferred over short-fiber composites, due to their being more efficient. Fibers can be of different dimensions and create planar or unidirectional reinforcement. The volume section of fibers impact the physical properties in the same direction. Planar and unidirectional reinforced composites are anisotropic. An isotropic composite has equal properties in all directions. Laminate composites are fiber-reinforced and anisotropic. Anisotropic layers of laminate composites differ in fiber orientation and volume fraction.

Collagen fibers are used in skin and ligament tissue engineering. Absorbable biogas fibers made up with calcium phosphate are also used in the biomedical field. Carbon fiber is lighter, stiffer, and stronger than glass fiber. Carbon fiber consists of a fine structure of axially-aligned graphite crystals and is used for the fabrication of lightweight composites. Carbon fibers are used in prosthetic structural components. Their fatigue resistance is considerable. Carbon fibers are anisotropic in thermal properties. Carbon fibers enhance the electrical conductivity of composites and cause corrosive effects to metals. Polylactide and polyglycolide and their copolymers imbue composites with greater adsorption ability than mechanical properties.

5.5.3 Particles

Particles, when added to composites, improve mechanical strength (hardness, toughness). Particles improve the dimensional stability, thermal conductivity, and insulation when added to a polymer matrix. Composites with particles randomly distributed in the matrix are called isotropic composites. Depending upon their shape, strength, or stiffness, particles will strengthen or weaken a matrix. Platelet particles are more effective than spherical particles for improving stiffness. Hard particles also increase stiffness if added to low modulus polymers. To make a comparatively softer composite, compliant particles (silicon rubber) are added to a stiff polymer matrix. Fillers are non-reinforcing particles. Particle-reinforced composites are used in ceramic matrices which are used in dental and bone applications, such as HAp. HAp and collagen form natural composite structures of bone. HAp has poor mechanical properties and acts as bioactive component.

5.5.4 Interface

The interface separates fibers/particles and the matrix. The interface acts in the transmission and dispersal of stress. Interfacial bonding and area contribute to a composite's properties. If the interface area is small, it represents poor saturation of fibers on the matrix. Coupling agents are sometimes used as a coating on fibers to improve chemical compatibility with the matrix, in case mechanical coupling is not sufficient. The failure of a composite depends on interfacial shear strength, as this controls debonding of the fibers and the matrix. When the polymer-matrix interface is strong, this makes the matrix stiff. Fracture toughness is lower for polymer matrices. For ceramic matrices, the interface is weak and brittle, which promotes matrix cracking, so again strength and stiffness is lower. Table 5.2 lists different constituents of composites.

5.6 POLYMER MATRIX COMPOSITE PROCESSING

Fiber composites that are continuous in nature can be fabricated by pultrusion, filament winding, and the resin transfer moulding method, the manual layering of laminates. The manual process consists of stacking and layering parallel fiber filament sheets together by a thermoplastic or thermosetting resin, followed by

TABLE 5.2
Constituents of Biomedical Composites

Particles	Fibers	Matrix
Inorganic	*Polymers*	*Thermosets*
Glass	Aromatic polyamides (aramids)	Epoxy
Aluminum (III) oxide (Alumina)	Ultra High Molecular Weight Polyethylene (UHMWPE)	Polyacrylates
	Polyesters	Polymethacrylates
	Polyolefins	Polyesters
	Polytetrafluoroethylene (PTFE)	Silicones
Organic	*Resorbable Polymers*	*Thermoplastics*
Polyacrylate	Polylactide (PLA), and its copolymers with polyglyocolide (PGA)	Polyolefins (Polyethylene (PE), Polypropylene (PP),
Polymethacrylate	Collagen	Ultra High Molecular Weight Polyethylene (UHMWPE))
	Silk	Polycarbonate
		Polysulfones
		Polyether ether ketones (PEEK)
		Polyesters
	Inorganic	*Inorganic*
	Carbon	Hydroxyapatite (HAp)
	Glass	Glass ceramics
	Hydroxyapatite (HAp)	Calcium phosphate ceramics
	Tricalcium phosphate (TCP)	Carbon
		Steel
		Titanium
		Resorbable Polymers
		Polylactide (PLA), polyglycolide (PGA), and their copolymers
		Polydioxanone (PDO, PDS)
		Polyhydroxy butyrate (PHB)
		Alginate
		Chitosan
		Collagen

Source: Iftekhar (2003).

autoclaving. Through filament winding, fiber and properties for tubular forms can be controlled. Pultrusion is good for preparing composite rods such as orthodontic arch wires. Very complex shapes can be made through resin transfer moulding and short cycle times requiring well-designed moulds. Thermosets and thermoplastics are prepared by compression moulding. Injection moulding is used to prepare short-fiber thermoplastic composites with rapid production, but tools are highly expensive. Again, volume fraction and length of the fibers are inadequate when made by this process. Distribution and alignment of fibers is tough to control.

5.7 PROCESSING OF CERAMIC MATRIX COMPOSITES

Infiltration or pressing methods are used to fabricate CMC. Hot pressing or the hot isostatic pressing (HIP) method is used to make dense reinforcement mixed-matrix powder. Through pressing methods, it is possible to achieve near-zero porosity. High pressure and temperature if used simultaneously may degrade fibers or can create strong interfacial bonds for fracture toughness reduction, which will reduce fracture toughness.

High pressure and temperature, if given time, poses the risk to degrade fibers create a robust interfacial bond which will reduce fracture toughness. In infiltration methods, fibrous or paniculate preform is filled by the matrix through glass melt infiltration, chemical vapour deposition, and pre-ceramic polymer or sol infiltration. Low pressure, complex shape, and flexible fiber construction of the composite are the advantages of this method. Alumina-glass dental composites are fabricated through this process. High matrix porosity and elongated fabrication cycles are considered disadvantages of this method. Tape casting and slip casting are also used for particulate reinforcement.

5.8 PHYSICAL PROPERTIES

Composite materials show improved physical and biochemical properties. The basic mechanical properties of a composite can be discovered through the mechanics of the materials, the theory of elasticity model, and semi-empirical models. (Iftekhar, 2003)

5.8.1 Mechanical Properties

Hardness, toughness, yield strength, ultimate strength, fatigue strength, wear resistance, ductility, and load-bearing capacity are important mechanical properties of biocomposite implants.

Assessment of mechanical behaviour of HAp-reinforced polycarbonate (PC) composite was studied (Mehar, 2016). A biocomposite is a compound having two or more chemically distinct phases separated by an interface. Composites have high strength, stiffness, and low weights. Metals and ceramics, being much stiffer, cause mismatch of the mechanical properties between the implant and adhered bone tissue. As metals are very stiff and lacking in biocompatibility and ceramics are very brittle, polymers are preferred due to being feeble and due to their flexibility to adjust its mechanical strength. Polymers on the other hand have better mechanical strength, low density, and formability, but the low stiffness and hardness and high wear rate makes the use of polymers limited to several applications. However, a composite with a combination of ceramic (HAp) and polymer (PC) can eliminate this problem. A HAp reinforced with PC ceramic–polymer composite was fabricated and tested for various properties. The HAp–PC composite was used for biomedical engineering for bone tissue repair and replacement, internal fixation, and load-bearing orthopaedic applications. HAp was synthesized by the chemical precipitation method. HAp particles were combined into the polymer matrix by various processing phases like melt compounding, granulation, and microinjection moulding. Samples were then tested through in-vitro experiments to assess the mechanical and other properties. Samples were examined by Fourier transform infrared (FTIR), X-ray diffraction (XRD) and scanning electron microscopy (SEM). Composite tensile strength, compression, and hardness were tested with differing HAp percentage by volume polymers. Thus, it can be proved that the fabricated HAp PC ceramic-polymer composite has similar mechanical properties to those of bone. Mechanical properties were tested through measurement of the Young's modulus. Tensile strength, compression strength, impact strength, and flexuous strength were higher for HAp–PC composite. With increased volume of HAp, hardness was higher when measured with a Vickers hardness instrument. XRD analysis done by an X-ray diffractometer showed that HAp is non-crystalline and pure with a hexagonal structure, as dry HAp showed a broad peak. (Mehar, 2016). As the SEM micrograph of HAp/PC demonstrates, when the HAp content increases from 10vol.% to 40vol.%, the number of voids in the matrix increases (Mehar, 2016). FTIR analysis was done by an FTIR spectrometer in sintered HAp. HAp was mixed with KBr under high pressure with a hydraulic press in order to prepare a pellet for use in the FTIR test. The FTIR spectrum was transmitted through these pellets and a peak was recorded in the range 4000 to 400 cm^{-1}. Broad peaks between 3500 and 3300 cm^{-1} were noticed on the IR spectrum, showing vibration of absorbed water (OH^-) in the HAp lattice concurrent with the HAp chemical formula ($Ca_{10}(PO_4)_6(OH)_2$). (Mehar, 2016). Peaks of the phosphate functional group were characterized between 500 and 600 cm^{-1} and 1000 and 1100 cm^{-1} (Mehar AK, 2016).

5.8.2 Yield Strength

Composite load-bearing capability is determined by the yield strength. In the case of total joint replacement surgery, a high load-bearing ability of composites is mandatory while for alloys the required property is high yield strength. Composites used for orthopedic purposes should have ductility with high yield strength.

5.8.3 Elastic Property

A composite implant has a higher modulus than bone. If a load transmitted from an implant composite to the bone on which it is adhered is insufficient, this might result in bone reabsorption, and subsequently the prosthetic devices will be loosened (Dujovne, 1993). Tensile strength, compressive load, or the bending moment of bone is reduced, thickness of bone is decreased, loss of bone mass, and an increased in osteoporosis will occur. The total incident is termed as the stress shielding effect. The stress shielding effect is caused by differences in stiffness and flexibility, while stiffness and flexibility depend on the difference in elastic moduli between the bones and the implant (Sumner, 1992). If lower modulus material is used in the composite it improves the stress distribution to adjacent bone and reduces the stiffness. Thus, stress shielding is minimized and the lifetime of the implant device is increased (Long, 1998).

5.8.4 Fatigue

While developing a composite, the fatigue resistance of metal implants is an important concern. During body movement, orthopaedic implants experience cyclic loading. This results in plastic deformation of small stress concentration zones. Notches and structural inhomogeneities create these stress zones. Tension or compression, torsion, and rotation bending are considered standard fatigue tests (Long, 1998).

5.8.5 Corrosion

Corrosion might occur in composite materials if metal is present, such as in hip joint implants due to chemical or electrochemical redox reactions in the salinized body environment. Interaction of composite materials with cell tissue also affects corrosion. Three different types of corrosion might occur in the presence of metal in composites, such as intergranular corrosion, galvanic corrosion, pitting corrosion, crevice corrosion, etc. Galvanic corrosion occurs mostly in metals, when two metals are coupled inside the body and physiological fluid acts as salt bridge to complete the cycle and corrosion occurs. Metal composition, processing, and handling techniques are also factors in cases of corrosion. Anything affecting the microstructure can change localized ion concentration, resulting in corrosion and thereby degradation. Due to motion and stress changes in orthopedic implants, crevice corrosion may occur in narrow deep cracks. A drop in pH increases metal dissolution. Pitting corrosion occurs if a small area of metal has a flaw in the passivation film due to processing or handling. The anode oxidizes and the cathode grows, which can lead to device failure after small amounts of material loss. Intergranular corrosion occurs in grain boundaries with high energy that act as a high energy cathode with the rest of the metal acting as the anode, facilitating redox reaction and the metal will crumble into pieces.

Implant materials are exposed to the harsh aqueous body environment which contains anions such as Cl^-, HCO_3^-, and HPO_4^{--}, cations such as Na^+, K^+, Ca^{++}, and Mg^{++}, organic matters, and dissolved oxygen. The incorporation of a second reinforcing phase into the metal matrix improves the physical and mechanical properties along with the corrosion behaviour. Addition of a filler to a Co-Cr-Mo-based composite can also improve corrosion resistance. (Adzali, 2012) Although, Co-Cr-based alloys demonstrate a low level of corrosion, after implantation metal-metal joints will experience corrosion, that is, metal degradation, and will release metal ions into the body (Bellefontaine, 2010). Addition of Ca and phosphate from HAp reduces the vulnerability of Co-Cr-Mo alloys to corrosion (Oksiutaa, 2009). Thus bioceramics as well as HAp play a combined role both in preventing the release of metal ions, accomplishing more corrosion resistance and converting the metal surface into bioactive material (Sathish 2010).

The interfacial reaction of orthopedic implants is important. Composite implants with metals and alloys undergo corrosion due to the complex environment in the body. Metals and alloys with surface passive film might have corrosion at a slow finite rate, resulting in device failure. Corrosion resistance can be produced by surface modification techniques such as ion beam. Material selection and handling, implant design, and quality control can help to improve corrosion resistance. Corrosion impact can be reduced by ion coating but materials cannot be totally resistant.

Corrosion has two impacts. One is implant weakening and permanent failure, and the other is tissue reaction and the release of corrosion foreign particles into the body. After implantation, the body might show several side effects: the implant can inhibit defence mechanisms of the body that will lead to an infection, it can inhibit tissue responses that can lead to mild edema, or it can cause chronic inflammation or alteration of bone tissue structure.

5.8.6 Fracture and Fatigue Failure

Successive, slow, and gradual internal damage can disastrously result in unsafe composite structure and failure of FRC. Damage can be process induced (dental resin curing control) or service induced (water absorption). Common failure mechanisms in composites are fiber breaking, interfacial debonding, matrix cracking, fiber pull out, fiber bridging, etc. Load types in composites, interfacial shear strength and debonding, and constituent properties impact the sequence and interaction of different failure mechanisms. The maximum strain theories are the dominant theories of fracture mechanics and the failure analysis of composites. Energy adsorption and crack deflection result in enhancement of composite toughness. Debonding at the interface and fiber pull out are two important energy absorbing failure mechanisms. In case the bonds at the interface debonds easily, the interface can create obstacles that can lead to crack propagation. The crack does not move through the fiber, but rather bounces on the fiber surface and the fiber becomes able to carry high load. As fibers do not break on the crack plane, fiber pullout occurs. Fibers break randomly in separate locations and far from the crack plane. Instead of very high friction force or interfacial shear stress, the occurrence of fiber pull out may result in increment of fracture toughness. The same thing might occur for particle reinforcement. Crack deflection frequently occurs in particle bridging and causes the fiber pulls out because of the small aspect ratio. Figure 5.9 below shows hollow cracks in an alumina–glass matrix deflecting around granules of alumina (Chan, 1997). Here crack propagation occurs more through granules which indicates stronger interfacial bonding but lower toughness.

In the case of laminate composites, the delamination load between layers resulting from planar shear load dominates as a failure technique. This load is created at the free edges or in the screw holes of a carbon-reinforced bone plate. (Iftekhar, 2003)

Durability in the long term is still challenging in the case of composites. Like metals, most load-bearing composites do not show long-term performance in case of cyclic loading due to damage occurring by fatigue, creep in the matrix, stress, and the effect of the surrounding environment (Figure 5.10). Both implant life and fatigue damage are difficult to predict. Many composites do not have an endurance limit. Endurance limits are a stress level below which material can carry a load up to an infinite number of times without failure. Variable directional and magnitude loads and loads due to accidents cause damage. If water is absorbed into the polymer matrix it can cause puffiness and initiate a plasticizing effect, which creates

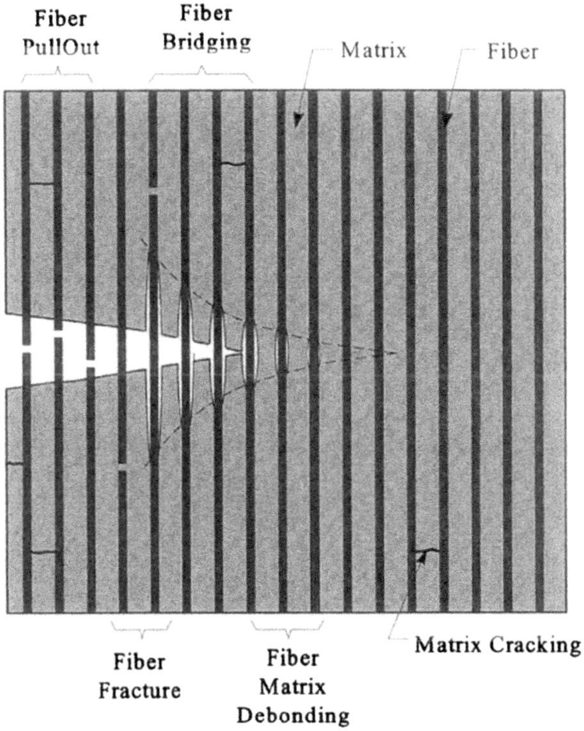

FIGURE 5.9 Failure mechanisms in a unidirectional fiber composite (Chan, 1997).

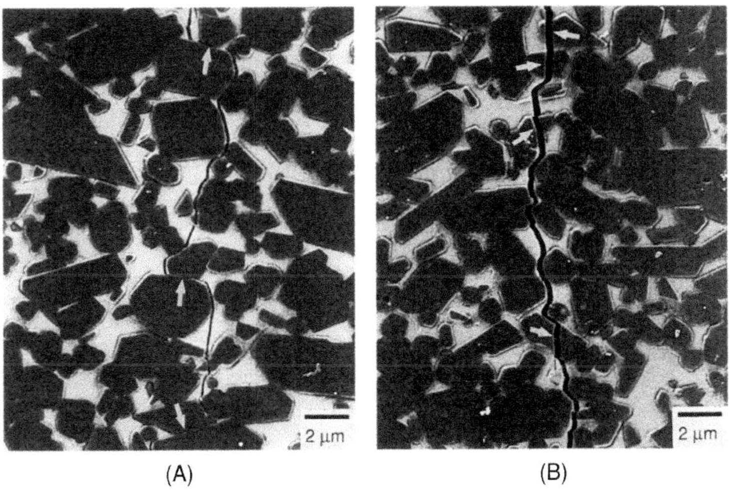

FIGURE 5.10 Surface traces of indentation cracks in alumina–glass dental composite by backscattered scanning electron microscopy (SEM). Grain bridging sites are near the arrows in (a) and trans granular fracture sites in (b) (Wolf, Vaidya, and Francis, 1996).

problems in bone cement and in dental composites. Viscoelastic effects like strain rate due to stiffness and creep proportional to the column fraction of the matrix are more dominant in fiber-reinforced continuous composites than in short fiber ones (Iftekhar, 2003).

5.9 BIOCOMPATIBILITY

Biocompatibility is related to the interaction of a biocomposite with the human body and its clinical success as a medical device, such as hip replacement or a pacemaker. The durability of a biomedical composite within the protein rich and warm salinized environment of the body and biocomposites' long-term consequences, such as their interaction with the body, characterize the biocompatibility feature of a biocomposite. An incompatible biocomposite may release toxic matter and may not integrate with the body environment, whereas biocompatible composites show the opposite, that is, they never release toxic substances and fully integrate within the body environment. Integration with tissue depends on the fibrous capsule formation between the composite and the tissue it is enclosed within. A moderate biocompatible composite can be strengthened through pairing with a matrix of high biocompatibility. Optimization of the structure of biocomposites can be done with advanced fabrication methods to reach high performance levels. Low atomic weight carbon-based biocomposites pose less problems after implantation while process of diagnostic implantation. Biocompatibility is the main criteria for selection and use of implants or materials. Figure 5.11 shows the interaction between an implant material and the body.

The long-term biocompatibility of implants refers to the capability of the device to accomplish its projected function with integration with the surrounding environment and without any unwanted defects in host body. For example, the

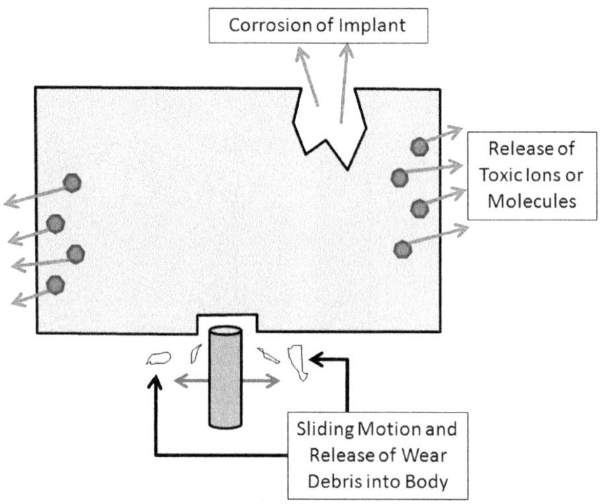

FIGURE 5.11 Interaction between the body and the implant material.

biocompatibility of an implant placed in the cardiovascular system for therapeutic or investigative purposes can be described as its ability to conduct its projected purpose in the bloodstream without any adverse interaction between blood and the device, while allowing any unrestrained activation for cellular or plasma protein cells (Ramakrishna, 2004).

Biocomposite implants interact with tissues at cellular and biochemical levels. When the implant is placed inside the body, it starts adsorption of proteins onto its surface. These proteins attract more cells and form an adaptive region around themselves. This bond is very important and suitable for bone implants where mechanical strength is dominant. The surrounding tissues of the implant are separated by a fibrous capsule formulated adjacent to the implant, but if the implant is biocompatible, the fibrous capsule will not be formed on its place, and there will be a required degree of integration between original tissue cell and implant. A high degree of biocompatibility shows bioactivity. Bioactivity refers to the integration and penetration of cells adjacent to the implants removing the implant boundary and forming a transition region, as shown in Figure 5.12 (Ramakrishna, 2004).

Osteogenesis, biocompatibility, and degradability of biocomposite scaffolds comprising nano-magnesium phosphate (nMP) and wheat protein were tested both in-vitro and in-vivo for suitability of application in bone regeneration (Xia et al., 2016).

Magnesium (Mg) is an important element in composites for bone formation and repair. Mg stimulates bone formation and healing due to bioactivity and osteostimulation. Wheat protein (WP) is a natural polymer having properties like biocompatibility and degradability. Scaffolds for bone tissue regeneration influence

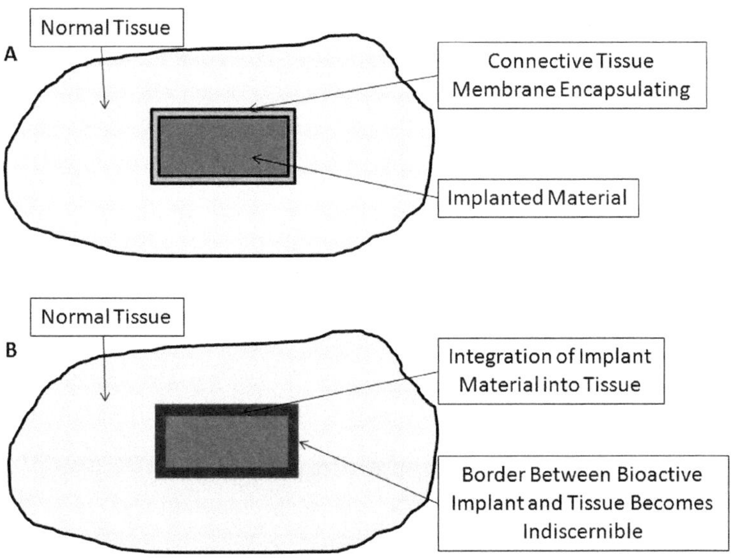

FIGURE 5.12 Bonding and nonbonding of implants inside the body for long-term basis with (a) a bioinert material; and (b) a biocompatible, bioactive material (Ramakrishna, 2004).

biological influences in the cell like cell attachment, new cell formation, and proliferation. A study was done on bioactive scaffolds with WP. nMP combined with WP was fabricated to form a nMP/WP composite (MWC) and tested for in-vitro degradability and cell growth. In-vivo tests were conducted to test the osteogenesis and degradation properties of the scaffolds. The nMP was synthesized through chemical precipitation. An MWC scaffold with 40wt% nMP was made up with solution casting and the particle leaching method. Experiments were conducted both in-vitro and in-vivo.

The degradability of the MWC composite scaffold and the WP scaffold were determined through in-vitro studies of samples by weight loss ratio in phosphate buffer saline (PBS) solution. Test results show that the weight loss ratio of the MWC scaffold increased in time, showing 46.34% after 12 weeks, while for the WP scaffold, the figure was 67.32%. Thus, the MWC scaffold is more suitable over WP scaffold because the weight loss ratio for WP is faster than the MWC scaffold. Cell proliferation and morphology of the scaffolds were tested with cells cultured in a chemical medium at 37°C for 2 days. The cells were then visualized under a laser scan microscope. The results show more cells with cytoplasmic extensions for the MWC scaffold with fibroblastic morphology and interconnected macrostructure. The MWC scaffold has higher compressive strength than WP (Xia et al., 2016).

In-vivo studies were conducted by attaching the MWC scaffold onto rabbits' femoral defects. The experiment was conducted using a total of 24 rabbits. 12 rabbits were used for MWC and WP scaffolding. Then, of the 4 rabbits were used for the MWC and WP scaffolding every 4 weeks (at 4 weeks, 8 weeks, and 12 weeks after the initial testing of 12 rabbits). A radial defect of 6 mm was created in the left thighbone of the rabbits through median incision. The defects were then fitted with the scaffolds. New bone formation was observed under a high power microscope. Samples were stained with Masson trichrome and observed under microscope. Masson trichrome staining proves the in-vivo effect of the MWC scaffold on new bone creation and degradation. Statistical study of histological sections was used to quantify newly formed bone. The same image was passed through analytical image software to find the new bone vs mature bone ratio. Synchrotron-microcomputed tomography (SRmCT) analysis was conducted using monochromatic light to detect bone defect repair after scaffold implantation. Samples were decalcified, washed, and dehydrated in alcohol for 30 days at room temperature. The results showed that defects were restored in line, without any complexity. The effectiveness and speed of bone defect repair was greater than that of WP. Histological elevation showed that the new bone vs mature bone ratio was higher in MWC scaffold. By 12 weeks, the MWC and WP scaffold degraded but the WP scaffold degraded faster than the MWC scaffold. Immunohistochemical analysis using XRD showed that stained MWC scaffold had a higher intensity than WP scaffold indicating better osteogenic properties. Histomorphometry analysis based on Masson trichrome staining suggested that the MWC scaffold stimulates effective new bone tissue growth with faster and improved bone regeneration and remodelling. So, Masson trichrome staining through immunohistochemical analysis showed that bone regeneration of the MWC scaffold is effective, improved, and faster, showing better osteogenic properties in the MWC scaffold (Xia et al., 2016).

A bioactive MWC scaffold was fabricated for this study. The results showed that the MWC scaffolds have a well interconnected microporous structure, improving the surface area which is suitable for tissue ingrowth. The MWC scaffold was degradable in PBS solution producing a weak alkaline environment and facilitating cell growth. Bioactive scaffolds of MWCs were made up. Experimental results showed that the MWC scaffolds were macroporous structures with proper interconnection. The micropore or macropore structures improved scaffold surface area, appropriate for bone tissue ingrowth. The MWC scaffolds were degradable in PBS and created a weak alkaline environment, which was beneficial for cell growth. Experimental results for the cell culture showed that MWC scaffolds extensively promoted cell growth and proliferation into the scaffolds. The experiment showed that the MWC scaffolds promoted new bone formation and degradation boosted new bone in growth in-vivo. Moreover, the immunohistochemistry test showed the indication of the stimulatory effects of MWC scaffolds on osteogenic cell differentiation and in-vivo osteogenesis. The MWC scaffolds thereby showed decent biocompatibility and degradability with improved bone rejuvenation (Xia et al., 2016).

5.10 STRUCTURAL BIOCOMPATIBILITY

Living tissue continuously regenerates. Aging of cells causes tissue to die, decompose, and be replaced by new cells. A similar process applied to bone tissue, which is comprised of osteoblasts and osteoclasts. Osteoblasts build up extracellular matrices and osteoclasts resorb them. The formation of bone by osteoblasts and resorption of bone by osteoclasts is balanced by hormones. The orientation, rearrangement, or density of trabecular architecture orientation of a bone if rearranged or density varies impacts the bone structural load-bearing capacity. Wolff's law of bone remodelling explains the influence of trabecular architecture on the main stress and the minimization of shear stress of composites. Placement of an orthopaedic implant inside the bone changes the local stress level. Loads on the bone may be same, but the mechanical stress of the bone adjacent to the implant will be impacted. As per Wolff's law, the insertion of a composite implant changes the bone density and architecture by remodelling. Thus, local stress on the bone becomes the dominant factor for bone material retention. When a high stress area of a bone observes a lack of stress, it creates a problem for the composite implants. The section of the bone where the implant is placed may atrophy due to reduction of mass in the bone and the decrease in mechanical strength known as "stress shielding". A biocomposite implant that minimizes this bone atrophy is structurally biocompatible. Structural bioincompatibility occurs due to composite implants that are stiffer than the original bone and implants. Again, composite implants which are very stiff and isotropic can cause a high extent of strain mismatch in the bone interface, resulting in the hindrance of implant bonding to the bone (Ramakrishna, 2004).

5.11 ADVERSE EFFECTS OF COMPOSITE IMPLANTS

A lack of biocompatibility may release toxins in the body, cause infection, and create adverse effects resulting in failure, rejection, and removal of the biocomposite. Some potential adverse impacts of biocomposite materials on a human host body can be:

- Coagulation, hemolysis, thrombosis
- Regular, acute, or chronic inflammation
- Toxicity and Immunogenicity
- Infection
- Corrosion, wear, or tear
- Loosening
- Fracture of implant and bone
- Rejection of Implant
- Tumorigenesis /Carcinogenesis.

5.12 THE ENVIRONMENT WITHIN THE BODY

The human body environment is aqueous, saline, warm, and cannot tolerate strong acidity or severe alkalinity. Living tissues regularly secrete protein, enzyme, and other chemicals. These proteins interact strongly with implanted devices and create problems. In the case of this interaction being adjustable and beneficial to body, the implanted device is called biocompatible. The environment within the body exposes a corrosive medium to the implants. The environment within the human body contains water, dissolved oxygen, calcium, magnesium, sodium, potassium, chloride, phosphate, bicarbonate, amino acid, protein, saliva, plasma, and organic compounds, and is highly oxygenated and saline with 7.4 pH at 37°C. These biological elements influence the electrochemical behaviour of implant constituents like metal, polymer, ceramic, or alloy. For example, protein adsorption on the implant surface can interact with oxygen and hydrogen particles present causing redox reaction. Again, human body bacteria influence the corrosiveness of the implanted composite. If electrochemical equilibrium within the body is imbalanced, this can result in infections or shifts to physiological pH. If metal ions are released from composite implants, it can influence electrochemical kinetics and interfere with equilibrium. The corrosion behaviours of implants are impacted by kinetic and thermodynamic factors. Oxygen, chloride content, and the pH of bodily fluids most highly influence the corrosion behaviour of implants. So, corrosiveness of a composite implants is tested through in-vivo studies. Appropriate material selection and design of a composite can be an effective measure to minimize and control the corrosion factor of composites.

Composites mainly engineered with polymers, ceramics, and metals, are used as implantable medical devices. Metallic composites undergo corrosion in the salty body environment and increase protein absorption. Ceramics and polymers are resistant to aqueous corrosion but the leaching of blood affects polymers. Some implant sites such as dental cavities, hip joints and knee joints are shown in Figure 5.13 (Alijani and Anvari, 2018).

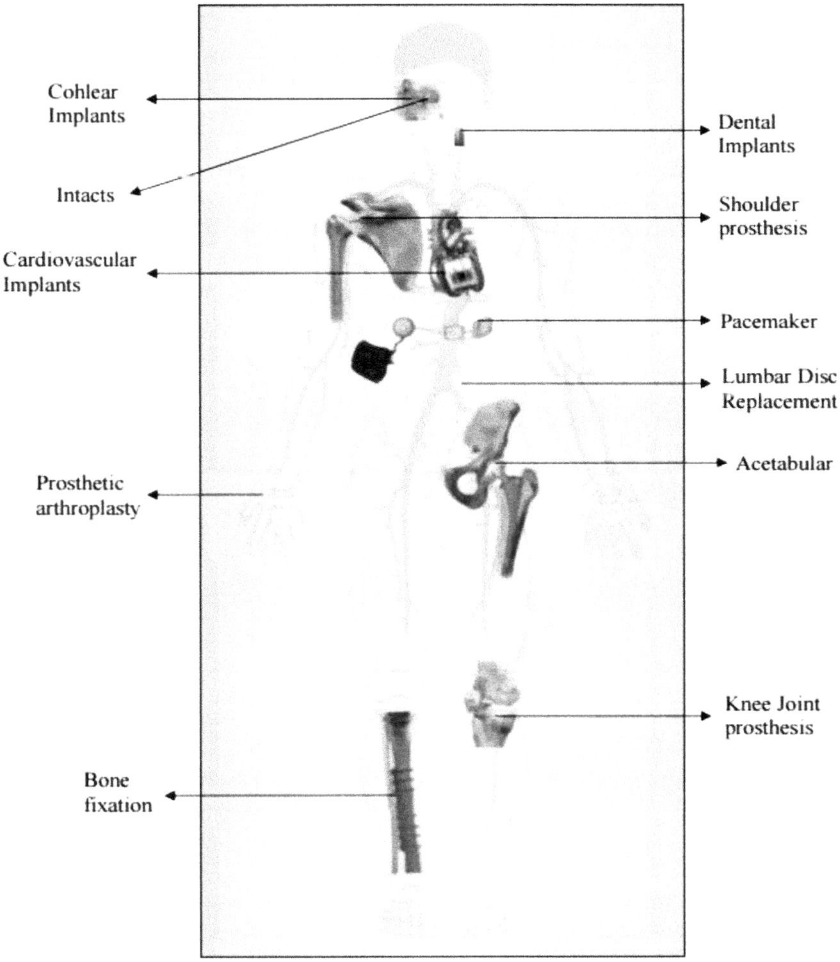

FIGURE 5.13 Different body implants (Alijani and Anvari, 2018).

5.13 STERILIZATION OF BIOCOMPOSITE IMPLANTS

The sterilization of a biocomposite implant is very important. Sterilization ensures patients' safety from bacteria, viruses, or other parasitic contamination during the implantation. Sterilization quality is measured by the Sterility Assurance Limit (SAL). The SAL shows the probability of an implant being non-sterile after sterilization. Some sterilization processes are listed below:

- Pressurised autoclave at 120–140°C, 2–3 bar pressure (Ramakrishna, 2004)
- Immersion in pure ethylene oxide or with carrier gases (freon or carbon dioxide) (Ramakrishna, 2004)
- Exposure to gamma radiation of cobalt-60 isotopes (Ramakrishna, 2004)

- Absorption in supercritical CO_2 at 20 MPa at ambient (or slightly higher) temperature (Ramakrishna, 2004)
- Advantages and disadvantages of each method are shown in Table 5.3

An application of antibacterial coating over biocomposites for craniofacial reconstruction was studied (Lazar, 2016). Hospitalization of patients or functional and structural impairment of the implant can happen after reconstructive surgery if infection occurs. Infection rates between 3% and 40% are reported for implants despite sterilization, even if antiseptic techniques are used and systemic antibiotics are prescribed before and after the craniofacial surgery. This large range of infections results in a high urgency to remove the implants. Reduction of the infection rate is possible if an antibacterial coating is applied to the surface of the biocomposite implants that are used for craniofacial reconstruction. Fiber-reinforced composite implants incorporated with antimicrobial properties were fabricated and tested. Gentamicin was applied as a coating on the external surface of the implant. Two bacterial strains that may cause infection (*Staphylococcus aureus* and *Pseudomonas aeruginosa*) were exposed to the implant, which was tested

TABLE 5.3
Different Sterilization Methods

Method	Principle	Advantage	Disadvantage
Steam sterilization	Saturated steam in the range of 121–134°C	Sterilises penetrable materials and exposable surfaces	Microcavitation in case of hydrophilic polymers Altered biocompatibility of heparinized surfaces
Dry heat sterilization	Carries out in electrically heated ovens at 160–180°C	Ability to penetrate solids. Lack of corrosion in case of non-stainless-steel metals	Rubber, plastic, etc. do not withstand high temperature
Ethylene oxide	Biocidal activity is achieved at 30% humidity, 45°C temperature levels with > 450 ml/l gas concentration	Suitable for heat sensitive implants	EO residues after sterilization cause Anaphylactoid reactions in dialysis patients Serious tissue reactions in pump oxygenators Mutagenicity and carcinogenicity
Radiation sterilisation	Primary biocidal action is via aqueous free radical formation following primary physical interaction of ionising radiation with the biological material	High-energy irradiation sterilisation	Degradation and/or crosslinking of polymers Gas evolution and free radical formation from polymers

Source: Matthews (1994).

for adherence of bacteria and efficiency against bacteria. The gentamicin coating repressed the growth of bacteria, leading to less adhesion of bacteria to the implant. The antibiotic coating of craniofacial implants reduced the infection rate. Gentamicin submission, bacterial strains and culture conditions, antibacterial efficiency tests, bacterial adherence, and data analysis were conducted. Data analysis was performed with SPSS 20 (Lazar, 2016).

An antimicrobial activity test for both *Staphylococcus aureus* and *Pseudomonas aeruginosa* proves the efficacy of gentamicin coating against bacteria strains as they create an inhibition zone on gentamicin coated FRC implants which inactivated bacteria with direct gentamicin contact. The inhibition zone for *Staphylococcus aureus* was between 7.21 mm to 20.13 mm and for *Pseudomonas aeruginosa* between 12.93 mm to 15.33 mm. A negative relation was detected for bacterial adhesion and gentamicin concentration, indicating inverse relationship between the bacterial adherence and the antibiotic dosage (Lazar, 2016).

In the bacterial adherence test, the interface between the material surface and the cells was measured to determine initial adhesion, which showed minimal *Staphylococcus aureus* and *Pseudomonas aeruginosa* accumulation to the gentamicin coated specimens than to the control discs. Ciprofloxacin and gentamicin are recommended for creating bacterial resistance in oral maxillofacial area as gentamicin has proven efficacy for activity against bacteria. This is an important design criterion while designing an antimicrobial coating for FRC implant surface to create an anti adhesive surface against bacterial inhibition. The statistical results of the bacterial adherence test were insignificant. The antimicrobial effectiveness test exhibited good results on both strains of bacteria due to the release of antibiotics from the implant coating (Lazar, 2016). A negative correlation was observed between bacterial strains and gentamicin concentration, indicating a reverse relation between antibiotic dosage and bacterial adherence (Lazar, 2016).

Inactivation of bacteria occurred with direct contact with gentamicin applied on the surface of FRC implant. The results of the microbiological study showed that gentamicin coating inhibited the growth of bacteria as well as lowered initial bacterial adhesion to the implant surface. So, an antibiotic coating of craniofacial composite implants for reconstructive surgery reduces the rate of infection (Lazar, 2016).

5.14 IMAGING OF BIOCOMPOSITES AFTER IMPLANTATION

After implantation, the correct implant location and attachment to adjacent tissue are confirmed through radiography or magnetic resonance imaging (MRI). Metallic components are opaque to X-rays. The presence of magnetic iron distorts MRI imaging. Sometimes composite implants can block the area being imaged. MRI magnetic currents may heat metallic implants, and human protein cells cannot tolerate even a small temperature increase. Thus, the imaging risks and hazards cause significant problems for the surgeon. Complications while taking a diagnostic image of composite implant, then, are as substantial as other long-term difficulties like corrosion or resorption of bone for implants. However, biocomposites are composed mostly of carbon fiber and organic materials. Their main elements are carbon and hydrogen, which are non-magnetic and have low atomic weight. Thus, the composites are transparent under X-rays and do not hinder MRI imaging. Sometimes a composite becomes so transparent to X-rays that 1wt.% tungsten metal as short fibers is added to biocomposites to create shadow.

5.15 BIOLOGICAL RESPONSE

When designing biocomposite implants and analysing their performance, the way they provide a biological response is very important. With the increase of the number of constituents in the composite, the body response will vary. Combinations of constituents need to be approved. Composition, interaction, arrangement, and biocompatibility of composites are considered while judging biological response. Again, composite biological response varies more in bulk form than in particulate or fibrous form. For fibrous or particulate form, the host response to composites varies. A constituent of the acetabular cup of a hip prosthesis might be biocompatible but show adverse reaction in fibrous form (Iftekhar, 2003). Inflammatory responses might include absorption by cells and transfer to other parts, thereby releasing enzymes and causing degradation, affecting the performance of the composite implant. While designing the composite, care should be taken so that the fibers and particles are not exposed to the body environment. For this, voids must be nullified at the interfaces of the fiber or particle matrix. The friction of dental and orthopedic composites can cause abrasion and create cavities at the interfacial area and the exposure of the implant to the human environment. The interaction of composite materials at the interface can be affected by tissue response. With the swelling of voids and the filling of voids with fluids and tissue, variance in the strength between the matrix and the reinforcement at the point of adhesion to the interface may occur. Such effects may cause the failure of composite implants in structural applications. In the case of thermoplastic and thermosetting polymer composites, residual solvents may leach from the matrix if not removed properly.

5.16 NANOCOMPOSITES

A nanocomposite is a composite material in which one of the components has at least one dimension that is around 10 nm. Elastin–collagen, nano-hydroxyapatite/chitosan–gelatin (nHAp/CS–Gel) biocomposites are examples of nanocomposites. Nanocomposites are applied as diagnostic tools, drug delivery mechanisms, and cancer treatments (Figure 5.14). Biphasic nanowires are very much useful for coating permanent transplants to improve osseointegrative properties. Nanocomposites have:

- High surface to volume ratio of the reinforcing
- Very high aspect ratio
- One matrix phase with one or more reinforcement material
- Increased electrical breakdown strength, melting temperature, and interacting zone
- Metal–ceramic nanocomposite

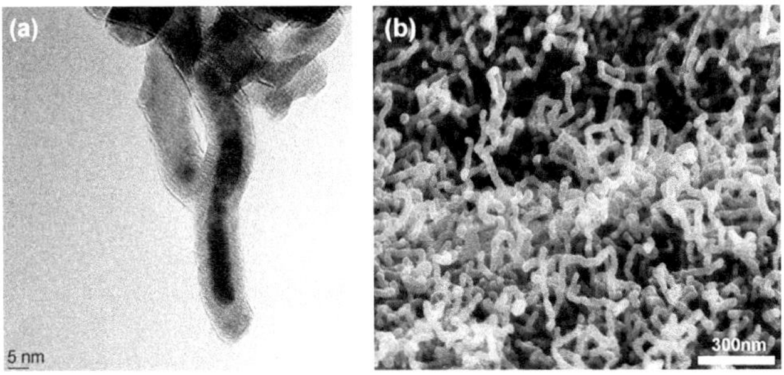

FIGURE 5.14 Nanocomposites (Pandya, 2013; Veith et al., 2008; Lee, 2014).

5.17 BIOMEDICAL APPLICATION OF BIOCOMPOSITE IMPLANTS

5.17.1 Hard Tissue Application

Bone is a natural composite with anisotropic properties. The anisotropy of the elastic properties of human tissue is considered an important criterion in composite design. Resorbable and porous ceramic–polymer composites have similar morphology and mechanical properties to bone. Again, this gives surgeons the advantage of cutting the graft at the surgery site, according to shape of the defect. Due to their low elastic modulus and high strength, ceramic–polymer composites are suggested for several orthopedic applications. The percentage of the reinforcement and continuous phases of composites can be controlled, therefore, the design of the composite can be altered to suit the physiological and mechanical properties of host tissue. If composites are used, they can release allergic metal ions, such as nickel or cadmium, and thereby can help with the removal of corrosion. Composites are low in weight and because of their better aesthetic characteristics, biopolymers are widely used as dental implants. A clinical application of biocomposites is in cages for spinal fusion, which gives the patient the benefits of faster bone healing, a cheaper and faster surgery, less pain, and no risk of pathogen transfer (Ramakrishna, 2004).

Composites are extensively used in orthopedic applications like total hip replacement, total knee replacement, bone fixation, and bone grafts. Cobalt–chromium alloys, stainless steel, and titanium alloys with high stiffness are used in hip replacement. The stiffness of titanium is 110 GPa and tensile strength is 800 MPa whereas the corresponding values of the cortical bone are 15 GPa and 90 MPa. For this reason, stress shielding and adverse bone remodelling occurs, resulting in the loss of bone and loosening of the implant in the long term. Fiber composites are fabricated matching the mechanical properties of bone. "For example, carbon fibers embedded in PEEK or Polysulfone, which are widely used polymers, can have strengths ranging from 70 MPa -900 MPa and a stiffness ranging from 1.0 GPa -170 GPa" (Yildiz et al., 1998). Femoral stem made of carbon fibers (unidirectional and laminar), liquid crystalline polymer (LCP), PEEK, polysulfone, or polyetherimide (PEI) are other examples of fiber composites. These composites are non-corrosive, flexible but not durable, and very difficult to fabricate. The biocompatibility problems of this composite caused by paniculate carbon debris can be addressed by application of HAp carbon–titanium alloy coating or polishing (Bacakova, 2001). Fully resorbable bone plate is desirable for fracture fixation, which eliminates the need for a second surgery for removal of the plate after healing. A low stiffness composite can serve this purpose, along with avoiding the stress shielding problem, which is very acute in the case of orthopaedic implants. The degradation rate must be controlled to maintain mechanical properties and prevent the loss of strength. The degradation by-products should be non-toxic. Continuous and laminated carbon fiber in a polylactide (PLA) matrix (partly resorbable) and calcium-phosphate glass fibers in PLA (fully resorbable) are examples of resorbable bone plates. However, though these composites degrade rapidly, they lack adequate mechanical properties. Bone cements fill up voids and enhance the bond. The cements are reinforced with fibers to prevent loosening and increase shear strength. PMMA powder mixed with a methacrylate-type monomer that is polymerized during fixation is an example of a composite bone cement. An increase in fatigue life and a reduction of creep deformation of the PMMA matrix is done through the addition of small volumes of carbon, graphite, and Kevlar fibers (Kelly, Cahn, and Bever, 1989). Some examples are

- Bone fracture repair – a composite of woven cotton fabrics (woven gauze) and a plaster of Paris matrix (calcium sulfate)
- Bone plates – CF/PEEK composite screw, carbon/PEEK composite plate
- Intramedullary nails – fixes long bone fractures (femoral neck), small fractures of knee, foot, or wrist – carbon fiber-reinforced (CFR) polyether ether ketone (GF/PEEK) composite
- Spine instrumentation – spondylolisthesis, metastasis of vertebral body and disc–CF/PEEK and CF/PS composite (Figure 5.15) (Hussain, 2006)
- Joint replacements – hip, knee, shoulder, or elbow due to osteoarthritis
    - Artificial joints are total hip replacement (THR) – cobalt–chromium–titanium alloys for the femoral shaft and neck, and cobalt–chromium alloys or ceramic (alumina and zirconia) materials for the head or ball of acetabular cups made with ultra-high molecular weight polyethylene (UHMWPE) with metal backing (Figure 5.16) (Katz, Lubovsky, and Yosibash, 2018; Hsu, 2018).
    - Total knee replacement (TKR) – fiber-reinforced UHMWPE polymer composite.

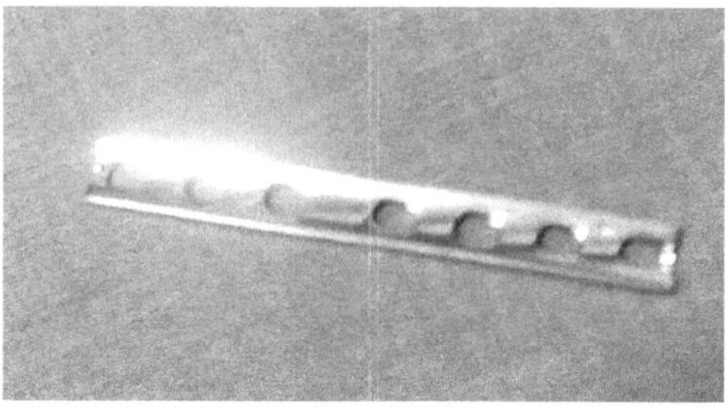

FIGURE 5.15 Spine instrumentation (Hussain, 2006).

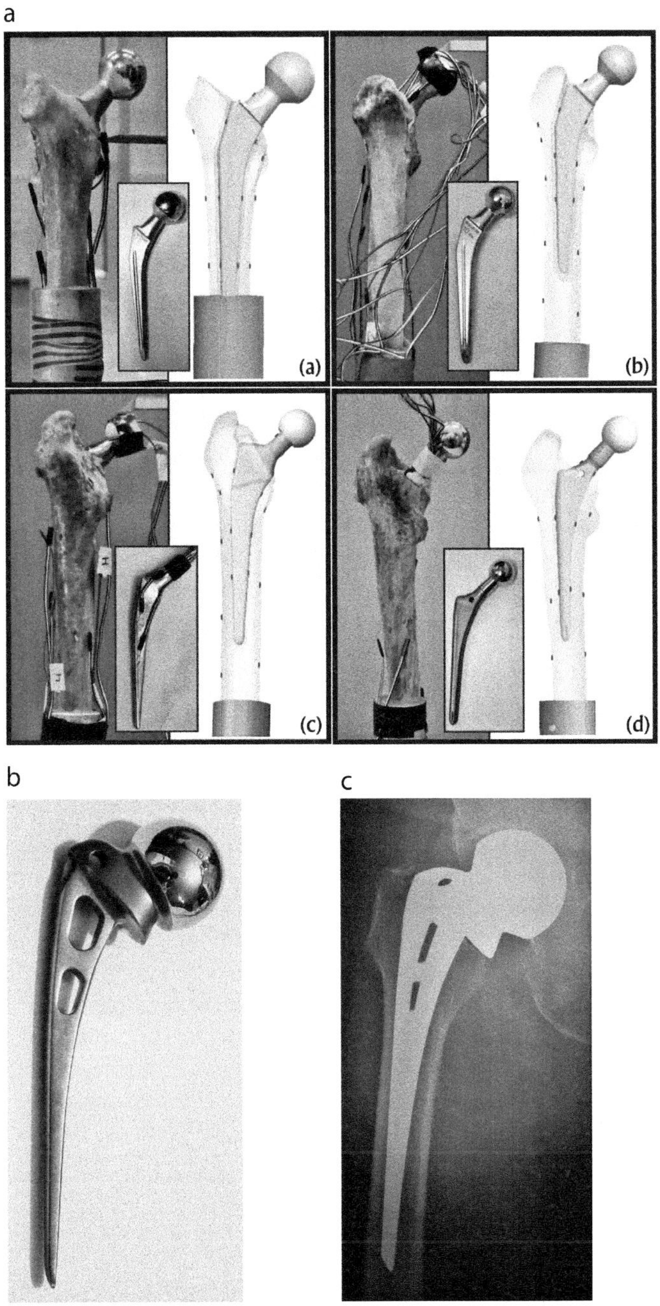

FIGURE 5.16 Hip joints (Katz, Lubovsky, and Yosibash, 2018; Hsu, 2018).

- Bone cement – acrylic bone cement (PMMA-based) – PMMA powder and liquid MMA monomer
- Bone graft (synthetic bone replacement) – polyethylene (PE) reinforced with HP and bioglass

5.17.2 Dental Application

The most successful applications of composites are in dental applications. Composites meet stringent design criteria which are not possible to achieve through single materials like ceramics or metal alloys. Dental crowns, cavity repairs, and dental implants are all applications whichc need to match in colour, gloss, and translucence with other teeth, along with hardness, wear resistance and fatigue fracture resistance. Dental composites must have stability in dimension and set shape and should withstand thermal stress in the mouth with a short processing time. Polymer resin matrices are used for cavity repair. Fillers are treated with saline or fused with silica to improve retention in the matrix, stress bearing, restoration, and to reduce water absorption. Fillers can occupy 50–80vol.% of the composite with a size variation from 20 nm to 50 μm. Microfilled dental resin polished with high gloss and fused silica particles of 20–40 nm can be incorporated up to 40vol.% to yield translucent composites. This is not mechanically strong for the posterior teeth and is difficult to handle due to low viscosity. Hybrid dental resins have particle sizes 0.1–10 μm, which allows 80 vol.% of filler and a high viscosity for ease of handling and minimization of water absorption. Dental composite resins have polymerization shrinkages of 1.6–2.5% and water absorption up to 1.5% with poor adhesion to dentin, which becomes bonding agents for fitting prevention of fitting and remedy for leakage (Iftekhar, 2003).

Ceramic dental composites are used as crowns as it repairs stress. Fracture toughness is addressed through crack deflection and bridging mechanisms in the composites. An alumina–glass composite with alumina particles is a composite of 75 vol.% α-alumina particles (3 μm size) and 25% glass with an aqueous tape-casting alumina core, and conforms well as tooth model. Dental composites are harder than the natural tooth enamel. Thus, they are coated with calcium–phosphate composites or alkali aluminosilicate dental porcelains to reduce the surface hardness (Kihara, 1984). Another application of dental composites is for orthodontic arch wires, for example, a unidirectional pultruded S2-glass-reinforced dimethacrylate thermoset resin. Fiber volume fraction varied from 30 to 75% based on the yarn of glass fiber used. Orthodontic brackets are also composites made up of a polyethylene matrix reinforced with ceramic HAp particles that demonstrate good isotropic properties and better bonding to enamel (Iftekhar, 2003). Some uses are listed below with examples (Figure 5.17).

- Dental composite resin – BIS–GMA as the matrix polymer and quartz, barium glass, and colloidal silica as fillers
- Dental post – root canal – zirconia, short glass fiber-reinforced polyester, unidirectional CFR epoxy composite posts

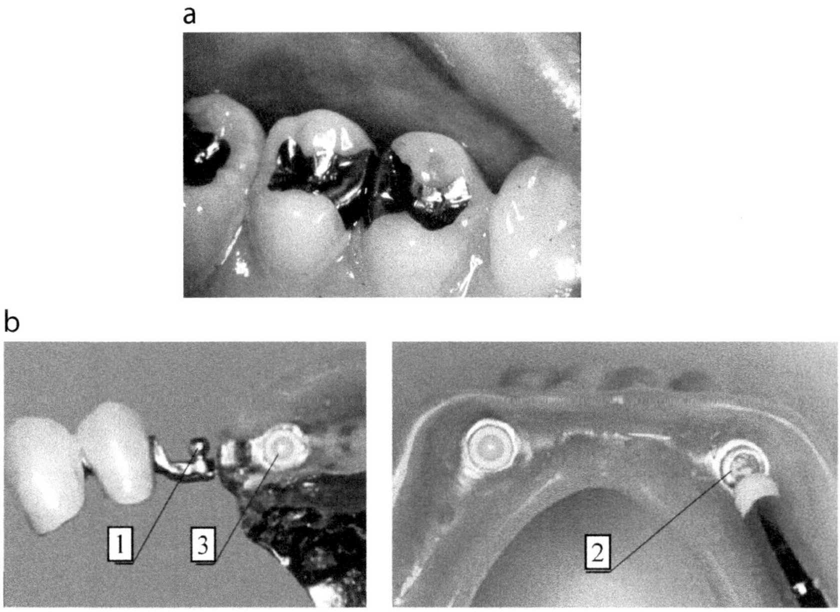

FIGURE 5.17 Composites in dentistry (Hayashi et al., 2018; Jałbrzykowski et al., 2018).

- Dental implant and bridge – the damaged tooth is extracted and replaced with a dental implant –SiC/carbon and CF/carbon composite CF/PMMA, GF/PMMA composite bridges, and dentures (false teeth)
- Bracket and arch wire - GF/PC, GF/Nylon, GF/PP, and GF/PMMA composite materials for arch wire application – polymer reinforced with chopped glass fiber composite as brackets

5.17.3 Soft Tissue Application

A crosslinked hydrogel network with more water content is suitable as a scaffold for skin regeneration. The break point and tensile strength of the hydrogel poly (2-hydroxyethyl) methacrylate (pHEMA) are improved by reinforcing spandex and gauze fibers which make gels capable to bear forces. Cartilage repair requires high compressive strength and shear properties. Improved shear properties and compressive properties are necessary in cartilage repair such as low-density polyethylene melt coated in three-dimensional woven fabric of UHMWPE fibers. The final composite shows compressive behaviour like natural cartilage (Seal, 2001). Fibrous polyglycolic acid (PGA) and poly(lactide-co-glycolide) (PLGA) fibers are applied as scaffolds and surfaces for tissue or cell regrowth for cartilage defect repair. Inter-fiber distance, fiber diameter, and degradation rate affect neocartilage quality. Woven fiber tubes made from polyester (Dacron) are used in vascular graft. These have high permeability during implantation causing blood leakage, which is avoided by the application of an impermeable material coating that results in composite formation on grafts. Some examples are listed below.

Bulk space fillers – re-establish cosmetic defects or atrophy to an aesthetically satisfactory condition – composites including PET or PTFE fabrics and PU treats cartilage defect.
Ureter prosthesis – polyester fiber-reinforced glycol methacrylate gel prostheses with a fabric backing
Catheters – Silicone rubber is reinforced with silica particles along with PU composites
Functional load – carrying and supporting implants, artificial tendons and ligaments, carbon–collagen fiber multifilament composite

Figure 5.18 shows some applications for soft tissue composites.
Tissue Engineering for Skin

5.17.4 Application as Drug Delivery and Scaffold

Drug delivery – composite implants are drug delivery media for thrombosis, osteomyelitis, periodontitis, infections, and inflammation.

Scaffold – scaffolds are produced using biocomposites. Scaffolds provide a cell's structural framework and mechanical strength and biochemical conditions, and cells proliferate in the presence of a scaffold and produce an extracellular matrix to form new tissue. An example can be seen below in Figure 5.19 (Department of Energy, 2009).

Left: Chitosan/hydroxyapatite-based ceramic polymer – Right: Metal–ceramic scaffold

5.17.5 Orthotics and Prosthetics

Old orthotic and prosthetic materials (aluminium, leather, or wood) are replaced by composites. Due to its low weight, reduced size, durability, fiber-reinforced plastic is widely attractive for this purpose. Composites are used for transtibial (TT) and transfemoral (TF) prostheses as the socket-frame transmits the load while interacting with the limb. CFR composites are ideal for TT and TF prostheses. CFR epoxy tubing is used instead of stainless steel in artificial arms. Hybrid composites used in shanks consisting of carbon and nylon fibers in polyester resins have better impact resistance than composites than those containing only carbon or nylon. Different composites used in prosthetic feet retain energy along the total length while moving. A flex foot unit with a long CFR epoxy composite beam stores flexural energy to entire prosthesis. Nylon-reinforced silicone elastomer is used as a hardwearing cover and encases the prosthesis. (Iftekhar, 2003). In orthotics, composites are used as a new bandage-form splinting material such as laminated fiberglass or polyester fabrics in polyurethane matrices, which are strong and have low water absorption and good X-ray transmission properties.

Prosthetic limbs, for example, artificial legs or arms, are thermoset polymer composites reinforced with glass, carbon, or Kevlar fibers. Polymer composites

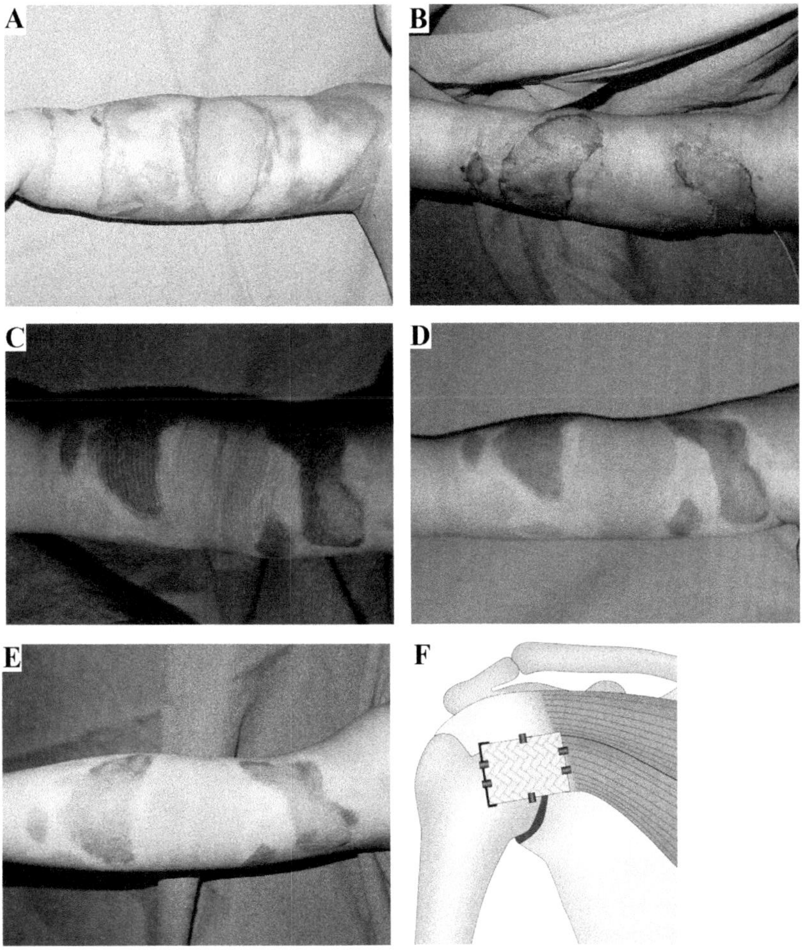

FIGURE 5.18 Soft tissue applications of composites (Hur, Seo, and Lee, 2014; Smith and Grande, 2015).

which are lightweight, corrosion resistant, fatigue resistant, easily fabricated, and have good aesthetics are ideal choices for modern limbs systems (Ramakrishna, 2004).

A biogenic HAp /CS/polyvinyl alcohol (PVA) biocomposite was fabricated and studied to test biocompatibility and as a drug delivering carrier for proneurogenic factor (Nellore, 2014).

HAp, CS, and PVA were combined to synthesize a 3D composite scaffold that favours craniofacial bone regeneration and which has increased mechanical properties. This composite was used for reconstruction of craniofacial deformities. HAp was synthesized through in-situ synthesis and the chemical precipitation method from biomolecules occluded in cucumber peel collected from kitchen waste. HAp was characterized through SEM, XRD and FTIR for purity and morphology of

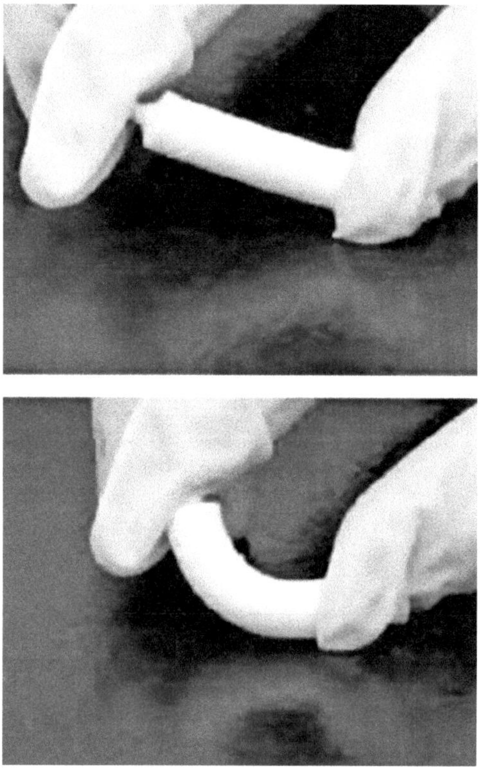

FIGURE 5.19 Biomimetic scaffold (Department of Energy, 2009).

CPHAp (cucumber peel-synthesized HAp). The biocomposite was fabricated using modified protocols.

PVA (7%) and CS (3%) mixed with methanol and HAp (2 w/v%) exposed to ultrasound to form CS-PVA-CPHAp-ATRA slurry. Slurry was then placed in a cylindrical mould, frozen, and dehydrated to form a CS-PVA-CPHAp composite encapsulated with all-trans retinoic acid (ATRA). A blank composite (CS-PVA-CPHAp) was made-up without ATRA. The prepared biocomposite was characterized by the analytical instruments SEM, FTIR, XRD through in-vitro release studies as well as cell activity studies. SEM analysis confirmed that the interconnected porous structure was uniformly spread with ATRA, indicating the morphology of biocomposite. Porous structures demonstrate the interconnectivity nature and cell attachment about spreading on scaffold. FTIR confirmed biocomposite formation without chemical modification of ATRA. XRD confirmed that the composite was amorphous in nature. It also confirmed that encapsulation of ATRA into the CS-PVA-CPHAp biocomposite expanded the dissolution of ATRA. The release of ATRA was measured at 55% after 10 days, which is within the range and proves the delivery of drug material into the scaffold (Figure 5.20). The ATRA release being in range also induces neurogenesis. Then the MTT assay was conducted to prove the biocompatibility features of the

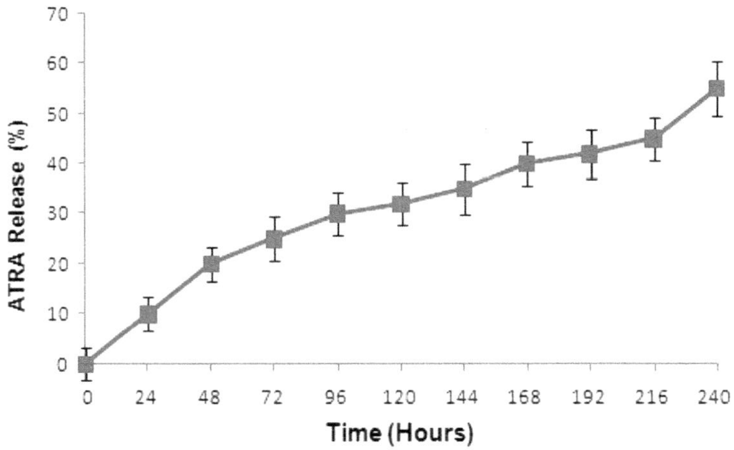

FIGURE 5.20 Release profile of all-trans-retinoic acid (ATRA) in chitosan-polyvinyl alcohol-CPHAp-ATRA bio composite (Nellore et al., 2014).

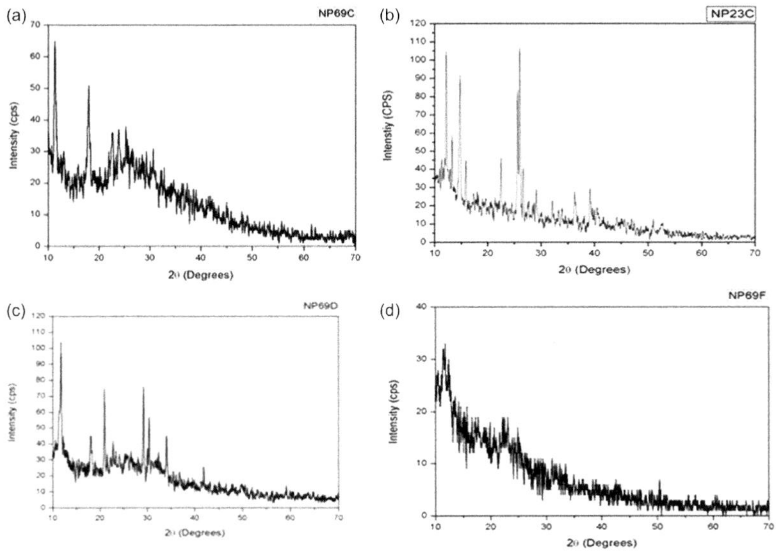

FIGURE 5.21 X-ray diffraction patterns of (a) Chitosan (CS), (b) CPHAp, (c) CS-CPHAp-all-trans-retinoic acid (ATRA) biocomposite, (d) CS-polyvinyl alcohol-CPHAp-ATRA biocomposite (Nellore, 2014).

composite and the non-cytotoxicity of the scaffolds. The MTT assay and phase contrast microscopic assessment proved that 30 nM ATRA release does not weaken cell attachment and spreading, rather, it improves them. The MTT assay or cell activity test proved that the composite is a better scaffold and is biocompatible (Figure 5.21) (Nellore, 2014).

5.18 ADVANCEMENT IN COMPOSITE IMPLANTS

Some recent advancement of biocomposite implants are as follows:

- Biodegradable biocomposite fabrication
- Biological and bioactive composite formation
- Antibiotic coating on composite implants (gentamicin coating reduces bacterial adhesion)
- Cosmetic or aesthetic composite implant formation
- Formation of microchip controlled bionic devices

The composite implant market is gradually expanding. The driving factors for the growth of the composite implant market are the increase in the ageing population, the increase in research and development investments, the increase in applications of biomaterials or composites, the growing demand for plastic surgery, and reimbursements offered by the Center for Medicare and Medicaid Services. (reportlinker.com, n.d.)

5.19 CONCLUSION

Two or more constituents which are physically and chemically different are combined, forming a composite. Constituents of composites do not blend or dissolve into each other but remain separated by an interface, and the resulting composite compound shows desired and improved mechanical and physical properties. Research shows that a PC-reinforced HAp composite has improved mechanical and thermal properties. Composites are widely used in the biomedical industry and are designed to interact with biological systems. Composites have low weight, low specific gravity, higher strength than metals or alloys, improved stiffness, improved corrosion and oxidation resistance, better fatigue strength, improved toughness, durability and design flexibility which put composites in an advantageous position. However, composites are heterogeneous, meaning that properties in the composite vary from point to point, and they are also anisotropic, meaning that their strength varies along with direction. Stiffness, thermal expansion, electrical and thermal conductivity, and creep resistance are anisotropic. Due to improved mechanical and thermal properties, composites are used for bone regeneration applications such as scaffolds, orthopedics, prosthetic limbs, and find success in dental application due to their aesthetic and cosmetic properties. Composites are replacing traditional homogenous materials like ceramics, metals, etc., and are being successfully used in forming biomedical devices for implants.

REFERENCES

Adzali, N. M. S. (2012). Mechanical properties, corrosion behavior, and bioactivity of composite metal alloys added with ceramic for biomedical applications. *Reviews on Advanced Materials Science*, 30(3), 262–266.

Alijani, S., and Anvari, A. (2018). Cycle numbers to failure for magnesium and its alloys in human body fluid. *Journal of Chemical Engineering and Materials Science*, 9(1), 1–8.

Bacakova, L. (2001). Polishing and coating carbon fiber-reinforced carbon composites with a carbon-titanium layer enhances adhesion and growth of osteoblast-like MG63 cells and vascular smooth muscle cells in vitro. *Journal of Biomedical Materials*, 54, 567–578.

Bellefontaine, G. (2010). *The Corrosion of CoCrMo Alloys For Biomedical Applications*. [MRes thesis]. School of Metallurgy and Materials, University of Birmingham.

Chan, W. S. (1997). Fracture and damage mechanics in laminated composites. In P. K. Mallick (Ed.), *Composites Engineering Handbook* (1st ed., pp. 309–370). Boca Raton, FL: CRC Press.

Department of Energy. (2009). *Hydrogel-HA*. [Online Image]. Retrieved from https://commons.wikimedia.org/wiki/File:Hydrogel-HA.jpg

Dujovne, A. R. (1993). Mechanical compatibility of noncemented hip prostheses with the human femur. *The Journal of Arthoplasty*, 8(1), 7–22.

Hayashi, M., Yamada, T., Lynch, C. D., and Wilson, N. H. F. (2018). Teaching of posterior composites in dental schools in japan - 30 years and beyond. *Journal of Dentistry*, 76, 19–23. doi:10.1016/j.jdent.2018.02.002

Hsu, A. Y. (2018). Changes in rehabilitation outcomes by new guidelines of hong kong hospital authority in implant choice for femoral neck fractures—Austin moore versus cemented exeter hemiarthroplasty. *Journal of Orthopaedics, Trauma and Rehabilitation*, 25, 37–48. doi:10.1016/j.jotr.2017.08.002

Hur, G., Seo, D., and Lee, J. (2014). Contracture of skin graft in human burns: Effect of artificial dermis. *Burns*, 40(8), 1497–1503. doi:10.1016/j.burns.2014.08.007

Hussain, N. (2006). *Dynamic Compression Plate*. [Online Image]. Retrieved from https://commons.wikimedia.org/wiki/File:Dynamic_compression_plate.jpg

Iftekhar, A. (2003). Biomedical composites. In M. Kutz, *Standard Handbook of Biomedical Engineering and Design*. McGraw-Hill.

Jałbrzykowski, M., Dąbrowski, J. R., and Minarowski, Ł. (2018). TiC-composite materials in applications for kinematic denture connections. *Advances in Medical Sciences*, 63(2), 374–381. doi:10.1016/j.advms.2018.04.012

John, M. J., and Thomas, S. (2008). Biofibres and biocomposites. *Carbohydrate Polymers*, 71(3), 343–364.

John, M. J., Anandjiwala, R. D., and Thomas, S. (2009). Hybrid composites. In. S. Thomas and L. A. Pothan (Eds.), *Natural Fibre Reinforced Polymer Composites: From Macro to Nanoscale* (pp. 315–328). Philadelphia, PA: Old City Publishing, Inc.

Katz, Y., Lubovsky, O., and Yosibash, Z. (2018). Patient-specific finite element analysis of femurs with cemented hip implants. *Clinical Biomechanics*, 58, 74–89. doi:10.1016/j.clinbiomech.2018.06.012

Kelly, A., Cahn, R. W., and Bever, M. B. (1989). *Concise Encyclopedia of Composite Materials*. Pergamon Press, MIT Press.

Kihara, S. (1984). Calcium Phosphate Glass–Ceramic Crown Prepared by Lost-Wax Technique. *Journal of the American Ceramic Society*, 67(6), C-100–C-101.

Kutz, M. (2002). *Standard Handbook of Biomedical Engineering and Design*. New York: McGraw-Hill.

Lazar, M. A. (2016). Antibacterial coating on biocomposites for cranio-facial reconstruction. *Clujul Medical*, 89(3), 430–434.

Lee, J. (2014). Composite Materials for Bio Medical Apllications. Retrieved from CVD/Biosurfaces, INM-Leibniz Institute for New Materials: http://training.pluscomposites.eu/sites/default/files/6_Lee.pdf.

Long, M. (1998). Titanium alloys in total joint replacement—a materials science perspective. *Biomaterials,* 19(18), 1621–1639.
Mallick, P.K. (1993). *Fiber-reinforced Composites: Materials, Manufacturing, and Design* (2nd ed.). CRC Press.
Matthews, I. P. (1994). Sterilisation of implantable devices. *Clinical Materials*, 15(3), 191–215.
Mehar A. K., Mahapatra, S. S., and Patel, S. K. (2016). Fabrication, processing and assessment of mechanical behaviour of Hydroxyapatite (HAp) reinforced Polycarbonate (PC) composite for bio-medical applications. *Indian Journal of Engineering*, 13(33), 386–393.
Mosallam, A., Elsanadedy, H. M., Almusallam, T. H., AL-Salloum, Y. A., and Alsayed, S. H. (2015). Structural evaluation of reinforced concrete beams strengthened with innovative bolted/bonded advanced frp composites sandwich panels. *Composite Structures*, 124, 421–440. doi:10.1016/j.compstruct.2015.01.020
Nellore, J., Vijayalakshmi, S., Priyanka, P., and Priyadarshini, S. (2014). Preparation and initial biocompatability evaluation of biogenic hydroxyapatite/Chitosan/PVA biocomposite as a drug delivery carrier of proneurogenic factor. *Asian Journal of Pharmaceutical and Clinical Research*, 8(1), 345–350. Retrieved from https://innovareacademics.in/journals/index.php/ajpcr/article/view/3618
Oksiuta, Z., Dabrowski, J. R., and Olszyna, A. (2009). Co–Cr–Mo-based composite reinforced with bioactive glass. *Journal of Materials Processing Technology*, 209(2), 978–985.
Pugliesi, D. (2010). *Fiber-reinforced Composites Tipology.* [Online Image]. Retrieved from https://commons.wikimedia.org/wiki/File:Fiber-reinforced_composites_tipology.svg
Ramakrishna, S. (2004). *Series on Biomaterials and Bioengineering Vol. 1: An Introduction to Biocomposites.* Imperial College Press.
Ramakrishna, S., Mayer, J., Wintermantel, E., and Leong, K. W. (2001). Biomedical applications of polymer-composite materials: A review. *Composites Science and Technology*, 61(9), 1189–1224.
Rao, P. D., Kiran, C. U., and Prasad, K. E. (2017). Effect of fiber loading and void content on tensile properties of keratin based randomly oriented human hair fiber composites. *International Journal of Composite Materials*, 7(5), 136–143. doi:10.5923/j.cmaterials.20170705.02
Sathish, S. (2010). Studies on the corrosion and wear behavior of the laser nitrided biomedical titanium and its alloys. *Materials Science and Engineering*, 30(3), 376–382 .
Seal, B. L., Otero, T. C., and Panitch, A. (2001). Polymeric biomaterials for tissue and organ regeneration. *Materials Science and Engineering,* 34(4-5), 147–230.
Smith, B. D., and Grande, D. A. (2015). The current state of scaffolds for musculoskeletal regenerative applications. *Nature Reviews. Rheumatology*, 11(4), 213–222. doi:10.1038/nrrheum.2015.27
Sumner D. R., (1992). Determinants of stress shielding: design versus materials versus interface. *Clinical Orthopaedics and Related Research*, 274, 202–212.
Swethaa, M., Sahithi, K., Moorthi, A., Srinivasan, N., Ramasamy, K., and Selvamurugan, N. (2010). Biocomposites containing natural polymers and hydroxyapatite for bone tissue engineering. *International Journal of Biological Macromolecules*, 47(1), 1–4. doi:10.1016/j.ijbiomac.2010.03.015
Veith, M., Sow, E., Werner, U., Petersen, C., and Aktas, O. C. (2008). The transformation of core/shell aluminium/alumina nanoparticles into nanowires. *European Journal of Inorganic Chemistry*, 2008(33), 5181–5184.
Wolf, W. D., Vaidya, K. J., and Francis, L. F. (1996). Mechanical properties and failure analysis of alumina-glass dental composites. *Journal of the American Ceramic Society*, 79(7), 1769–1776. doi:10.1111/j.1151-2916.1996.tb07994.x

Xia, Y., Zhou, P., Wang, F., Qiu, C., Wang, P., Zhang, Y., and Xu, S. (2016). Degradability, biocompatibility, and osteogenesis of biocomposite scaffolds containing nano-magnesium phosphate and wheat protein both in vitro and in vivo for bone regeneration . *International Journal of Nanomedicine*, 11, 3435–3449.

Yildiz, H., Ha, S., and Chang, F. (1998). Composite hip prosthesis design. I. Analysis. *Journal of Biomedical Materials Research*, 39(1), 92–101.

6 Biopolymers and Their Applications; with a Focus on Chitosan

6.1 INTRODUCTION

Biopolymers are polymers (macromolecules composed of many repeated subunits) that are produced by living organisms (Mohanty et al., 2005), which can include animals, plants, bacteria, and fungi (Kaplan, 1998). Biopolymers can be categorized as polypeptides, polysaccharides, polynucleotides, or a combination of the previously mentioned three (Abe et al., 2010). Polypeptides are long continuous chains consisting of amino acids, covalently bonded at the carboxyl group of one and the amine group of another (McNaught and Wilkinson, 1997) Polysaccharides are polymeric carbohydrate structures composed of monosaccharide units glycosidically bonded via hydrolysis (Varki et al., 1999). Polynucleotides consist of 13 or more nucleotide monomers (Kasianowicz et al., 1996), however, they are not currently used in biomaterial applications due to their size and properties (Quake and Volkmuth, 2001). Many applications look to utilize biopolymers for their sustainability, industrial ecology, eco-efficiency, renewable nature (Kaplan, 1998), antibacterial activity, and inherent properties (Rinaudo, 2006). As they are made from living organisms, they can be grown indefinitely, as opposed to polymers created from petrochemicals, which are not renewable and will deplete (Mohanty et al., 2002). Biopolymers often exhibit monodispersity, a property that makes their sequence identical throughout, whereas synthetic polymers tend to exhibit a polydispersity (Stupp and Braun, 1997). They have been utilized in a number of industries, some of which include; medicine, agriculture, textiles, automotive, and food (Kasapis et al., 2009). Some biopolymers, such as chitosan, exhibit natural antimicrobial properties that inhibit the growth of bacteria, which has been used in a wide range of applications including as an antimicrobial agent in automotive electrocoating (Heydarian et al., 2015). Chitosan is a particularly interesting biopolymer due to its abundance and it is very useful as a biomaterial due to the wide range of tailorable features.

6.2 CHITOSAN

Chitosan is a linear polysaccharide that is usually a semi-crystalline solid, composed of (1 to 4 attached)-2-acetamido-2-deoxy-B-D-glucan (N-acetyl D-glucosamine) and (1 to 4 attached)-2-amino-2-deoxy-B-D-glucan (D-glucosamine) units (Rinaudo, 2006). It is the only positively charged polysaccharide that is naturally occurring (Pavinatto

by Yaser Dahman and Chad Gorman Groves

et al., 2010). Figure 6.1 shows the structure of this biopolymer, although it is not generally present in nature in this structure. Instead Figure 6.2 shows the structure of chitin, which chitosan can be derived from through partial deacetylation (Acosta et al., 1993). The degree at which the deacetylation occurs alters various properties, but in general to be classified as chitosan, the chitin should contain at least 60% of D-glucosamine residues after deacetylation (Liu et al., 2011). The deacetylation degree is the ratio of D-glucosamine to the sum of both D-glucosamine and N-acetyl D-glucosamine residues (Acosta et al., 1993). Chemical hydrolysis under high alkaline conditions (Jayakumar et al., 2010) and enzymatic hydrolysis through specific enzymes including chitin deacetylase (Venkatesan and Kim, 2010) are the processes in which the deacetylation of chitin occurs. Chitin is the second most abundant biopolymer and is generally found in invertebrates such as insects or crustaceans (Shahidi et al., 1999). Chitosan has also been produced through the mycelia of various strains of fungi, including; *Absidia coerulea*, *Absidia blakesleeana*, *Gongronella butieri*, *Mucor rouxii*, and *Phycomyces blakesleeanus* (Rane and Hoover, 1993). Chitin is found in the cell walls of yeast (Kollár et al., 1997), and has been synthesized using the yeast *Candida albicans* (Braun and Calderon, 1978). Chlorella, a type of green algae, also has a chitin-like glycan found in its cellular wall (Kapaun and Reisser, 1995). The two main sources of chitosan for industrial purposes are fungal mycelia and crustaceans (Kannan et al., 2010). Sourcing from crustaceans can be particularly problematic due to being a seasonal product, a limited supply, and a variability among the animal populations which may lead to some inconsistent characteristics, physiochemically speaking (Kannan et al., 2010). Fungal sources allow for a year-round and controlled production environment that ensures better reproducibility of resulting chitosan (Wu et al., 2004). The fungus *A. coerulea* has been experimentally shown to

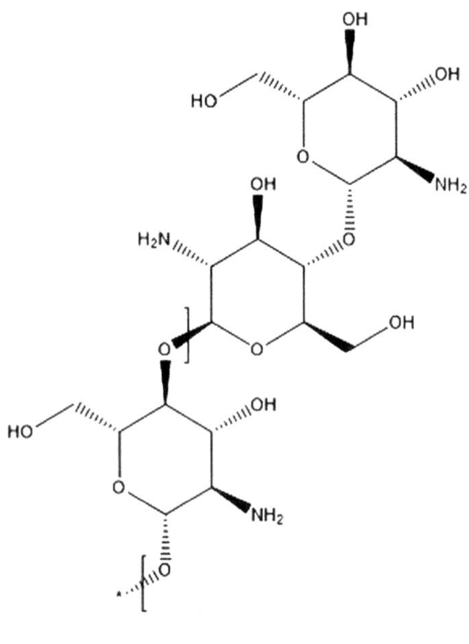

FIGURE 6.1 Chemical structure of chitosan, created using ChemDraw Professional.

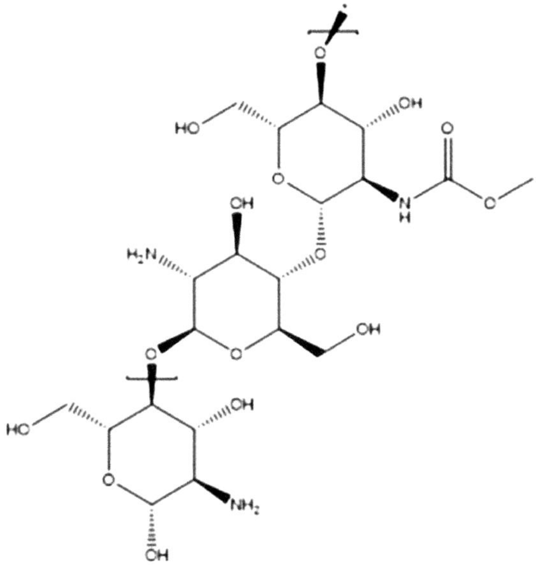

FIGURE 6.2 Chemical structure of chitin, created using ChemDraw Professional.

yield the most chitosan, where strains ATCC 14076 and NRRL 1315 were considered the best producers of *A. coerulea* (Rane and Hoover, 2009). This method of production is often preferred in medical applications due to potential allergies to crustaceans among human populations (Riva et al., 2011) Though the fungus-derived chitosan tends to have a narrower molecular mass distribution than the chitosan derived from crustaceans (Riva et al., 2011), and may also have a different deacetylation degree and molecular mass (Aranaz et al., 2009). The deacetylation degree can vary between 60 and 100%, where molecular weight can vary between 300 and 1000 kilodaltons (Liu et al., 2011). During production, it is difficult to predict the chitosan characteristics, so most approaches rely on analysing the resulting product for the composition and properties (Balázs and Sipos, 2007). Characterization becomes extremely important due to this but it is not easy, as chitosan can easily be contaminated through organic and inorganic impurities (Balázs and Sipos, 2007). Therefore, it has a broad polydispersity and often is referred to by the viscosity of the sample rather than the molecular mass (Balázs and Sipos, 2007). It is soluble mainly in acidic mediums, which limits analysis to tools and methods such as; infrared (IR) spectroscopy, hydrogen nuclear magnetic resonance (H NMR) spectroscopy, ultraviolet (UV) spectroscopy, pH-potentiometric titration, colloidal titration, enzymatic degradation, and both viscosimetry and size exclusion chromatography for the molecular mass (Balázs and Sipos, 2007).

6.2.1 Chitosan Structure to Property Relationship

Referring to Figures 6.1 and 6.2, it is shown that the amino groups in the chitosan structure differentiate it from chitin, which explains the properties of chitosan (Leedy et al., 2011). Acidic aqueous solutions with a pH of less than 6 will cause chitosan to be soluble while the D-glucosamine residues will become protonated (Leedy et al., 2011). This allows it to be processed under mild conditions and thus allows chitosan to be used in a number of applications (Madihally and Matthew, 1999). With protonated amino groups, chitosan can become a polycation, which can form ionic complexes (Madihally and Matthew, 1999). The amino groups also allow chitosan to complex various species (Rinaudo, 2006). Chitosan can be multilayered, through a layer-by-layer deposition, when it is in a polyelectrolyte form (Decher, 1997). This can be achieved through carboxyalkylation; which results in a carboxy-alkyl-chitosan (Kim and Choi, 1998), through phosphorylation; which results in a phosphonic-chitosan (Heras et al., 2001), or through sulphation; which results in a sulphonic-chitosan (Crini et al., 2008). The amino and alcohol groups along with the chitosan structure allow for stable covalent bonding with many other species, as well as allowing for etherification or esterification reactions (Rinaudo, 2006). The D-glucosamine residues can be reacted with aldehydes (Rinaudo, 2006), which has been shown to improve the antibacterial properties of chitosan (Jia and Xu, 2001). Depolymerization reactions can cause a lower molecular weight and result in lower viscosity and better solubility (Allon and Peyron, 1995). Table 6.1 summarizes many of the types of reactions chitosan can undergo and the effects they can produce. Although not an entirely exhaustive summary, it gives a good understanding of the diversity of chitosan. Chitosan also exhibits other properties related to its structure;

TABLE 6.1
Summary of Chitosan Reactions

Chitosan Type	Reaction Required	Use/Purpose	Source
N-alkyl chitosan	Alkylation	Improved water solubility, amphiphilic	Kim et al. (1997)
Acyl-chitosan	Acylation	Improved water solubility, fungicidal activity, amphiphilic	Badawy et al. (2004)
Carboxyalkyl-chitosan	Carboxyalkylation	Polyelectrolytes, multilayering	Kim and Choi (1998)
Phosphonic-chitosan	Phosphorilation	Polyelectrolytes, multilayering	Heras et al. (2001)
Sulfonic-chitosan	Sulfatation	Polyelectrolytes, multilayering	Crini et al. (2008)
Sugar-modified chitosans	Alkylation, cross linking	Cell targeting, drug delivery	Sashiwa and Aiba (2004)
Azidated-chitosan	Azidation	Photosensitivity	Jameela et al. (2002)
High DD chitosan	Deacetylation	More cationic properties, altered thermostability	Kiang et al. (2004)
Low molecular weight chitosan	Depolymerization	Lower viscosity, better solubility	Allan and Peyron (1995)

one previously mentioned is the antibacterial property, which will be examined later in this review (Sudarshan et al., 1992). Chitosan has also been shown to exhibit antifungal (Aranaz et al., 2009), antiviral (Davydova et al., 2011), mucoadhesive (Lehr et al., 1992), hemostatic (Yang et al., 2008), and analgesic properties (Aranaz et al., 2009). Antifungal properties have been explored usually with plants, as done by El Hassni et al. (2004), but studies exploring applications for antifungal treatments in humans or for use in biomaterials have not specifically been explored. Interestingly, Davydova et al. (2011) showed that the antiviral activity of chitosan actually increased significantly with lowering of the polymerization degree, which would also increase its solubility (Allan and Peyron, 1995). Studies have been done using chitosan for its hemostatic potential, as seen by Yang et al. (2008) and Rao and Sharma (1997). Okamoto et al. (2002) showed chitin and chitosan have an analgesic and anti-inflammatory effect in mice, which could have novel applications in implantable biomaterials. Chitosan has a structure that allows for easy functionalizing and tailoring, which makes chitosan a biopolymer that is particularly useful in a wide range of biomaterial applications.

6.2.2 Chitosan Physicochemical Properties

Chitosan is a biopolymer that has a huge variation in physicochemical properties, which include molecular weight and deacetylation degree, due to it being derived from multiple organisms including animals, fungi, and plants (Aranaz et al., 2010). The difficulty of chitosan characterization previously outlined has caused an issue in direct comparisons of chitosan samples used by various studies, which makes understanding the various physicochemical properties so important (Aranaz et al., 2010). Samples of n-alkyl-chitosan, discussed briefly in Table 6.1, tend to have a molecular weight in the range of 150 to 200 kilodaltons (Dos Santos Alves et al., 2009). Dos Santos Alves et al. used n-alkyl-chitosan in rheological studies, which showed a non-Newtonian behaviour; likely due to the presence of the hydrophobic groups present in this chitosan derivative. It was also observed that viscosity increased at higher concentrations but decreased with higher temperatures (Dos Santos Alves et al., 2009). Li et al. (2007) observed that chitosan molecular weight affected both morphology and viscosity of oleoyl-chitosan samples; at lower molecular weights spherical nanoparticles were observed, which were not present at higher molecular weights. Likely this observation was due to increasing viscosity of the higher molecular weight chitosan (Li et al., 2007). This spherical morphology was also observed in a study by Mincheva et al. (2008) at molecular weights of 45 kilodaltons, although increasing concentration had no effect on their presence. Mincheva et al. (2008) were able to find the presence of spherical morphologies in higher molecular weights of 670 kilodaltons, although this was only at pH 5.0 and 5.6, whereas the lower molecular weight had spherical morphologies at all pH levels. Increasing the concentration of the 670 kilodalton chitosan would result in worm-like or rod-type structures (Mincheva et al., 2008). Another physicochemical property of chitosan, deacetylation degree (DD), has been found to increase micelle size in water when increasing the DD (Opanasopit et al., 2007a). As deacetylation degree increases, cytotoxicity was found to increase as well (Hu et al., 2008). Opanasopit et al. (2007a) loaded a drug called camptothecin into chitosan micelle cores through dialysis and found that as the deacetylation degree ratio increased, the micelle stability also increased. Similar results were seen in another study by Opanasopit et al. (2007b) when using all-trans retinoic acid as the loaded drug.

6.3 ANTIBIOTIC TYPE PROPERTIES OF BIOPOLYMERS

As biopolymers are derived from living organisms, some exhibit antibacterial activity (Rinaudo, 2006), essentially the ability to kill or inhibit bacterial growth (Dorman and Dean 2000). This is of importance as pharmacological industries have produced a number of antibiotic drugs designed to treat bacterial infections; misuse of these has led to a number of antibiotic resistant strains of bacteria (Goossens et al., 2005). Antibacterial properties of many phytochemicals produced by plants have been evaluated, along with their synergistic effects when added with antibiotics developed pharmaceutically (Lee et al., 2001). Yeast and other yeast-like fungi have been found to produce antibacterial compounds including polypeptide structures called enzymes, which have been experimentally isolated from the phylloplane they produce (McCormack et al., 1994). Biopolymer based packaging has been developed using materials such as oregano oil, polylysine, sodium lactate, sorbitol-plasticized whey protein isolate films, etc. (Zinoviadou et al., 2011). This is likely the next area for development in packaging technology, due to concerns of food-borne microorganisms, as these materials have natural antimicrobial properties (Cha and Chinnan, 2004). Biopolymers have also been studied for use in enhancing drug treatments or therapies. Although not entirely an antibiotic treatment, a recent example involves a study utilizing an alginate scaffold to load tumour-specific T cells which act with antibiotic properties (Stephan et al., 2015). This was designed as an implantable scaffold aimed at destroying inoperable tumours or cancer cells left behind post-operation (Stephan et al., 2015). The alginate scaffold was functionalized with collagen-mimetic molecules to promote T cell migration, and soluble factor interleukin 15 and other signalling molecules were added to increase T cell growth (Stephan et al., 2015). These functionalized scaffolds showed a 22-times increase in cell proliferation and 8-times increase in migration, over a 7 day period, when compared to scaffolds without the modification (Stephan et al., 2015). In a breast cancer mouse model, tumours were removed surgically, with roughly 1% of tumour tissue left over (Stephan et al., 2015). Groups were treated with the cell-loaded fully modified scaffold, intravenous T cell injection, intracavitary T cell injection, and no treatment (Stephan et al., 2015). All mice treated with the scaffold survived at least 80 days, which was the last follow up for the scaffold treatment, and an average of 21, 25 and 19 days respectively for the remaining treatments (Stephan et al., 2015). The inoperable tumour model was a stage 3 ovarian carcinoma mouse; 6 of the 10 animals with the implanted scaffold saw survival rates of at least 80 days (Stephan et al., 2015).

6.3.1 Chitosan Antimicrobial Properties

Chitosan has been identified as bacteriostatic, essentially hindering the growth of bacteria, but it has also been considered to be bactericidal, where it kills live bacteria

(Coma et al., 2002). Although the mechanisms behind the antibacterial properties, both bacteriostatic and bactericidal, are not fully understood, the molecular weight and degree of acetylation have both independently been shown to influence the antibacterial properties of chitosan (Sekiguchi et al., 1994). Some proposed models of chitosan's antibacterial activity include the ionic surface interaction causing cell wall leakage in bacteria, the penetration of chitosan into bacterial nuclei causing an inhibition of mRNA and protein synthesis, and the formation of an external barrier chelates metals from bacteria and limits essential nutrients needed for growth (Goy et al., 2009). Similar to the antimicrobial properties of chitosan, it has also been found that chitosan exhibits both antifungal properties (Aranaz et al., 2009) and antiviral properties (Davydova et al., 2011). Antifungal properties essentially hinder the growth of, or actively kill fungus; chitosan has been found to exhibit both properties (Aranaz et al., 2009). Where antiviral properties actively prevent viral infections through similar mechanisms (Davydova et al., 2011). These aspects aren't as explored as the antibacterial properties of chitosan and the mechanisms behind them are still not fully understood (Aranaz et al., 2009).

A recent study by Heydarian et al. (2015) explored the use of chitosan as an antibacterial agent used for a water base coating formulation. This formulation was used in cathodic electrodeposition coatings, a process that is used in coating alloys for biomaterial applications (Song et al., 2008). Since these substances contain water and biodegradable substances, bacterial growth is possible, although it is unwanted as it can cause disruptions of the coatings (Huang et al., 2014), and lead to biocompatibility concerns (Song et al., 2008). A minimum inhibitory concentration (MIC) value was determined for chitosan by dissolving the chitosan biopolymer in lactic acid, at concentrations of 0.06, 0.125, 0.25, 0.05, and 1%weight by volume and testing against *E. Coli* bacterial growth (Heydarian et al., 2015). They determined that complete inhibition was achieved at 0.125%, but 80% of bacteria was inhibited at 0.06%, so these values were used, along with increased values, to determine their effect on the cathodic electrodeposition coating. Changing the values of the chitosan concentration caused the appearance of the films to change significantly, as seen by Figure 6.3. Using a previous growth model developed by Pierce et al. (1978), it was determined that increasing the chitosan concentration would dramatically increase the initial and residual current density, as outlined in Figure 6.4. Thus, an increase of chitosan biopolymer concentration would lead to a higher density of positive charge on the biopolymer chain leading to an accumulation of entanglements. Heydarian et al. (2015) determined that the tendency of coil-structured chains was the reason for the rise in the deposited layer on the surface. At MIC, the chitosan solution of

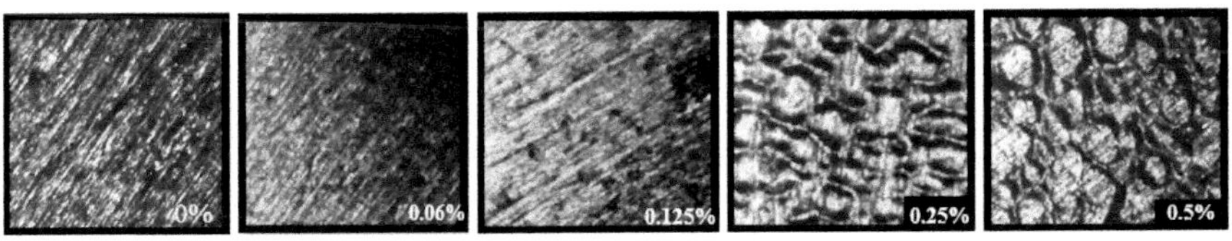

FIGURE 6.3 Morphology of the final films, observed at varying concentrations of chitosan biopolymer (Heydarian, Ranjbar, and Rastegar, 2015).

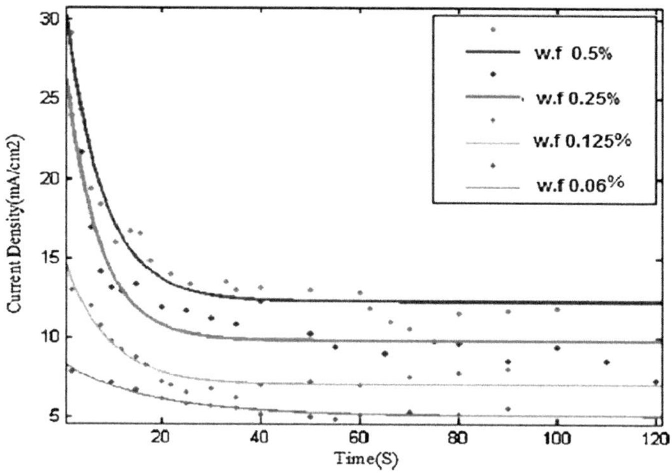

FIGURE 6.4 Pierce-modelled current density–time for varying chitosan solutions, w.f; weight fraction of chitosan biopolymer (Heydarian, Ranjbar, and Rastegar, 2015).

0.125%, no porosity or defect showed on the surface, thus this would be the optimal concentration as increased levels would show defects, and decreased levels would find bacterial growth. Heydarian et al. (2015) demonstrated a novel use of chitosan that allows for a sustainable and environmentally friendly method for cathodic electrodeposition baths. Through tailoring concentrations, optimal conditions were achieved, and this is an excellent example of using biopolymers to replace environmentally toxic or non-renewable methods of production.

Plant pathogens, bacteria that damage plants, are important for economic considerations as they can induce decay on agricultural crops in both growing and harvest seasons (Casida and Quistad, 1998). Conventional means to control and eliminate these pathogens involve the use of many chemical pesticides, which has had recent backlash due to health concerns as well as pesticide-resistant bacteria as a result of overuse (Houeto et al., 1995). A safer alternative to these chemical pesticides involves the use of chitosan, which has negligible risk to human health and to environmental health (Muzzarelli, 1983). Again, the exact mechanisms behind chitosan's ability to reduce plant disease are not fully understood, though evidence suggests that it is due to direct toxicity or a chelation of nutrients which pathogens are dependant upon (El Hadrami et al., 2010). Due to their ability to adhere to biological membranes or molecules, chitin dressings have been made to be placed on injured or damaged tree-bark, resulting in a faster wound healing process (El Hadrami et al., 2010). More research is needed in developing chitosan formulations for different stages of plant growth, and to be implemented in pest management strategies for major crops such as potatoes or canola (El Hadrami et al., 2010). Quintana-Obregon et al. (2011) released a study that showed the inhibitory properties chitosan of middle viscosity has on *Ramularia cercosporelloides*. *R. cercosporelloides* is a fungus that has led to several yield losses of Mexico-grown safflower, forming necrotic lesions on the leaves of the plant (Huerta-Espino et al., 2006).

Skin is a barrier used to block pathogens, and generally should be covered with a dressing immediately after it is damaged to help prevent infection by bacteria or other pathogens (Bruin et al., 1990). Using the antibacterial properties demonstrated by chitosan, Lu et al. (2008) designed a novel wound dressing, which was composed both of chitosan and nano-silver particles. A chitosan film was prepared using 1.5 grams of chitosan dissolved into a 100 mL acetic acid (1% volume by volume) solution, filtered and then coated onto plastic plates. These plates were oven-dried at a constant 40°C, and then later sterilized through alcohol and UV light. The nano-silver was prepared through a process previously outlined by Kneipp et al. (1997). Lu et al. (2008) used self-assembly technology to fabricate the wound dressingsy immersing the chitosan films in the nano-silver solutions for 12 hours and washing extensively with sterile water. Figure 6.5 shows the surface morphology of these prepared wound dressings, both with atomic force and electron scanning microscopy (Lu, Gao, and Gu, 2008). Wound dressings composed entirely of nano-silver and dressings composed entirely of chitosan were also generated to test the efficacy of the combination. Figure 6.6 shows an increase in healing rate among the combined chitosan and nano-silver dressings; Figure 6.7 compares the wound healing of each of these dressings on rabbits (Lu, Gao, and Gu, 2008). The combined approach showed the best healing; no bacterial growth was found in the nano-silver/chitosan wound dressings after incubation for 14 days. Potential concerns were identified by Lu et al. (2008), as silver concentrations increased over 100 times in the liver for both nano-silver/chitosan combined dressings, and nano-silver only dressings, although lesser in the combinatory dressing. The use of the biomaterial chitosan along with the nano-silver has made a dressing that is capable of promoting wound healing, combating infection, and reducing the amount of silver absorption. The simple processing and use of readily available materials, including renewable chitosan, shows that this dressing could have wide clinical applications (Lu et al., 2008).

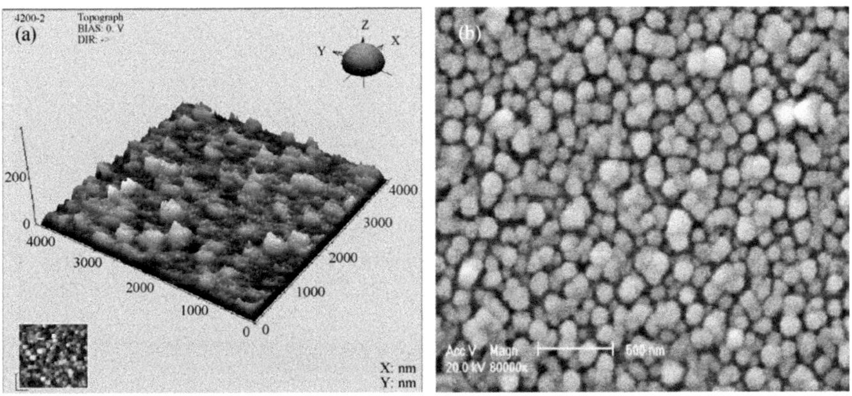

FIGURE 6.5 (a) Atomic force microscopy of chitosan wound dressing with nano-silver. (b) Scanning electron microscopy of chitosan wound dressing with nano-silver (Lu, Gao, and Gu, 2008).

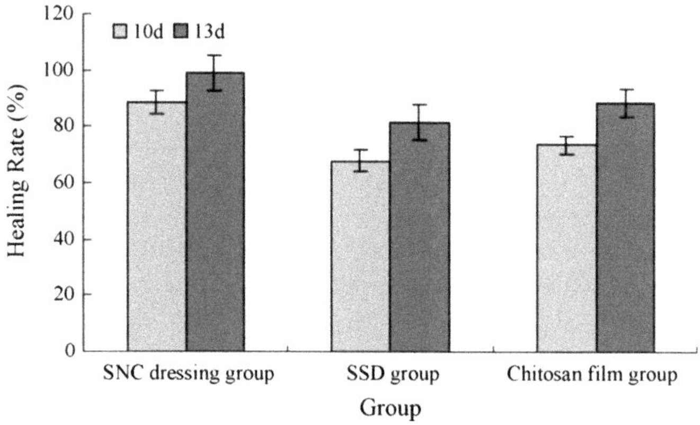

FIGURE 6.6 Healing rates among nano-silver/chitosan dressing (SNC), nano-silver only dressing (SSD) and chitosan film dressing. 10 and 13 day bars present, data is mean of trials with standard deviation bars present, $n = 14$ (Lu, Gao, and Gu, 2008).

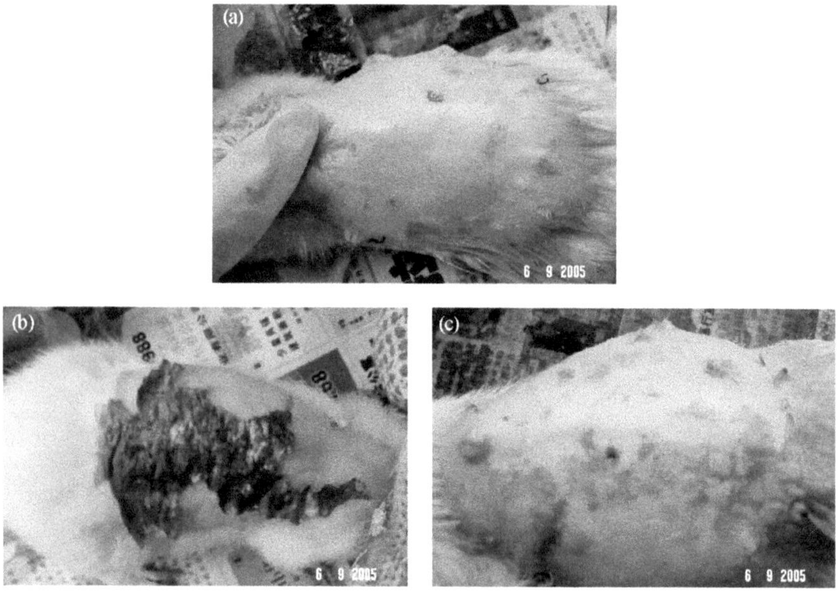

FIGURE 6.7 Ten days post-operation, healing conditions of (a) nano-silver/chitosan dressing (SNC), (b) nano-silver only dressing (SSD) and (c) chitosan film dressing (Lu, Gao, and Gu, 2008).

6.4 BIOPOLYMERS FOR TISSUE ENGINEERING

Traditionally, three main components of tissue engineering include the cells to be sourced and used; the scaffolds, which allow for attachment, migration, growth, and proliferation; and the growth factors and signals, which cause cell differentiation, proliferation, survival, growth, and stimulation (Langer et al., 2000). Biopolymers are used, though not exclusively, in designing scaffolds in tissue engineering. A large variety of scaffolds have been investigated for tissue engineering, with the goal often is to mimic the extracellular matrix of the desired tissue (He et al., 2012). For purposes of this review, however, discussion of scaffolds and the chosen biomaterials will be made using biopolymer examples, but it is important to note that scaffolds have been produced using a wide range of materials including synthetic polymers, ceramics, and other biomaterials (Karp et al., 2003). The choice of biopolymer for application is crucial due to the physical contact between cells used; thus, the chosen biopolymer can be crucial in directing the behaviour of the cell and ultimately the fate and efficacy of the tissue engineering process (He et al., 2012). Since the cell can detect various chemical cues from the biomaterials, cellular changes in the morphology, behaviour, and dynamics are all dependent on the type of material chosen (He et al., 2012). Various chemical functional groups have been extensively studied in this field as they have been shown to influence the behaviour of human fibroblasts, adipose stem cells, and mesenchymal stem cells (He et al., 2012). The mechanism behind this influence is the interaction of receptors on cell membranes and extracellular matrix proteins, which can direct cell fate without the presence of growth factors (Ren et al., 2009). A study by Ren et al. (2009) showed that neuronal stem cells would have increased migration on amidogen and carboxylic acid functionalized surfaces when compared to alcohol and methyl group functionalized surfaces. Studies such as this show that surface interaction can be quite complex, and further studies are needed in this field to better understand, and control for, cellular behaviour with biopolymer or biomaterial surface choice. Another important consideration in biopolymer selection as scaffolds is the modulus of elasticity, as stem cells can commit to various lineages in response to different modifications of the substrates they are grown on (He et al., 2012). Gel system cultures can be altered with various stiffness; stiffer matrices would influence mesenchymal stem cells into myogenic or muscle tissue differentiation, where soft matrices mimicked brain tissue and would induce mesenchymal stem cells into neurogenic lineages (He et al., 2012). Glial cells, used as support cells in neuronal tissue engineering, have been shown to suppress adhesion and proliferation on soft gels, and support adhesion and proliferation on hard gels (Georges et al., 2006). Optimal elastic modulus of scaffolds should most often resemble the desired tissue modulus, for example, optimal neuronal tissue engineering modulus and brain tissue modulus are both at around several hundred Pascals (Georges et al., 2006). Therefore, optimizing the modulus of elasticity is a way to ensure unwanted cellular lineages are suppressed (He et al., 2012).

6.4.1 Chitosan in Tissue Engineering

Chitosan is a polysaccharide which has proved useful in applications of tissue engineering. It has proved useful in spinal tissue engineering for injuries caused to the spinal cords of both rats and dogs (He et al., 2012). It was also found to produce an inflammatory response and become ingested by macrophages in these brain tissue engineering studies, which limits its application in this area (He et al., 2012).

Funakoshi et al. (2005) utilized a chitosan-based scaffold which proved useful in ligament tissue engineering. In this application, chitosan saw exception cellular adhesive properties, showed biocompatibility, excellent wound healing, and helped to repair the tissue. Though Funakoshi et al. (2005) utilized a chitosan-based scaffold, they also emphasized the importance of the hyaluronan, a major component of glycosaminoglycans, in its ability to stimulate the tissue regeneration. Though these scaffolds displayed a similar stress–strain shaped curve as the ligaments they were replacing, they displayed a lower magnitude in stress endured when compared to a native ligament. Another use of chitosan is in utilizing electrospun nanofibers in creating a bridge graft, as seen in Figure 6.8 (Wang et al., 2009). These are created using a process called electrospinning, wherein nanoscale fibers can be made into a mesh which has a high porosity, surface to volume ratio, and can exhibit the hierarchal structure found in the extracellular matrix in tissues (He et al., 2012). Wang et al. (2009) utilized these chitosan nanofibers to bridge graft a sciatic nerve gap in rats; with the use of C17.2 neural stem cells, regeneration and new blood vessel formation was observed. Chitosan-based biopolymer materials have been found to be quite suitable for orthopedic tissue engineering due to the antibacterial nature, its ability to be moulded and shaped, tailorable porosity, and its capability of cellular ingrowth and osteoconduction (Di Martino et al., 2005). As well the ability of chitosan to make into a gel allows for it to deliver various pharmaceutically active agents, growth factors, and other signalling molecules in a controlled manner (Di Martino et al., 2005). It has even been utilized as a bioactive coating for various orthopedic implants, which include craniofacial and dental implants (Bumgardner et al., 2003).

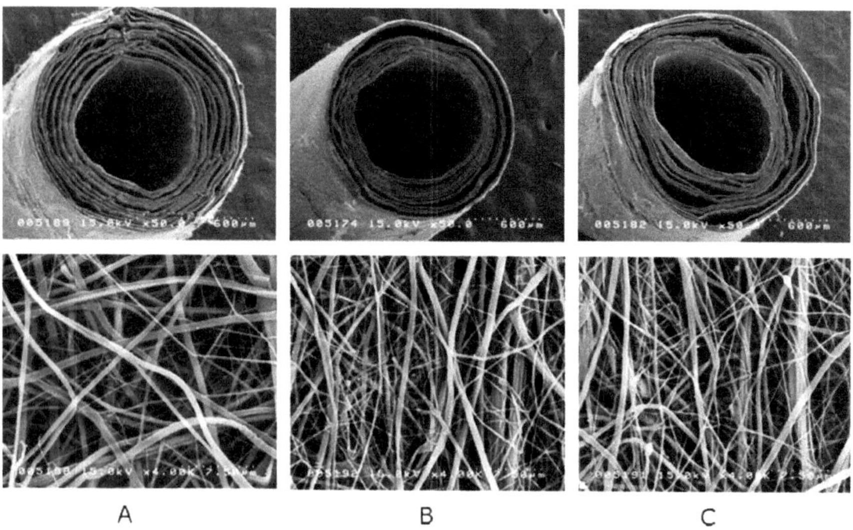

FIGURE 6.8 Scanning electron microscopy images of chitosan electrospun nanofiber mesh tubes. (a) Nonoriented chitosan mesh, (b) oriented chitosan mesh, (c) bilayered chitosan mesh (Wang et al., 2009).

6.5 SUSTAINABLE AND ENVIRONMENTAL IMPACTS OF BIOPOLYMERS

Biopolymers can be sustainable, as they tend to be carbon neutral and renewable resources (Kaplan, 1998). Their carbon neutrality comes from the fact that any carbon emissions released through production, use, or degradation of these biopolymers would likely be returned or reabsorbed by the crops or organisms that are grown to replace them (Kaplan, 1998). In recent years, biotechnology has focused on producing biodegradable polymers to replace non-biodegradable counterparts, which has proven particularly useful in biomaterial applications (Steinbüchel, 2005). Green chemistry is guiding the next generation of materials and products, which has seen an increase in biopolymer studies (Mohanty et al., 2002). This trend is even seen within biopolymer research as some studies are focusing on improving sustainability and reducing their impact on the environment (Lichtenthaler, 2007). A study by Fei et al. (2016) sought to find an environmentally friendly solvent system in order to recover Poly-β-hydroxybutyrate (PHB); PHB is a biopolymer that is synthesized with bacterial fermentation of sugars by *Cupriavidus necator*. To obtain the PHB from the bacteria, an extraction process is required. Previous extraction methods utilize organic solvents such as methylene chloride, 1,2-dichloroethane and chloroform (Sayyed et al., 2009). These can be toxic and have harmful effects on the environment, thus this study produced a novel extraction method that utilized a 1:1:1 solvent mixture of acetone, ethanol and propylene carbonate. Significantly lower PHB recovery amounts were seen with acetone, butanol, acetone/ethanol, acetone/isopropanol when compared to chloroform (Fei et al., 2016). This suggested that polarity of the solvents was not as important for extraction as previously thought. This is an observation that future experiments, aiming to design more environmentally friendly extraction methods, may wish to factor when choosing solvents, as solvent selection may not be as intuitive using polarities as previously. Another interesting observation from Fei et al. (2016) is that residual biomass from butanol extraction was waxier at higher temperatures, thus likely indicating elevated temperatures will improve the release of PHB. This again is a trend that should be explored in future studies developing "greener" methods for extraction of biopolymers. This trend of "green" technologies and "green" chemistries is likely the cause of there being a scarcity of work done on non-biodegradable biopolymers; as it is thought to contradict these "green" paradigms (Steinbüchel, 2005). Though most biopolymers are biodegradable, which means they are brown down into carbon dioxide by microorganisms, some of these biopolymers are compostable, which allows them to be broken down by at least 90% within six months (Kale et al., 2007). Polythioesters are a relatively new type of biopolymer that are not considered to be biodegradable as they are resistant to microbial degradation, likely due to the fact they are produced by bacteria themselves (Steinbüchel, 2005). Chitosan is a biopolymer which is considered to be biodegradable (Bourtoom and Chinnan, 2008). Should biopolymers replace their synthetic, fossil fuel derived counterparts, studies on both biodegradable and non-biodegradable biopolymers will be required. Though in terms of biomaterials and medical applications, biodegradable biopolymers are preferred due to biocompatibility issues (Novikova et al., 2003)

6.5.1 Chitosan Bioplastic

Chitin is second to cellulose as the most abundant organic compound found on Earth, which has industry seeking it out for manufacturing strategies (Fernandez and Ingber, 2014). Fernandez and Ingber (2014) performed X-ray diffraction and Fourier transform IR spectroscopy to characterize chitosan films prepared by solution evaporation. Using birefringence microscopy, the team discovered the existence of millimeter-sized liquid crystal domains which underwent molecular rearrangement when stretched. Using these chitosan liquid crystals, Fernandez and Ingber (2014) created three-dimensional objects through casting and injection moulding. When adding wood flour to this chitosan bioplastic, a solid object was formed; adding natural colours would produce coloured and transparent plastics (Fernandez and Ingber, 2014). Fernandez and Ingber (2014) also gave the plastic water resistance by coating them in parylene, which gives this wide potential in industrial applications. This chitosan bioplastic also has biomaterial applications; Fernandez et al. (2017) demonstrated that this chitosan bioplastic could bond to bodily tissues for wound repair and is capable of holding implanted medical devices in place. This approach is thought to be capable of sealing tissue tears or other injuries to prevent infection prior to the patient seeking hospital attention. Fernandez et al. (2017) used transglutaminase, an enzyme that occurs in the human body that strengthens blood clots and epithelium tissue, to make the chitosan bioplastic adhere to living tissues. Transglutaminase was applied to the chitosan in many ways to test efficacy; including, foams, sprays, and sheet bonding (Fernandez et al., 2017). Sprays utilized liquid streams of chitosan plastic and transglutaminase that could be applied to wounds (Fernandez et al., 2017). These methods were compared with surgical glues currently being used with similar mechanical strength adhesive properties (Fernandez et al., 2017).

6.6 CONCLUSION

Biopolymers offer an effective and environmentally friendly solution to synthetic polymers created by fossil fuels. More importantly, the diverse properties and applications of biopolymers allow for them to be useful as a biomaterial in medical applications. Through exploring a single biopolymer of chitosan, this diversity of properties can be easily seen. As these polymers are derived from living organisms, their biodegradability makes them great for implantation, but they may not benefit from the inertness some synthetic materials have. This can negatively affect the biocompatibility, but the wide range of tailorable properties can help ensure this along with little to no toxicity. Interactions between living tissues and biopolymers need to be explored more in detail, as current studies show how varying certain properties can have profound impacts on cellular lineage differentiation. With more detailed studies in this area, it is possible that biopolymers will replace synthetic materials or ceramics altogether, as scientists will better be able to integrate the materials with host tissues. Chitosan has been shown to be effective in many forms of tissue engineering as its natural antimicrobial activity and biodegradability make it an effective scaffold. Chitosan is also capable of being implemented in many media such as hydrogels or electrospun nanofibers, which furthers its diverse capability. The overlap of biomaterial studies with other industries is seen in studies being done on chitosan to produce a bioplastic capable of replacing petrochemical plastics. In making a product capable of being used as plastic bags or plastic egg cartons, it was found that it also integrated with living tissue quite easily. It is now being explored in more detail as a potential treatment for damaged tissues, better surgical glue, and even as a material designed to hold implantable devices inside the host body. More studies should also be done on further developing non-biodegradable biopolymers, which admittedly do go against the "green" push in science and technology. However, the fact that a bioplastic was so easily taken from a manufacturing industry into a medical industry should be an indicator of how research in this field may open new avenues for biomaterials. Though this review primarily focused on chitosan examples, with some exceptions, there are a huge number of biopolymers which can be tailored, functionalized, combined with synthetic materials, and much more to produce a seemingly unlimited number of biomaterials. Future studies will likely focus on improving our current understanding of biopolymer interactions with living and nonliving materials, novel types of biopolymer production and finding more environmentally friendly, "greener", methods of production and extraction.

REFERENCES

Abe, A., Dusek, K., and Kobayashi, S. (2010). *Biopolymers: Lignin, Proteins, Bioactive Nanocomposites*. Springer.

Acosta, N., Jimenez, C., Borau, V., and Heras, A. (1993). Extraction and characterization of chitin from crustaceans. *Biomass and Bioenergy*, 5(2), 145–153.

Allan, G. G., and Peyron, M. (1995). Molecular weight manipulation of chitosan I: kinetics of depolymerization by nitrous acid. *Carbohydrate Research*, 277(2), 257–272.

Aranaz, I., Harris, R., and Heras, A. (2010). Chitosan amphiphilic derivatives. Chemistry and applications. *Current Organic Chemistry*, 14(3), 308.

Aranaz, I., Mengíbar, M., Harris, R., Paños, I., Miralles, B., Acosta, N., and Heras, Á. (2009). Functional characterization of chitin and chitosan. *Current Chemical Biology*, 3(2), 203–230.

Badawy, M. E., Rabea, E. I., Rogge, T. M., Stevens, C. V., Smagghe, G., Steurbaut, W., and Höfte, M. (2004). Synthesis and fungicidal activity of new N, O-acyl chitosan derivatives. *Biomacromolecules*, 5(2), 589–595.

Balázs, N., and Sipos, P. (2007). Limitations of pH-potentiometric titration for the determination of the degree of deacetylation of chitosan. *Carbohydrate Research*, 342(1), 124–130.

Bourtoom, T., and Chinnan, M. S. (2008). Preparation and properties of rice starch-chitosan blend biodegradable film. *LWT—Food Science and Technology*, 41(9), 1633–1641. doi:10.1016/j.lwt.2007.10.014

Braun, P. C., and Calderone, R. A. (1978). Chitin synthesis in *Candida albicans*: comparison of yeast and hyphal forms. *Journal of Bacteriology*, 133(3), 1472–1477.

Bruin, P., Jonkman, M. F., Meijer, H. J., and Pennings, A. J. (1990). A new porous polyetherurethane wound covering. *Journal of Biomedical Materials Research Part A*, 24(2), 217–226.

Bumgardner, J. D., Wiser, R., Gerard, P. D., Bergin, P., Chestnutt, B., Marini, M., and Gilbert, J. A. (2003). Chitosan: potential use as a bioactive coating for orthopaedic and craniofacial/dental implants. *Journal of Biomaterials Science, Polymer Edition*, 14(5), 423–438.

Casida, J. E., and Quistad, G. B. (1998). Golden age of insecticide research: past, present, or future? *Annual Review of Entomology*, 43(1), 1–16.

Cha, D. S., and Chinnan, M. S. (2004). Biopolymer-based antimicrobial packaging: a review. *Critical Reviews in Food Science and Nutrition*, 44(4), 223–237.

Coma, V., Martial-Gros, A., Garreau, S., Copinet, A., Salin, F., and Deschamps, A. (2002). Edible antimicrobial films based on chitosan matrix. *Journal of Food Science*, 67(3), 1162–1169.

Crini, G., Martel, B., and Torri, G. (2008). Adsorption of CI basic blue 9 on chitosan-based materials. *International Journal of Environment and Pollution*, 34(1–4), 451–465.

Davydova, V. N., Nagorskaya, V. P., Gorbach, V. I., Kalitnik, A. A., Reunov, A. V., Solov'eva, T. F., and Ermak, I. M. (2011). Chitosan antiviral activity: dependence on structure and depolymerization method. *Applied Biochemistry and Microbiology*, 47(1), 103–108.

Decher, G. (1997). Fuzzy nanoassemblies: toward layered polymeric multicomposites. *Science*, 277(5330), 1232–1237.

Di Martino, A., Sittinger, M., and Risbud, M. V. (2005). Chitosan: a versatile biopolymer for orthopaedic tissue-engineering. *Biomaterials*, 26(30), 5983–5990.

Dorman, H. J. D., and Deans, S. G. (2000). Antimicrobial agents from plants: antibacterial activity of plant volatile oils. *Journal of Applied Microbiology*, 88(2), 308–316.

Dos Santos Alves, K., Vidal, R. R. L., and De Carvalho Balaban, R. (2009). Chitosan derivatives with thickening properties obtained by reductive alkylation. *Materials Science and Engineering: C*, 29(2), 641–646.

El Hadrami, A., Adam, L. R., El Hadrami, I., and Daayf, F. (2010). Chitosan in plant protection. *Marine Drugs*, 8(4), 968–987.

El Hassni, M., El Hadrami, A., Daayf, F., Barka, E. A., and El Hadrami, I. (2004). Chitosan, antifungal product against fusarium oxysporum f. sp. albedinis and elicitor of defence reactions in date palm roots. *Phytopathologia Mediterranea*, 43(2), 195–204.

Fei, T., Cazeneuve, S., Wen, Z., Wu, L., and Wang, T. (2016). Effective recovery of poly-β-hydroxybutyrate (PHB) biopolymer from *Cupriavidus necator* using a novel and environmentally friendly solvent system. *Biotechnology Progress*, 32(3), 678–685.

Fernandez, J. G., and Ingber, D. E. (2014). Manufacturing of large-scale functional objects using biodegradable Chitosan bioplastic. *Macromolecular Materials and Engineering*, 299(8), 932–938.

Fernandez, J. G., Seetharam, S., Ding, C., Feliz, J., Doherty, E., and Ingber, D. E. (2017). Direct bonding of chitosan biomaterials to tissues using transglutaminase for surgical repair or device implantation. *Tissue Engineering Part A*, 23(3–4), 135–142.

Funakoshi, T., Majima, T., Iwasaki, N., Suenaga, N., Sawaguchi, N., Shimode, K., and Nishimura, S. I. (2005). Application of tissue engineering techniques for rotator cuff regeneration using a chitosan-based hyaluronan hybrid fiber scaffold. *The American Journal of Sports Medicine*, 33(8), 1193–1201.

Georges, P. C., Miller, W. J., Meaney, D. F., Sawyer, E. S., and Janmey, P. A. (2006). Matrices with compliance comparable to that of brain tissue select neuronal over glial growth in mixed cortical cultures. *Biophysical Journal*, 90(8), 3012–3018.

Goossens, H., Ferech, M., Vander Stichele, R., Elseviers, M., and ESAC Project Group. (2005). Outpatient antibiotic use in Europe and association with resistance: a cross-national database study. *The Lancet*, 365(9459), 579–587.

Goy, R. C., Britto, D. D., and Assis, O. B. (2009). A review of the antimicrobial activity of chitosan. *Polímeros*, 19(3), 241–247.

He, J., Wang, X. M., Spector, M., and Cui, F. Z. (2012). Scaffolds for central nervous system tissue engineering. *Frontiers of Materials Science*, 6(1), 1–25.

Heras, A., Rodriguez, N. M., Ramos, V. M., and Agullo, E. (2001). N-methylene phosphonic chitosan: a novel soluble derivative. *Carbohydrate Polymers*, 44(1), 1–8.

Heydarian, S., Ranjbar, Z., and Rastegar, S. (2015). Chitosan biopolymer as an antimicrobial agent in automotive electrocoatings: the effect of solid content. *Progress in Organic Coatings*, 89, 344–347.

Houeto, P., Bindoula, G., and Hoffman, J. R. (1995). Ethylenebisdithiocarbamates and ethylenethiourea: possible human health hazards. *Environmental Health Perspectives*, 103(6), 568.

Hu, F. Q., Wu, X. L., Du, Y. Z., You, J., and Yuan, H. (2008). Cellular uptake and cytotoxicity of shell crosslinked stearic acid-grafted chitosan oligosaccharide micelles encapsulating doxorubicin. *European Journal of Pharmaceutics and Biopharmaceutics*, 69(1), 117–125.

Huang, Y., Wu, P., Wang, Y., Wang, W., Yuana, D., and Yaob, J. (2014). Materials Chemistry A. *Journal of Materials Chemistry*, 100(2), 4057–4062.

Huerta-Espino, J., Constantinescu, O., Velasquez, C., Herrera-Foessel, S. A., and Figueroa-López, P. (2006). First report of *Ramularia cercosporelloides* on *Carthamus tinctorius* in Northwestern México. *Plant Disease*, 90(12), 1552–1552.

Jameela, S. R., Lakshmi, S., James, N. R., and Jayakrishnan, A. (2002). Preparation and evaluation of photocrosslinkable chitosan as a drug delivery matrix. *Journal of Applied Polymer Science*, 86(8), 1873–1877.

Jayakumar, R., Menon, D., Manzoor, K., Nair, S. V., and Tamura, H. (2010). Biomedical applications of chitin and chitosan based nanomaterials—A short review. *Carbohydrate Polymers*, 82(2), 227–232.

Jia, Z., and Xu, W. (2001). Synthesis and antibacterial activities of quaternary ammonium salt of chitosan. *Carbohydrate Research*, 333(1), 1–6.

Kale, G., Kijchavengkul, T., Auras, R., Rubino, M., Selke, S. E., and Singh, S. P. (2007). Compostability of bioplastic packaging materials: an overview. *Macromolecular Bioscience*, 7(3), 255–277.

Kannan, M., Nesakumari, M., Rajarathinam, K., and Singh, A. J. A. R. (2010). Production and characterization of mushroom chitosan under solid-state fermentation conditions. *Advances in Biological Research*, 4(1), 10–13.

Kapaun, E., and Reisser, W. (1995). A chitin-like glycan in the cell wall of a Chlorella sp.(*Chlorococcales, Chlorophyceae*). *Planta*, 197(4), 577–582.

Kaplan, D. L. (Ed.) (1998). Introduction to biopolymers from renewable resources. In *Biopolymers from Renewable Resources*. Springer.

Kasapis, S., Norton, I. T., and Johan, B. (Eds.). (2009). *Modern Biopolymer Science: Bridging the Divide Between Fundamental Treatise and Industrial Application*. Academic Press.

Kasianowicz, J. J., Brandin, E., Branton, D., and Deamer, D. W. (1996). Characterization of individual polynucleotide molecules using a membrane channel. *Proceedings of the National Academy of Sciences*, 93(24), 13770–13773.

Kiang, T., Wen, J., Lim, H. W., and Leong, K. W. (2004). The effect of the degree of chitosan deacetylation on the efficiency of gene transfection. *Biomaterials*, 25(22), 5293–5301.

Kim, C. H., Choi, J. W., Chun, H. J., and Choi, K. S. (1997). Synthesis of chitosan derivatives with quaternary ammonium salt and their antibacterial activity. *Polymer Bulletin*, 38(4), 387–393.

Kim, C. H., and Choi, K. S. (1998). Synthesis and properties of carboxyalkyl chitosan derivatives. *Journal of Industrial and Engineering Chemistry*, 4(1), 19–25.

Kneipp, K., Wang, Y., Kneipp, H., Perelman, L. T., Itzkan, I., Dasari, R. R., and Feld, M. S. (1997). Single molecule detection using surface-enhanced Raman scattering (SERS). *Physical Review Letters*, 78(9), 1667.

Kollár, R., Reinhold, B. B., Petráková, E., Yeh, H. J., Ashwell, G., Drgonová, J., and Cabib, E. (1997). Architecture of the yeast cell wall β (1→ 6)-glucan interconnects mannoprotein, β (1→ 3)-glucan, and chitin. *Journal of Biological Chemistry*, 272(28), 17762–17775.

Langer, R. S., Lanza, R. P., and Vacanti, J. (2000). *Principles of Tissue Engineering* (2nd ed.). Academic Press.

Lee, M. H., Lee, H. J., and Ryu, P. D. (2001). Public health risks: chemical and antibiotic residues-review. *Asian-Australasian Journal of Animal Sciences*, 14(3), 402–413.

Leedy, M. R., Martin, H. J., Norowski, P. A., Jennings, J. A., Haggard, W. O., and Bumgardner, J. D. (2011). Use of chitosan as a bioactive implant coating for bone-implant applications. In R. Jayakumar, M. Prabaharan, and R. A. A. Muzzarelli (Eds.), *Chitosan for Biomaterials II*. Springer.

Lehr, C. M., Bouwstra, J. A., Schacht, E. H., and Junginger, H. E. (1992). In vitro evaluation of mucoadhesive properties of chitosan and some other natural polymers. *International journal of Pharmaceutics*, 78(1-3), 43–48.

Li, Y. Y., Chen, X. G., Liu, C. S., Cha, D. S., Park, H. J., and Lee, C. M. (2007). Effect of the molecular mass and degree of substitution of oleoylchitosan on the structure, rheological properties, and formation of nanoparticles. *Journal of Agricultural and Food Chemistry*, 55(12), 4842–4847.

Lichtenthaler, F. W. (2007). Carbohydrates as renewable raw materials: a major challenge of green chemistry. In P. Tundo, A. Perosa, and F. Zecchini (Eds.), *Methods and Reagents for Green Chemistry: An Introduction*. John Wiley & Sons.

Liu, X., Ma, L., Mao, Z., and Gao, C. (2011). Chitosan-based biomaterials for tissue repair and regeneration. In R. Jayakumar, M. Prabaharan, and R. A. A. Muzzarelli (Eds.), *Chitosan for Biomaterials II*. Springer.

Lu, S., Gao, W., and Gu, H. Y. (2008). Construction, application and biosafety of silver nanocrystalline chitosan wound dressing. *Burns*, 34(5), 623–628.

Madihally, S. V., and Matthew, H. W. (1999). Porous chitosan scaffolds for tissue engineering. *Biomaterials*, 20(12), 1133–1142.

McCormack, P. J., Wildman, H. G., and Jeffries, P. (1994). Production of antibacterial compounds by phylloplane-inhabiting yeasts and yeastlike fungi. *Applied and Environmental Microbiology*, 60(3), 927–931.

McNaught, A. D., and Wilkinson, A. (1997). *Compendium of Chemical Terminology* (Vol. 1669). Blackwell Science.

Mohanty, A. K., Misra, M., and Drzal, L. T. (2002). Sustainable bio-composites from renewable resources: opportunities and challenges in the green materials world. *Journal of Polymers and the Environment*, 10(1), 19–26.

Mohanty, A. K., Misra, M., and Drzal, L. T. (Eds.). (2005). *Natural Fibers, Biopolymers, and Biocomposites*. CRC Press.

Muzzarelli, R. A. (1983). Chitin and its derivatives: new trends of applied research. *Carbohydrate Polymers*, 3(1), 53–75.

Muzzarelli, R. A. A., Aiba, S., Fujiwara, Y., Hideshima, T., Hwang, C., Kakizaki, M., and Sinskey, A. J. (1986). Filmogenic properties of chitin/chitosan. In Gooday, G.W., Jeuniaux, C., and Muzzarelli, R. (Eds.), *Chitin in Nature and Technology*. Springer US.

Novikova, L. N., Novikov, L. N., and Kellerth, J. O. (2003). Biopolymers and biodegradable smart implants for tissue regeneration after spinal cord injury. *Current Opinion in Neurology*, 16(6), 711–715.

Okamoto, Y., Kawakami, K., Miyatake, K., Morimoto, M., Shigemasa, Y., and Minami, S. (2002). Analgesic effects of chitin and chitosan. *Carbohydrate Polymers*, 49(3), 249–252. doi:10.1016/S0144-8617(01)00316-2

Opanasopit, P., Ngawhirunpat, T., Rojanarata, T., Choochottiros, C., and Chirachanchai, S. (2007). Camptothecin-incorporating N-phthaloylchitosan-g-mPEG self-assembly micellar system: effect of degree of deacetylation. *Colloids and Surfaces B: Biointerfaces*, 60(1), 117–124.

Opanasopit, P., Ngawhirunpat, T., Rojanarata, T., Choochottiros, C., and Chirachanchai, S. (2007). N-Phthaloylchitosan-g-mPEG design for all-trans retinoic acid-loaded polymeric micelles. *European Journal of Pharmaceutical Sciences*, 30(5), 424–431.

Pavinatto, F. J., Caseli, L., and Oliveira Junior, O. N. D. (2010). Chitosan in nanostructured thin films. *Biomacromolecules*, 11(8), 1897–1908.

Pierce, P. E., Kovac, Z., and Higginbotham, C. (1978). The kinetics and mechanism of film growth during the electrodeposition process. *Industrial & Engineering Chemistry Product Research and Development*, 17(4), 317–322.

Quake, S., and Volkmuth, W. D. (2001). U.S. Patent No. 6,221,654. Washington, DC: U.S. Patent and Trademark Office.

Quintana-Obregón, E. A., Plascencia-Jatomea, M., Sánchez-Mariñez, R. I., Burgos-Hernandez, A., González-Aguilar, G. A., Lizardi-Mendoza, J., and Cortez-Rocha, M. O. (2011). Effects of middle-viscosity chitosan on *Ramularia cercosporelloides*. *Crop Protection*, 30(1), 88–90.

Rane, K. D., and Hoover, D. G. (1993). Production of chitosan by fungi. *Food Biotechnology*, 7(1), 11–33.

Rao, S. B., and Sharma, C. P. (1997). Use of chitosan as a biomaterial: studies on its safety and hemostatic potential. *Journal of Biomedical Materials Research*, 34(1), 21–28.

Ren, Y. J., Zhang, H., Huang, H., Wang, X. M., Zhou, Z. Y., Cui, F. Z., and An, Y. H. (2009). In vitro behavior of neural stem cells in response to different chemical functional groups. *Biomaterials*, 30(6), 1036–1044.

Rinaudo, M. (2006). Chitin and chitosan: properties and applications. *Progress in Polymer Science*, 31(7), 603–632.

Riva, R., Ragelle, H., des Rieux, A., Duhem, N., Jérôme, C., and Préat, V. (2011). Chitosan and chitosan derivatives in drug delivery and tissue engineering. In R. Jayakumar, M. Prabaharan, and R. A. A. Muzzarelli (Eds.), *Chitosan for Biomaterials II*. Springer.

Sashiwa, H., and Aiba, S. I. (2004). Chemically modified chitin and chitosan as biomaterials. *Progress in Polymer Science*, 29(9), 887–908.

Sayyed, R. Z., Gangurde, N. S., and Chincholkar, S. B. (2009). Hypochlorite digestion method for efficient recovery of PHB from *Alcaligenes faecalis*. *Indian Journal of Microbiology*, 49(3), 230–232.

Sekiguchi, S., Miura, Y., Kaneko, H., Nishimura, S. I., Nishi, N., Iwase, M., and Tokura, S. (1994). Molecular weight dependency of antimicrobial activity by chitosan oligomers. In K. Nishinari, and E. Doi (Eds.), *Food Hydrocolloids*. Springer US.

Shahidi, F., Arachchi, J. K. V., and Jeon, Y. J. (1999). Food applications of chitin and chitosans. *Trends in Food Science & Technology*, 10(2), 37–51.

Song, Y. W., Shan, D. Y., and Han, E. H. (2008). Electrodeposition of hydroxyapatite coating on AZ91D magnesium alloy for biomaterial application. *Materials Letters*, 62(17), 3276–3279.

Steinbüchel, A. (2005). Non-biodegradable biopolymers from renewable resources: perspectives and impacts. *Current Opinion in Biotechnology*, 16(6), 607–613.

Stephan, S. B., Taber, A. M., Jileaeva, I., Pegues, E. P., Sentman, C. L., and Stephan, M. T. (2015). Biopolymer implants enhance the efficacy of adoptive T-cell therapy. *Nature Biotechnology*, *33*(1), 97–101.

Stupp, S. I., and Braun, P. V. (1997). Molecular manipulation of microstructures: biomaterials, ceramics, and semiconductors. *Science*, 277(5330), 1242–1248.

Sudarshan, N. R., Hoover, D. G., and Knorr, D. (1992). Antibacterial action of chitosan. *Food Biotechnology*, 6(3), 257–272.

Varki, A., Cummings, R., Esko, J., Freeze, H., Hart, G., and Marth, J. (1999). *Essentials of Glycobiology*. Cold Spring Harbor Laboratory Press.

Venkatesan, J., and Kim, S. K. (2010). Chitosan composites for bone tissue engineering—an overview. *Marine Drugs*, *8*(8), 2252–2266.

Wu, T., Zivanovic, S., Draughon, F. A., and Sams, C. E. (2004). Chitin and chitosan value-added products from mushroom waste. *Journal of Agricultural and Food Chemistry*, 52(26), 7905–7910.

Wang, W., Itoh, S., Konno, K., Kikkawa, T., Ichinose, S., Sakai, K., and Watabe, K. (2009). Effects of Schwann cell alignment along the oriented electrospun chitosan nanofibers on nerve regeneration. *Journal of Biomedical Materials Research Part A*, 91(4), 994–1005.

Yang, J., Tian, F., Wang, Z., Wang, Q., Zeng, Y. J., and Chen, S. Q. (2008). Effect of chitosan molecular weight and deacetylation degree on hemostasis. *Journal of Biomedical Materials Research Part B: Applied Biomaterials*, 84(1), 131–137.

Zinoviadou, K. G., Koutsoumanis, K. P., and Biliaderis, C. G. (2011). Biopolymer-based films as carriers of antimicrobial agents. *Procedia Food Science*, 1, 190-196.

7 Surface Modification of Polymer Biomaterials

7.1 INTRODUCTION

There are many variables that are taken into consideration when designing a biomedical device for implantation into the human body. The most prominent of these are: (1) bioactivity with the host site; (2) biocompatibility; and (3) protein adsorption. When a biomaterial is introduced into the human body, the first contact occurs between the surface of the biomaterial and the tissues at the host site (Ikada, 1994a). It is because of this that the surface of the biomaterial must be adequate in the sense that it can produce a favourable response. If the response towards the biomaterial is strong then the implanted device will not work as intended. For example, contact lenses will ruin the cornea if not wettable by tears. The effectiveness of an immunoadsorbent, which is determined by the surface moiety to bind pathogenic substances for removal, will be significantly altered if the binding sites are blocked because of cell and protein adhesion to said sites (Ikada, 1994b). It is clear that the surface interaction plays a very crucial role, and this becomes an issue because common materials used for designing biomedical devices don't always possess the desired surface properties for their given application. A host of biomaterials are chosen because of their mechanical and chemical properties being robust and non-toxic, such as polylactide (PLA) and poly(lactide-co-glycotide) (PLGA), but they do not have the necessary surface properties to induce cellular absorption and regeneration (Ma et al., 2007). It is through this desire, to use polymers and materials with specific mechanical and chemical bulk properties, that the study of material surface modification started. Through the application of surface modification, the interacting surface of the biomaterial can be altered in order to undergo a positive reaction from the human body once implanted, without sacrificing the bulk properties of the material itself (Ikada, 1994a; Elbert and Hubbell, 1996; Ikada, 1994b, "Interfacial Biocompatibility"). Figure 7.1 shows a general representation of how surface modification techniques alter the surface composition of a polymer (Goddard and Hotchkiss, 2007).

For over half a century, polymers have been undergoing surface modifications to allow for better biocompatibility and interactions with the host tissues and bodily fluids at the implantation site. There is a list of methods that are currently used, and as technology increases so will the ability to apply more complex surface modifications for specific purposes. These different techniques aim to alter the three major aspects of the polymer through four distinct methods. These methods include altering the (1) hydrophobicity/hydrophilicity of the surface; (2) altering the surface energy of the material; (3) changing the chemical topography of the material; and

by Yaser Dahman and Stefan Balko

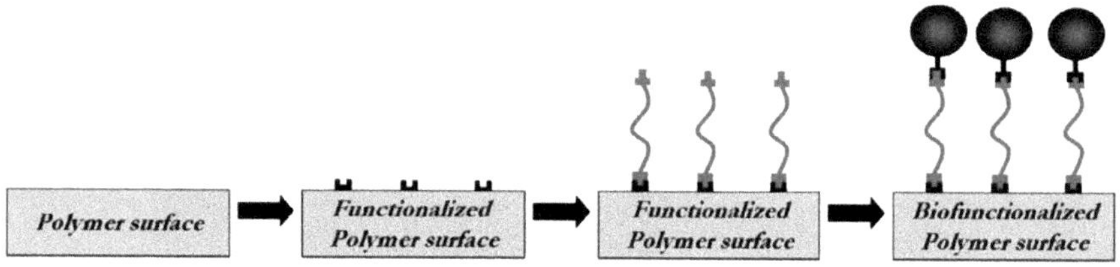

FIGURE 7.1 Concept of surface modification (Goddard and Hotchkiss, 2007).

(4) changing the physical topography of the material. By tuning these specific properties of the surface, the correct combination of bioactivity, biocompatibility, and protein adsorption can be achieved while maintaining the overall bulk properties of the original polymer.

7.1.1 Bioactivity

Bioactivity refers to the ability of a polymer to elicit a biological response from the host site that it is implanted in. Many factors are taken into account when trying to tune common polymers to be bioactive. Some of these factors include (1) cellular adhesion; (2) scaffold degradation; (3) growth factor inclusion; (4) matrix protein inclusion; and (5) immune response. As result, there have been many approaches that aim to specifically target these issues. In terms of surface modification, it has been shown that small peptide sequences integrated into the surface of polymers show the most promise as they continue to be stable for modification, are tunable for cell binding, and are easy to synthesize in large-scale applications (Ma et al., 2007; Zhu, 2010; Teramura and Iwata, 2010). Generally speaking, specific peptide chains or proteins are bound to the surface of inert or non-biologically active polymers to address the specific issues stated earlier. For example, ECM protein derived peptides, specifically collagen, elastin, fibronectin, and laminin, have been used to increase cellular adhesion of poly(ethylene glycol) (PEG) hydrogels and increase the bioactivity by 240 times (Zhu, 2010).

7.1.2 Biocompatibility

The biocompatibility of an implanted biomaterial plays a crucial role in the successfulness of the implant itself. Biocompatibility refers to the ability of a device to perform with an appropriate host response in a given application. What this means, in regard to a polymeric implant, is that there can't be any cytotoxicity which might cause cell death, no chronic inflammation, and no impairment of regular cellular or tissue function (Bauer et al., 2012; Zhou et al., 2016). In terms of design, biocompatibility not only refers to the actual implant itself but even stems to the process by which the polymer is manufactured as well as any post-processing that occurs. A proper host response not only relates to the initial response of the surface-tissue interaction, but it also includes the response during the entire time of implantation. This could mean the biodegradable particulates, as a tissue-engineered scaffold degrades over a certain time and new tissue replaces it, must not be harmful to the tissue as well as the host environment.

When any implant is introduced into the body, a foreign body reaction is induced which usually consists of three steps. The first step is the adsorption of proteins onto the surface of the implant. These non-specific proteins signal cells to bind to the surface of the implant, which is the second phase. These cells release cytokines and subsequently create an inflammatory response. The release of these cytokines causes macrophages to migrate to the implant site that, in a negative scenario, will bind with the adsorbed proteins and begin to fuse to form a giant cell which then releases degradative enzymes onto the surface of the biomaterial. These degradative enzymes have been known to lead to the degradation of the implant material, which leads to the failure of the device. In order to elicit a positive response, the surface of a biomaterial must be tailored such that it can mediate and direct the response (Zhou et al., 2016; Stunkard, 2009).

Many different surface modification techniques are used to mediate the natural foreign body response as the surface-cell interaction is what leads to potential degradation. A plasma surface treatment of polymers is a common technique used as it is simple to use and can be repeated quite easily, with the ability to tailor exposure times (Goddard and Hotchkiss, 2007). Plasma is essentially a gas that is partially ionized into charged particles, electrons, and neutral molecules. Plasma has the ability to penetrate the first nanometer of a polymer surface, meaning the bulk of the polymer goes unchanged. This is important, as many applications require the strength and bulk properties of the polymer in order to work properly. Many different types of gases can be used to plasma treat the surface of polymers, but the most common are ammonia, nitrogen, oxygen, carbon dioxide, or carbon monoxide. Through plasma treatment, polymer surfaces can be tuned to be more hydrophilic, reducing foreign body responses (De Valence et al., 2013).

7.1.3 Protein Adsorption

Protein adsorption has been shown to play a crucial role in cellular adhesion to polymers. Depending on the specific design and function of the polymer being implanted in the human body, protein adsorption can either lead to it failing or succeeding. When a polymer is introduced into the human body, nonspecific protein adsorption occurs immediately. This can lead to further cascading of other biological responses depending on the proteins adsorbed (Wei et al., 2014; Jo and Park, 2000). This can lead to potential foreign body responses that would elicit the degradation of the polymer and the overall failure of the device. Because of this, non-specific protein adsorption is something that is not desired. By inhibiting nonspecific protein adsorption, the biocompatibility of the polymer is directly increased in a favourable manner.

That being said, by eliciting specific protein adsorption to occur, the potential for specific cellular attachment and proliferation can be increased. Van Wachem et al. (1987) have shown a direct increase in human endothelial cell adhesion when surfaces were coated with specific proteins found in blood plasma. Polymers used as scaffolds for tissue growth, repair, and replacement require cellular growth and proliferation to occur in order to repair the damaged tissue. If the polymer scaffold does not allow for any protein adsorption, cellular adhesion would not occur and ultimately the device would fail. Therefore, protein adsorption is a difficult subject and one of great concern, because although indirect and undesired protein adsorption should be avoided to increase biocompatibility of the device, allowing specific and selective protein adsorption to occur will ultimately allow for certain devices to succeed.

Protein adsorption has been directly linked to the wettability, or hydrophilicity, of a polymer's surface. It is generally believed and accepted that materials that exhibit a hydrophilic surface will not adsorb proteins. Therefore, by altering the surface hydrophilicity, the undesired protein adsorption can be avoided, allowing for materials to be more biocompatible. The surface energy of the material also plays a role. Generally speaking, higher surface energies allow the binding of proteins to reduce the surface energy to a more favourable state (Wei et al., 2014; Jo and Park, 2000; Liu et al., 2010).

Different techniques have been developed, over time, that allow for the modification of these distinct surface properties. The most commonly used surface modification techniques are: (1) grafting; (2) polyelectrolyte layer addition; and (3) surface topographical changes (Ikada, 1994a). These techniques will be looked at in more detail as to how they alter the characteristics mentioned above, and in what situations they are useful.

7.2 GRAFTING

The general idea behind grafting is the addition of a monomer or polymer to the surface of a material through the creation of functionalized groups on the surface of the polymer. There are two broad techniques that are used. Graft polymerization is the synthesis of polymer chains from monomers by initiating chain growth directly on the substrate surface via free radicals and binding sites previously created, and polymer grafting is the attachment of previously synthesized polymer chains to the substrate surface, also through the use of free radicals (Penn and Wang, 1994). When looking at the specific techniques used to achieve this binding of polymers and creation of free radicals and binding sites, the three most used techniques are: (1) Direct chemical modification; (2) Ozone treatment and; (3) Plasma treatment. In essence, each method attaches a monomer or a second polymer to the surface of the polymer to tune the hydrophilicity, the non-specific protein adsorption, the biocompatibility, and the polymer's overall ability to promote cellular growth and proliferation. These techniques add functionality to the surface of the polymer, which tunes it one way or another depending on the desired result (Uyama et al., 1998). These bound monomers or polymers have specific tail properties that elicit a response from the environment they are implanted into.

7.2.1 Direct Chemical Modification

Direct chemical modification involves altering the surface through the use of other chemicals in a stepwise fashion, first to achieve functionalization and second to apply the monomer or polymer to the surface. This is also known as a wet-chemical technique. These were some of the earliest techniques for treating the surfaces of polymers, and traditionally involve an aqueous solution containing oxidizing chemicals (Goddard and Hotchkiss, 2007; Penn and Wang, 1994). Usually applied to polyethylenes, polypropylenes, and polyesters, the oxidation once applied introduces functional groups that increase the polarity potential for hydrogen bonding which in turn increase the hydrophilicity of the surface (Penn and Wang, 1994; Uyama et al., 1998; Peez et al., 1998). This technique requires no special equipment and therefore can be conducted in any laboratory, with the added bonus of its having a better ability to penetrate 3D substrates (Goddard and Hotchkiss, 2007). Bamford and Al-Lamee (1994) used peroxydisulfate and peroxymonosulfate in aqueous solution to oxidize the surface of different polymers and applied a second solution to graft a vinyl monomer to the surface for enhanced properties. Rasmussen et al. (1977) also investigated wet chemistry surface modifications of polymers, more specifically low-density polyethylene. Initially, they oxidized the surface through the introduction of chromic acid solution at temperatures between 25°C and 75°C and followed up with further oxidation with the introduction of a 70% nitric acid solution at 50°C. This introduced functional groups to the surface, tuning it to become hydrophilic. In general, wet chemical grafting involves multiple steps in order to achieve the final result. Because of the ease of the technique, a multitude of polymers and monomers can be used to graft onto surfaces, showing how robust and versatile this technique is. The general outcome is to achieve a hydrophilic surface for better biocompatibility and reduce protein adsorption. That being said, oxidized polymer surfaces my actually exhibit a phenomenon called surface reconstruction, where the oxygen functionalized group diffuses and reorients away from the surface, leaving the lower-energy hydrophobic portion exposed at the surface. This will result in an increase in hydrophobicity of the surface rather than an increase in hydrophilicity (Penn and Wang, 1994). Another drawback of this method is that it can lead to hazardous waste, which can lead to irregular surface etching. On top of this, the degree of surface functionalization may not be repeatable between polymers of different molecular weight, crystallinity, or tacticity (Goddard and Hotchkiss, 2007).

7.2.2 Ozone Treatment

Ozone treatment plays a similar role in that it oxidizes the surface of the polymer in order to create functionalized groups for further grafting. The polymer surface is treated with ozone gas either on its own, or in conjunction with another form of stimulation such as UV, which creates peroxides on the surface of the polymer, which in turn can then be used for further grafting of other polymers or left on its own (Tu et al., 2005; Oláh et al., 2005). The advantage that ozone treatment has is that it is clean and is considered a "dry" grafting approach in that no liquids or aqueous solutions are used to initiate the oxidizing process or functionality of the surface. In addition to this, it is inexpensive and can be carried out in atmospheric pressures (Tu et al., 2005). Mathieson and Bradley (1996) explored the use of combined ozone and UV on polyethylene and polyether ether ketone noting an increase in surface wettability after treatment. Tu et al. (2005) examined the effects of ozone treatment on polytetrafluoroethylene (PTFE) substrates, noting the formation of peroxide groups on the surface, which were then used to graft acrylamide, acrylic acid, glycidyl methacrylate, and 2-(2-bromoisobutyryloxy) ethyl acrylate as monomers to the surface. It was noted that the surface showed a hydrophilic response after treatment, exhibiting a 21° water contact angle.

Although ozone treatment is a simple and clean technique to use, there are some drawbacks. The first of these drawbacks is that ozone will not cause functionalization on all surfaces. Fully fluorinated PTFE will not be affected by ozone treatment because of the strong bonding energy of the C-F bonds and would require a pretreatment to allow for ozone treatment to work (Tu et al., 2005). Another drawback is that after exposure to ozone, depending on the intensity of exposure, the polymer reverts back to its hydrophobic nature. The hydrophilic surfaces of certain polymers (e.g. poly(dimethylsiloxane)) are unstable and a hydrophobic recovery will occur. At low degrees of oxidation, the process by which it recovers is called the reversible relaxation process, whereas at a higher degree of oxidation, the free low molar mass molecules diffuse to the surface, reducing the free surface energy (Oláh et al., 2005).

7.2.3 Plasma Treatment

Plasma treatment is a field that is getting more attention of late because of its robust ability to graft polymers and monomers onto the desired surface. Similar again to both the direct chemical method and the ozone method, plasma treatment involves applying plasma to the surface of a polymer to create free radicals on the surface, which in turn can be used for further grafting of polymers and monomers. Plasma itself is a partially ionized gas in the form of a gaseous mixture of electrons, ions, radicals, and neutral particles. For plasma treatment of polymers, nonthermal plasma is generated by an electrical gas discharge. Generally speaking, there are four different methods by which the plasma can be used for grafting. These four methods are (1) plasma treatment; (2) plasma postirradiation grafting; (3) plasma syn-irradiation; and (4) plasma polymerization deposition. Each method is different in its own right and is generally used in its own manner (Jacobs et al., 2012; Desmet et al., 2009).

Plasma treatment applies an inert gas, which creates functional groups or free radicals on the surface, depending on the type of plasma used. Typically, Ar, He, O_2, N_2, NH_3, and CF_4 are the gases that are used (Desmet et al., 2009). Figure 7.2 below is a schematic representation of how the surface is treated.

This method is very similar to the ozone treatment method mentioned earlier and has a similar drawback, in the sense that some time after exposure the polymer surface will exhibit a hydrophobic recovery, depending on the original structure of the polymer used. Cui and Brown investigated the modification of polypropylene films

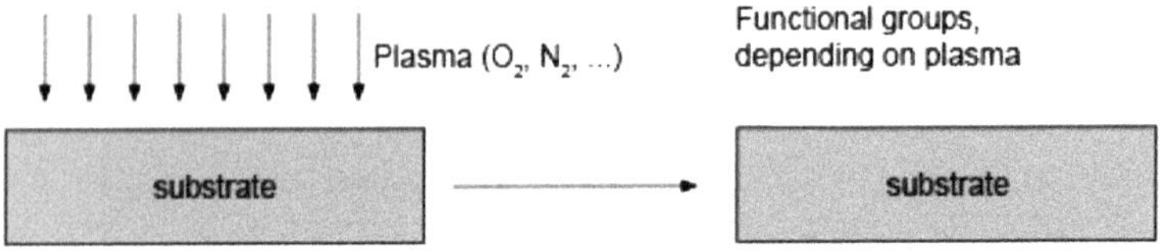

FIGURE 7.2 Schematic of plasma treatment (Desmet et al., 2009).

using an air dielectric barrier discharge plasma and found that the surface was tuned to hydrophilic directly post-treatment, but after long-term storage, the hydrophobic nature recovered (Cui and Brown, 2002). Generally speaking, plasma treatment is used mainly for increasing hydrophilicity of the surface, but it can also be used to immobilize biologically active ligands (Jacobs et al., 2012, Desmet et al., 2009, Cui and Brown, 2002).

7.2.3.1 Plasma Post-Irradiation Grafting

Plasma post-irradiation grafting involves the specific use of He and Ar plasmas, which introduce free radicals on the surface without actually functionalizing the surface or introducing plasma-deposited polymers (Desmet et al., 2009). When exposed to atmospheric gas, or to O_2, specific functional groups are formed. These groups are peroxides and hydroperoxides, which themselves can then be used for grafting of other polymers and monomers to the surface. This method is referred to as the "grafting-from" approach. In this approach, the monomer or polymer that is grafted onto the surface is not subjected to the initial plasma application (Desmet et al., 2010; Siow et al., 2006). Figure 7.3 has a schematic representation of the process by which plasma post-irradiation occurs.

Vasilets et al. (1997) have demonstrated the ability to bind human thrombomodulin to the surface of polytetrafluoroethylene by treating with CO_2 plasma to functionalize the surface and then subsequently grafting acrylic acid onto the surface.

7.2.3.2 Plasma Syn-Irradiation

Unlike post-irradiation, plasma syn-irradiation involves a monomer that is subjected to the plasma alongside the surface of the polymer that is trying to be tuned. Syn-irradiation involves the introduction of a monomer in aqueous or gaseous form to the

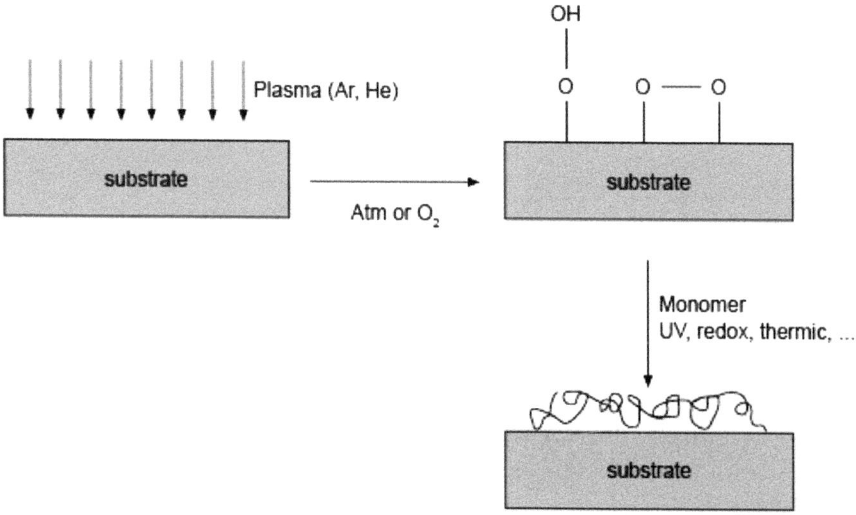

FIGURE 7.3 Schematic of plasma post-irradiation (Desmet et al., 2009).

surface, which is then subjected to plasma. The plasma creates free radicals on the surface and the monomer allows for crosslinking to occur. Typically, the monomer is already adsorbed onto the surface of the polymer being tuned beforehand (Jacobs et al., 2012; Desmet et al., 2009). The top set of images in Figure 7.4 show the schematic representation of this technique.

The last method is similar to this one but involves the introduction of a monomer to the plasma gas itself.

The final plasma technique is plasma polymerization deposition, which involves a monomer being introduced into the plasma in vapour phase. The monomer is converted into reactive fragments which combine to create polymers in gas phase in the plasma itself. These polymers can then bind to the surface of the polymer being tuned to create a polymer coating on the desired surface. This technique is known as a "grafting-to" approach. The bottom set of images in Figure 7.4 show the schematic representation of plasma polymerization deposition (Jacobs et al., 2012; Desmet et al., 2009; Tendero et al., 2006).

There are a few different sources from which the plasma can be generated, each following the principle of creating an electrical discharge in a partially ionized gas. Essentially, each technique causes electrons, which are generated by applying a strong electric field to a natural gas in order to ionize it, to move in the direction of the electrical field towards the surface of the polymer that is being treated. Traditionally, the sources operate at low pressures because lower pressure plasma discharge is more stable (Jacobs et al., 2012; Desmet et al., 2009). The main methods for applying plasma are as follows: (1) corona discharge, which involves a low current discharge caused by a breakdown of a gas gap with a strong electric field at atmospheric pressure; (2) dielectric barrier discharges, where one of two electrodes in the apparatus is covered in a dielectric layer which accumulates charge after ionization and generates

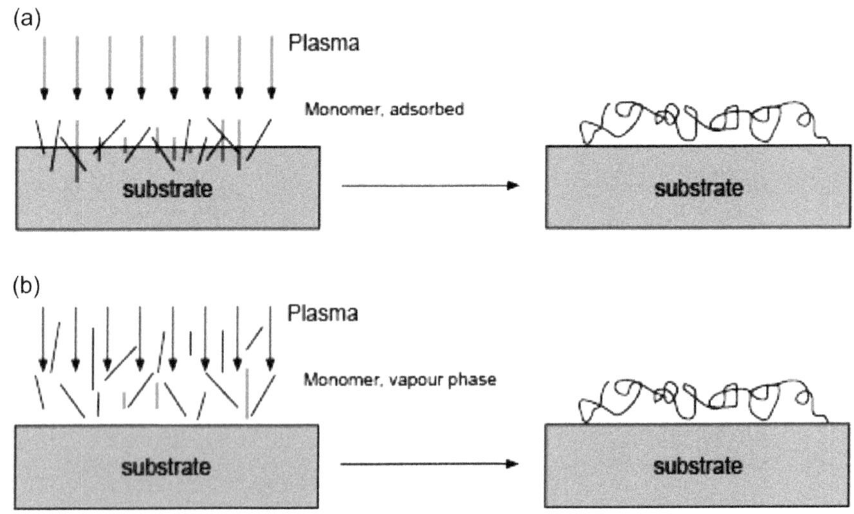

FIGURE 7.4 Schematic of plasma syn-irradiation (a) and plasma polymerization deposition (b) (Desmet et al., 2009).

an electrical field which reduces the field in the gap and causes a breakdown; and lastly, (3) radio frequency discharge, which involves high frequency electromagnetic waves which excite the gas and cause ions and electrons to move along the electromagnetic field as the frequency of the field increases (Desmet et al., 2009).

Grafting as a technique can be used in a vast amount of different applications. Generally speaking, grafting is used to alter the surface of a polymer by adding a monomer or polymer which contains specific properties to the surface of another polymer. This secondary polymer is usually of hydrophilic nature in order to tune the overall surface of the first polymer to be hydrophilic. Although this is the most-used application of grafting, grafting can be used to add specific bioactive molecules onto the surface of another polymer, which would allow for better cellular proliferation and adhesion. Enhanced cellular adhesion and proliferation can also be accomplished through the grafting of specific biopolymers which have been known to promote cellular growth and adhesion, such as collagen, to the surface. Specifically, after graft polymerization Gupta et al. (2002) immobilized collagen onto the surface of poly(ethylene terephthalate), which in turn increased the growth rate of smooth human muscle cells. In terms of manufacturing applications, grafting has been used in drug delivery systems, tissue engineering scaffolds, and even macroscale polymeric implants for bone replacement. Although grafting allows the specific tunability of the surface hydrophobicity and surface composition, it does have one major drawback. Depending on the shelf life or wait time before use, grafted polymers tend to reorient themselves and revert back to their hydrophobic nature, depending on the intensity of the grafting procedure and the technique used.

7.3 POLYELECTROLYTE LAYER-BY-LAYER DEPOSITION

Another form of surface modification that is simple to use and has grown in popularity recently is the application of polyelectrolytes to the surface of a material in a stepwise fashion to create a coating that adheres to the surface through electrostatic forces. The method is based on the assembly of a thin film coating through the subsequent deposition of polyanions and polycations onto a surface. As long as there exists a charge difference between the materials being deposited, essentially any form and combination of materials can be used, including linear polyelectrolytes, enzymes, antibodies, viruses, and inorganic nanoparticles (Lvov et al., 2003). Figure 7.5 shows a schematic of the general step-by-step procedure as to how the thin film coating is built up.

The multilayer thin film is self-assembling and can have its thickness tuned depending on the number of layers that are wanted. Because of this, films with nanometer thickness can be created. Depending on the terminating layer of the thin film, the surface free energy, hydrophobicity and surface chemistry can be altered to match what is desired. There are three techniques that are widely used to create a thin film polyelectrolyte coating on polymer surfaces: (1) dip method; (2) spray coating; and (3) electrospinning. Each method has its advantages and disadvantages, but in general, they all apply the same principle of depositing oppositely charged polyelectrolytes to the surface of a material to build up the thin film coating.

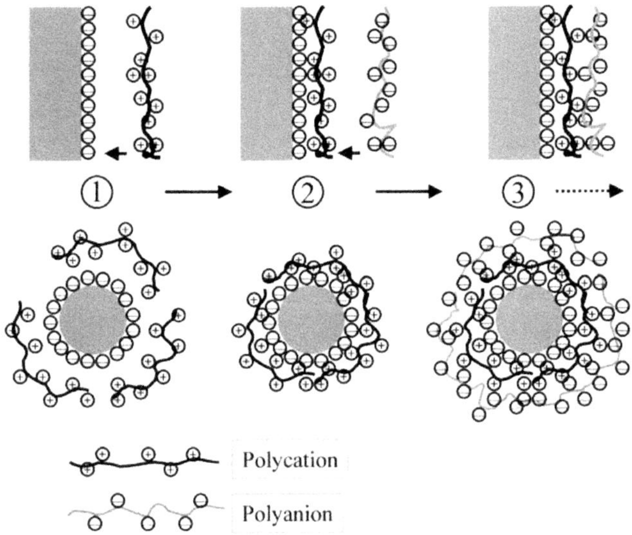

FIGURE 7.5 Schematic of electrostatic layer-by-layer deposition in 2D and 3D (Ai, Jones, and Lvov, 2003).

7.3.1 Dip Method

The most basic of the methods, and subsequently requiring the least amount of equipment to implement, is the dip method. Figure 7.6 has a representation of the step-by-step procedures involved in the dip method of layer-by-layer thin film coating.

Essentially the surface of a polymer is given an initial charge and dipped into a polycation solution, followed by a wash, then dipped into a polyanion solution, which is then followed by a subsequent wash. This process of dipping into polycation and polyanion solutions is repeated until the desired number of layers is achieved. As the original layer is adsorbed onto the surface, the surface charge reverses because of the high concentration of polyelectrolytes in solution, which is what allows for the subsequent deposition of the next layer (Lvov et al., 2003; Ariga et al., 2007; Picart, 2001). The advantages of the dip method are that it is extremely simple and requires no extra lab equipment except beakers and tweezers, and at the same time, it still maintains the versatility of the technique in terms of what materials can be used. The downsides of this method are that, because the solutions are aqueous in form, they require a washing step between each layer deposition leading to a longer process time overall. That being said, with the introduction of a spinning platform, Lee et al. (2003) were able to decrease the process time and increase the homogeneous deposition of the polyelectrolytes to the surface.

7.3.2 Spray Coating

The spray coating technique is very similar to the dip method, except instead of dipping the surface into the aqueous solution, the solution is sprayed onto the

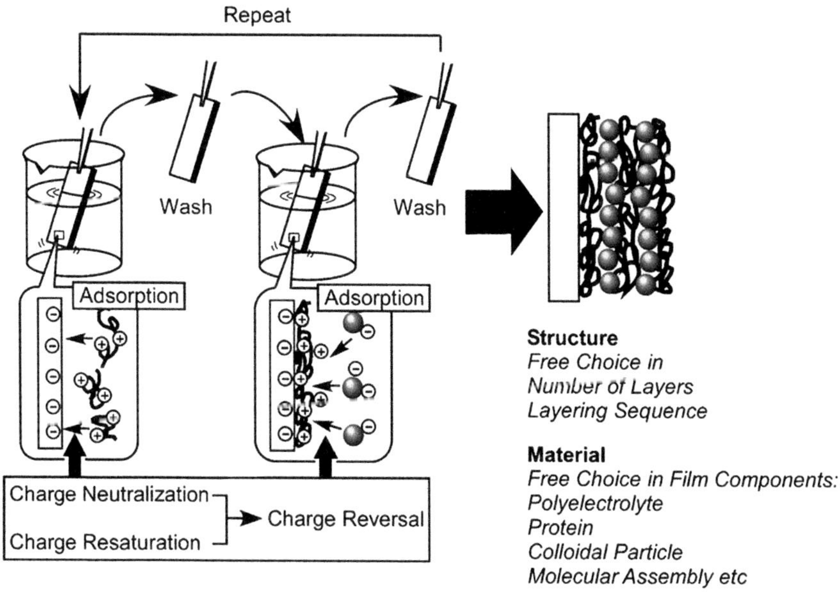

FIGURE 7.6 Dip method of layer-by-layer thin film coating (Ariga, Hill, and Ji, 2007).

surface. Figure 7.7 shows a step-by-step representation of how the spray method is carried out.

The polyanion and polycation solution are sprayed onto the surface, with the nozzle an optimum distance away from the surface to allow for full coverage of the polymer surface. Although this technique requires the ability to bring the polycation and polyanion solution into spray form, it has many benefits over that traditional dipping method technique. The largest benefit is that it can cut the time of the process down significantly. Izquierdo et al. (2005) investigated the total time savings by applying a spray method and showed that the spray method could be up to 250 times faster than the traditional dipping method. The major reason for this is because by applying the polyions through a spray you are generating a more homogeneous layer deposition in a faster amount of time. This, combined with the fact that the drainage of the excess spray essentially constantly removes excess material which essentially eliminates the need for a washing stage, dramatically decreases the overall process time of the procedure. That being said, dipping methods tend to allow for a thicker layer to be deposited than the spraying method (Izquierdo et al., 2005; Ariga et al., 2014). The last layer-by-layer technique that is widely used and has seen a lot of attention recently is the electrospinning technique.

7.3.3 Electrospinning

Electrospinning requires a much more complicated setup than the spray or dip methods listed above. The apparatus involves a polymer solution reservoir with a flow-controlled pump, which pumps the polymer solution through a needle or capillary with a

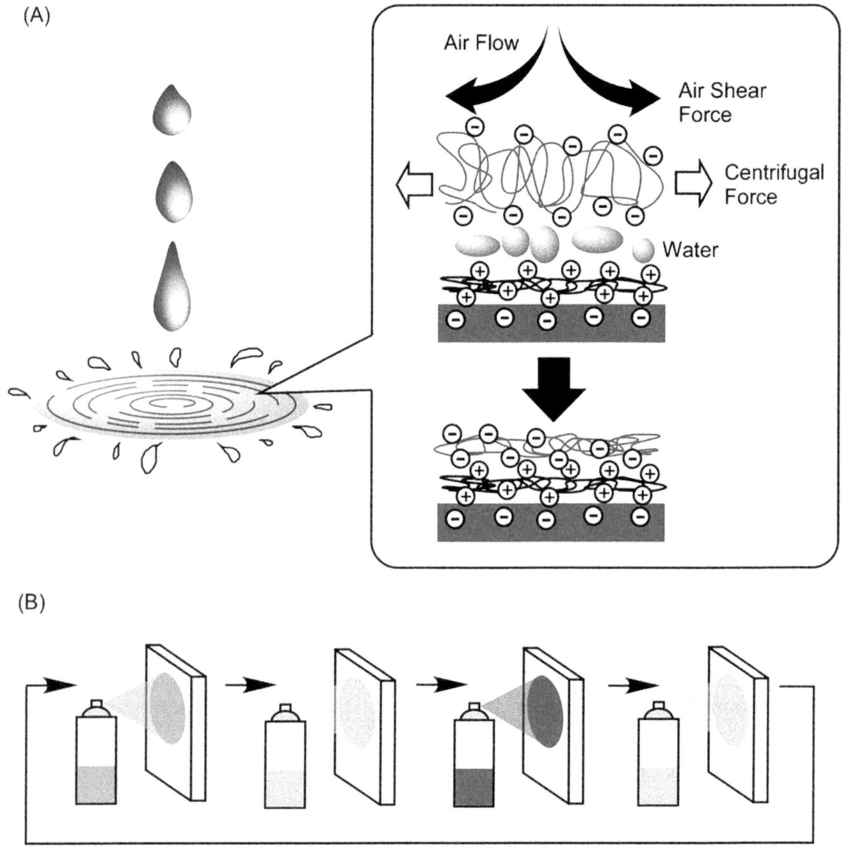

FIGURE 7.7 Layer-by-layer method via (a) spin-coating and (b) spray coating (Ariga, Hill, and Ji, 2007).

metal tip which is connected to a high voltage source. As the polymer travels through the needle or capillary it becomes charged. The surface that you want to alter resides on a separate target surface which is conductive, and as the polymer is pumped out it becomes charged and is bonded to the surface via electrochemical charges (Li Shanti, and Tuan, 2007). An example of this apparatus can be seen in Figure 7.8.

As the polymer is pumped out of the reservoir, it forms a droplet on the end of the needle or capillary which over time elongates and forms a cone as the voltage applied increases. Once the surface charge overcomes the surface tension of the droplet, the droplet releases from the tip and a polymer jet is initiated, which causes the solvent to evaporate and the polymer itself to divide into micro- and nanofibers. With multiple needle heads, the deposition of a multilayered thin film can be achieved. The nano- and microstructure of the membrane can be finely tuned with electrospinning, leading to more or less porosity as well as the immobilization of specific biomolecules if desired. Electrospinning also allows the tunability of surface roughness, which can lead to better cellular adhesion.

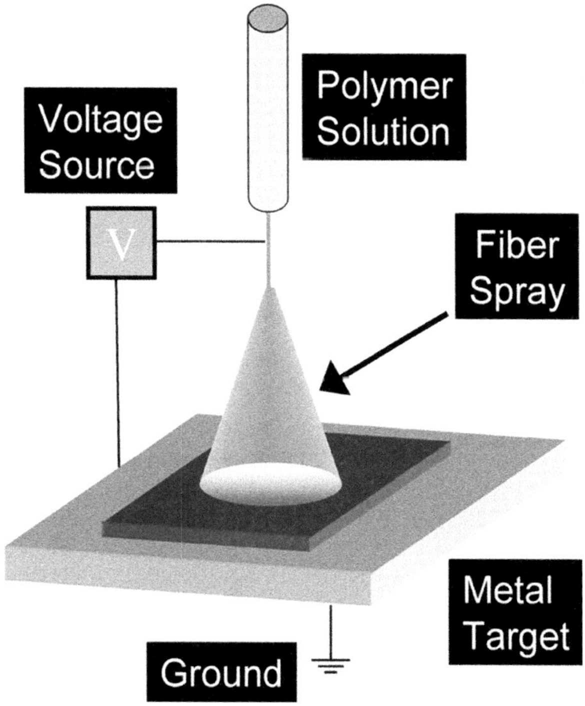

FIGURE 7.8 Basic electrospinning apparatus (Li, Shanti, and Tuan, 2006).

Similar to the grafting technique, the polyelectrolyte layer deposition technique has a multitude of applications. Through the layer-by-layer technique, biomolecules and specific proteins can be embedded into the thin film coatings, increasing cellular adhesion properties as well as cellular proliferation. In regard to this, the layer-by-layer system can be used for drug delivery techniques, embedding specific drugs or viruses into a microencapsulated device which only releases its drug once the specific protein has attached to the surface and allowed for the degradation of the layer-by-layer membrane. Zhang et al. developed such a system for a cancer therapy, releasing the model drug doxorubicin once the microcapsules were exposed to the acidic environment found around cancer cells (Zhang et al., 2012). The layer-by-layer technique can also be used to tune the specific surface architecture of the surface, increasing surface roughness or even creating microchannels that allow for specific orientation of cells as they grow. In terms of protein adsorption, by including hydrophilic polymers to the surface and altering the surface charge, the layer-by-layer technique is a readily available technique.

7.4 SURFACE TOPOGRAPHY

Another key aspect that is responsible for how polymers will react once implanted into the human body is the surface topography. Specifically speaking, the surface roughness and the overall surface architecture, both physical and chemical, play a crucial role in how cells will interact with that surface (Ma et al., 2007). The surface topography can be divided into two distinct groups based on how they are able to affect the interaction with the host tissues. The first set is related to the surface features at a microscale, which involve ridges, cracks, and microchannels (Tan and Saltzman, 2004). Cracks and ridges have the ability to act as a grafting surface by entrapping polymers in the microchannels and ridges of the surface. Grafting and etching techniques that increase the surface roughness at a microscale allow for the entrapment of polymers to occur (Ito, 1999). These microfeatures also have the ability to control the direction and proliferation of cells, directing cells to proliferate and grow along the microchannels. This is very crucial in specific tissue architectures where the direction of cell growth is important to the strength and function of the tissue (Ito, 1999; Tsuda et al., 2007). There are a few different methods which are used to specifically tune the surface architecture and topography, such as hot-embossing imprint lithography, microcontact printing, and block copolymers but the most readily used technique is photolithography (Charest et al., 2006; Nie and Kumacheva, 2008).

7.4.1 Photolithography

Photolithography involves the patterning of a polymer surface by exposing it to photoirradiation through a mask, which allows only parts of the surface to be affected. Depending on the surface, the light will either render the surface more soluble or less soluble to a solvent. After exposure to the photoirradiation, a solvent is applied to the surface which removes areas either affected or unaffected by the light, depending on the original surface composition. This creates a micropattern, which in itself can be used as a template for further functionalization and surface modification (Nie and Kumacheva, 2008). Figure 7.9 shows a schematic representation of photolithography and how it is applied.

The photoirradiation causes the polymer to also become crosslinked and functionalized for potential further use. Through the use of photolithography, specific architectures can be created on the polymer surface. The resolution of patterns ranges from micrometers to nanometers in scale, and depending on the mask that is used, a vast amount of different and unique architectures are possible. A specific example of this micropatterning can be seen in Figure 7.10.

There are many different applications that can be achieved by altering the surface topography of a polymer. As mentioned previously, through the creation of microchannels on the surface, the directional growth and proliferation of cells can be controlled. Specific tissues require directional growth in order to operate as intended, such as neurons and specific muscle cells (Tai and Buettner, 1998). Revzin et al. (2003) explored this idea, creating wells which had PEG walls which were highly cell repelling, essentially forcing cells to migrate into the wells and grow. They noticed cell occupancy of 96.7% inside the wells when culturing 3T3 fibroblasts and primary rat hepatocytes. Dowel-Mesfin et al. (2004) investigated the effects of topography on rat hippocampi neuron growth and orientation. Using photolithography, they created micropillars and channels and cultured the hippocampi neurons on

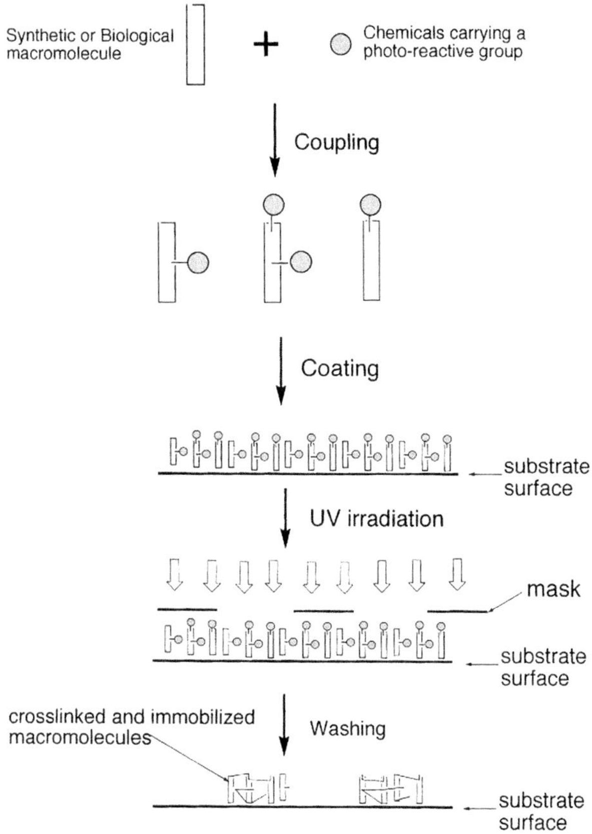

FIGURE 7.9 Schematic of photolithography (Ito, 1999).

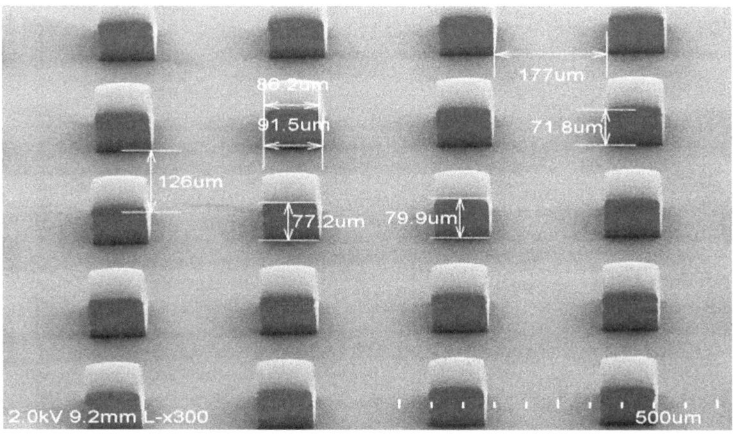

FIGURE 7.10 Scanning Electron Microscopy (SEM) of pillars and channels created through photolithography (Kim and Goulbourne, 2014).

these surfaces and noted the growth followed the orientation and alignment of the pillars and subsequent channels. Another major area of application is the creation of topographically specific architectures for future processing. Photolithography can be used as an imitator for further grafting, which in turn creates a 3D architecture with regions of high hydrophobicity and regions of low hydrophobicity. Wang et al. explored the use of photolithography combined with surface vapour polymer deposition. They created a specific crisscross pattern of poly(γ-benzyl-L-glutamate) and poly(γ-methyl-L-glutamate) which was initiated with traditional photolithography and the application of a photoresistor to block out unwanted deposition sites (Penn and Wang, 1994). As is the general goal with surface modifications, surface topography alterations can also change the biocompatibility of the surface. Surface roughness has been shown to have a direct link to surface wettability, leading to localized packing of certain proteins, depending on the geometry (Lord et al., 2010).

7.5 DRAWBACKS

As with many other techniques used in polymer modification for biomaterials, surface modification techniques still have drawbacks associated with them. The largest drawback is the fact that over time, polymers with altered surface properties tend to recover back to their original surface chemistry. This is an issue for many reasons, the first of which is that if a polymer is implanted into a human and the surface starts to revert back to its original form, it will elicit a negative response from the body and will end up failing. If a polymer with a surface tuned to be hydrophilic so as to reduce nonspecific protein adsorption reverts back to its hydrophobic nature, the polymer will attract unwanted proteins, which will lead to a foreign body response and ultimately cause the polymer to fail. This is also an issue when the polymer is used in tissue engineering applications, specifically if the polymer is meant to act as a scaffold. If the polymer loses its surface properties over time, cellular proliferation and adhesion will lower, causing the tissue engineered construct to ultimately fail. Even if the scaffold is meant to degrade over time, if the surface loses its biocompatible properties and starts to degrade at a very high rate it can cause cleaves and fractures in the structure, which will lead to its eventual failure. Although the recovery of the surface properties does occur, the severity at which it occurs depends on the technique used.

Another potential drawback is the price associated with different techniques. Specific techniques, such as plasma grafting and electrospinning, require precise instruments, which can be costly. When looking at it from a manufacturing perspective, if surface modifications had to be scaled up in order to be applied to multiple surfaces, the cost would increase even more.

In relation to manufacturing, another issue is the scaling up factor of these techniques. Although these techniques are applicable in small batch settings, if surface modification for polymers in biomaterials is to be seen as a method for allowing polymers in regenerative medicine to be mass produced, these techniques would need to be scaled up to be able to handle hundreds of surfaces at the same time. Not only would the techniques and apparatuses need to be adjusted to handle a higher quantity, but control of the devices will need to be regulated such that each surface is getting the exact same treatment for the exact same amount of time.

7.6 CONCLUSION AND FUTURE PROSPECTS

The use of surface modification for polymers intended for biomedical applications is quite vast and plays a major role in ensuring that these devices and designs succeed when they are implanted into the human body. There are many different types of surface modification techniques, but generally speaking, they can be broken down into the three main categories: grafting, polyelectrolyte layer deposition, and surface topography alteration. Each technique uniquely alters the biocompatibility, bioactivity, and protein adsorption of the polymer surface to elicit a specific response from the human body. Although the field is showing great promise already, it is still relatively new and will need to continue to move forward as technology increases. In regard to where the field is moving, there are definitely a few directions in which it can go. Firstly, as technology increases, so does the ability to scale down the surface modification procedures. With further scaling down of surface modification techniques, very specific surface chemistries and topographies can be created to elicit the exact response required from the human body for the given application. Although techniques currently allow for better-lowered protein adsorption and increased biocompatibility, there is still an unknown as to what the exact specific response from the human body will be. With further scaling down of surface modification treatments, specific and complex surface chemistries can be created which will eliminate this unknown factor and allow for an exact response to be known.

In conjunction with this, a combination of techniques will need to start being investigated in order to create more complex surfaces. Although each technique works on its own, combining different surface techniques together will be required in order to tune the surface to be more complex and specific. This is already starting to happen, but as technologies increase the ability to further combine techniques will be readily available which will lead to better surfaces overall.

As mentioned earlier, one of the major drawbacks of surface modifications is how long they last after application. In order for regenerative medicine and tissue engineering applications to make it into actual manufacturing, the issue with the post-application recovery of the surfaces will have to be addressed.

As more researchers start investigating the combination of different surface techniques, new surface modification principles will arise which will allow for a "drug release" type mechanism to be possible on surfaces of polymers and biomaterials; specifically, polymers used for tissue engineering and regenerative medicine. In applications of regenerative medicine, as surface modification techniques move forward, the ability to encapsulate specific enzymes and biomolecules into the surface will be able to be explored. Essentially it will allow these enzymes to follow a release profile, which will elicit regeneration and healing in the body. This would be a huge step forward in wound healing, specifically with patients who have reduced ability for their own body to elicit the proper healing response, and also could act to accelerate regular cellular regeneration.

Another issue that needs to be looked into is that of optimization. There are many different surface techniques, each with their own set of variables involved. Something that is overlooked a lot in this field is the optimization of these variables in order to produce one clear-cut "best" form of surface modification for a given technique. As an example, in a given category such as grafting, there are many different techniques that can be used to graft polymers onto surfaces but there is no clear-cut "best" technique. Each has its pros and cons depending on the availability of equipment, lab space, money, etc. Even going further, if you look at plasma grafting, there are a multitude of different methods of plasma grafting as well as a vast array of different gases that can be used. In order to move the field forward, the optimization of these techniques will be required. Furthermore, as techniques become more involved and start to include multiple types of surface modifications, optimization will play an even bigger role in ensuring that the best outcomes are achieved.

Lastly, as the field of surface modification moves forward, the application of these surface modification techniques will start to be investigated on not just polymers but also cells. As technology increases and the ability to manipulate individual polymers becomes possible, cellular surfaces will start to be investigated. This application would be huge in the field of regenerative medicine, as you would have the ability to tune a specific cellular surface to elicit in-vitro responses in-vivo through binding specific enzymes and proteins directly to the surface of the cells.

REFERENCES

Ai, H., Jones, S. A., and Lvov, Y. M. (2003). Biomedical applications of electrostatic layer-by-layer nano-assembly of polymers, enzymes, and nanoparticles. *Cell Biochemistry and Biophysics*, 39(1), 23–43. doi:10.1385/CBB:39:1:23

Ariga, K., Yamauchi, Y., Rydzek, G., Ji, Q., Yonamine, Y., Wu, K. C.-W., and Hill, J. P. (2014). Layer-by-layer nanoarchitectonics: invention, innovation, and evolution. *Chemistry Letters*, 43(1), 36–68.

Ariga, K., Hill, J.P., and Ji, Q. (2007). Layer-by-layer assembly as a versatile bottom-up nanofabrication technique for exploratory research and realistic application. *Physical Chemistry Chemical Physics*, 9(19), 2319–2340.

Bamford, C., and Al-Lamee, K. (1994). Studies in polymer surface functionalization and grafting for biomedical and other applications. *Polymer*, 35(13), 2844–2852.

Bauer, S., Schmuki, P. von der Mark, K., and Park, J. (2012). "Engineering biocompatible implant surfaces. Part I: Materials and surfaces. *Progress in Materials Science*, 58(3), 261–326.

Charest, J. L., Eliason, M. T., García, A. J., and King, W. P. (2006). Combined microscale mechanical topography and chemical patterns on polymer cell culture substrates. *Biomaterials*, 27(11), 2487–2494.

Cui, N. Y., and Brown, N. M. D. (2002). Modification of the surface properties of a polypropylene (PP) film using an air dielectric barrier discharge plasma. *Applied Surface Science*, 189(1–2), 31–38.

De Valence, S., Tille, J. C., Chaabane, C., Gurny, R., Bochaton-Piallat, M. L. Walpoth, B. H., and Möller, M. (2013). Plasma treatment for improving cell biocompatibility of a biodegradable polymer scaffold for vascular graft applications. *European Journal of Pharmaceutics and Biopharmaceutics*, 85(1), 78–86.

Desmet, T., Morent, R., De Geyter, N., Leys, C., Schacht, E., and Dubruel, P. (2009) Nonthermal plasma technology as a versatile strategy for polymeric biomaterials surface modification: a review. *Journal of the American Chemical Society*, 10(9), 2351–2378.

Desmet, T., Billiet, T., Berneel, E., Cornelissen, R., Schaubroeck, D., Schacht, E., and Dubruel, P. (2010). Post-plasma grafting of AEMA as a versatile tool to biofunctionalise polyesters for tissue engineering. *Macromolecular Bioscience*, 10(12), 1484–1494.

Dowell-Mesfin, N. M., Abdul-Karim, M. A., Turner, A. M., Schanz, S., Craighead, H. G., Roysam, B., and Shain, W. (2004). Topographically modified surfaces affect orientation and growth of hippocampal neurons. *Journal of Neural Engineering*, 1(2), 78–90.

Elbert, D. L., and Hubbell, J. A. (1996). Surface treatments of polymers for biocompatibility. *Annual Review of Materials Science*, 26(1), 365–394.

Goddard, J. M., and Hotchkiss, J. H. (2007). Polymer surface modification for the attachment of bioactive compounds. *Progress in Polymer Science*, 32(7), 698–725.

Gupta, B., Plummer, C., Bisson, I., Frey, P., and Hilborn, J. (2002). Plasma-induced graft polymerization of acrylic acid onto poly(ethylene terephthalate) films: characterization and human smooth muscle cell growth on grafted films. *Biomaterials*, 23(3), 863–871.

Ikada, Y. (1994a). Interfacial Biocompatibility. *Polymers in Biology Biomedical Significance*, 35–48.

Ikada, Y. (1994b). Surface modification of polymers for medical applications. *Biomaterials*, 15(10), 725–736.

Ito, Y. (1999). Surface micropatterning to regulate cell functions. *Biomaterials*, 20(23–24), 2333–2342.

Izquierdo, A., Ono, S. S., Voegel, J. C., Schaaf, P., and Decher, G. (2005). Dipping versus spraying: exploring the deposition conditions for speeding up layer-by-layer assembly. *Langmuir*, 21(16), 7558–7567.

Jacobs, T., Morent, R., De Geyter, N., Dubruel, P., and Leys, C. (2012). Plasma surface modification of biomedical polymers: influence on cell-material interaction. *Plasma Chemistry Plasma Process*, 32(5), 1039–1073.

Jo, S. and Park, K. (2000). Surface modification using silanated poly(ethylene glycol)s. *Biomaterials*, 21(6), 605–616.

Kim, Y., and Goulbourne, N. (2014). Micropillar arrays for high sensitivity sensors. Paper presented at the *Materials Research Society Symposium Proceedings, 1685*.

Lee, S. S., Lee, K. B., and Hong, J. D. (2003). Evidence for spin coating electrostatic self-assembly of polyelectrolytes. *Langmuir*, 19(18), 7592–7596.

Li, W., Shanti, R. M., and Tuan, R. S. (2007). Electrospinning technology for nanofibrous scaffolds in tissue engineering. In C. S. Kumar (Ed.). *Nanotechnologies for the Life Sciences*, doi:10.1002/9783527610419.ntls0097

Liu, P.S., Chen, Q., Wu, S. S., Shen, J., and Lin, S. C. (2010). Surface modification of cellulose membranes with zwitterionic polymers for resistance to protein adsorption and platelet adhesion. *Journal of Membrane Science*, 350(1–2), 387–394.

Lord, M. S., Foss, M., and Besenbacher, F. (2010). Influence of nanoscale surface topography on protein adsorption and cellular response. *Nano Today*, 5(1), 66–78.

Lvov, Y. M., Ai, H., and Jones, S. A. (2003). Biomedical applications of electrostatic layer-by-layer nano-assembly of polymers, enzymes, and nanoparticles. *Cell Biochemistry and Biophysics*, 45(February 2004), 59–70.

Ma, Z., Mao, Z., and Gao, C. (2007). Surface modification and property analysis of biomedical polymers used for tissue engineering. *Colloids and Surfaces B: Biointerfaces*, 60(2), 137–157.

Mathieson, I., and Bradley, R. H. (1996). Improved adhesion to polymers by UV/ozone surface oxidation. *The International Journal of Adhesion and Adhesives*, 16(1), 29–31.

Nie, Z., and Kumacheva, E. (2008). Patterning surfaces with functional polymers. *Nature Materials*, 7(4), 277–290.

Ogawa, T., Ding, B., Sone, Y., and Shiratori, S. (2007). Super-hydrophobic surfaces of layer-by-layer structured film-coated electrospun nanofibrous membranes. *Nanotechnology*, 18(16), 165607.

Oláh, A., Hillborg, H., and Vancso, G. J. (2005). Hydrophobic recovery of UV/ozone treated poly(dimethylsiloxane): adhesion studies by contact mechanics and mechanism of surface modification. *Applied Surface Science*, 239(3–4), 410–423.

Peez R. F., Dermody D. L., Franchina J. G., Jones S. J., Bruening, M. L., Bergbreiter D. E., and Crooks, R. M. (1998). Aqueous solvation and functionalization of weak-acid polyelectrolyte thin films. *Langmuir*, 14(7), 4232–4237.

Penn, L. S., and Wang, H. (1994). Chemical modification of polymer surfaces: a review. *Polymers for Advanced Technologies*, 5(12), 809–817.

Picart, C., Lavalle, P., Hubert, P., Cuisinier, F. J. G., Decher, G., Schaaf, P., and Voegel, J. (2001). Buildup mechanism for poly(l -lysine)/hyaluronic acid films onto a solid surface. *Langmuir*, 17(23), 7414–7424.

Rasmussen, J. R., Stedronsky, E. R., and Whitesides, G. M. (1977). Introduction, modification, and characterization of functional groups on the surface of low-density polyethylene film. *Journal of the American Chemical Society*, 99(14), 4736–4745.

Revzin, A., Tompkins, R. G., and Toner, M., "Surface engineering with poly(ethylene glycol) photolithography to create high-density cell arrays on glass," *Langmuir*, 19(23), 9855–9862, 2003.

Siow, K. S., Britcher, L., Kumar, S., and Griesser, H. J. (2006). Plasma methods for the generation of chemically reactive surfaces for biomolecule immobilization and cell colonization—a review. *Plasma Processes Polymers*, 3(6–7), 392–418.

Stunkard, A. J. (2009). NIH Public Access. *Psychiatry: Interpersonal and Biological Processes*, 162(3), 214–220.

Tai, H.C., and Buettner, H. M. (1998). Neurite outgrowth and growth cone morphology on micropatterned surfaces. *Biotechnology Progress*, 14(3), 364–370.

Tan, J. and Saltzman, W. M. (2004). Biomaterials with hierarchically defined micro- and nanoscale structure. *Biomaterials*, 25(17), 3593–3601.

Tendero, C., Tixier, C., Tristant, P., Desmaison, J., and Leprince, P. (2006). Atmospheric pressure plasmas: a review. *Spectrochimica Acta Part B: Atomic Spectroscopy*, 61(1), 2–30.

Teramura, Y., and Iwata, H. (2010). Cell surface modification with polymers for biomedical studies. *Soft Matter*, 6(6), 1081.

Tsuda, Y., Shimizu, T., Yamato, M., Kikuchi, A., Sasagawa, T., Sekiya, S., and Okano, T. (2007). Cellular control of tissue architectures using a three-dimensional tissue fabrication technique. *Biomaterials*, 28(33), 4939–4946.

Tu, C. Y., Liu, Y. L., Lee, K. R., and Lai, J. Y. (2005). Surface grafting polymerization and modification on poly(tetrafluoroethylene) films by means of ozone treatment. *Polymer*, 46(18), 6976–6985.

Uyama, Y., Kato, K., and Ikada, Y. (1998). Surface Modification of Polymers by Grafting. In *Grafting/Characterization Techniques/Kinetic Modeling* (Vol. 137, pp. 1–39). Retrieved from https://archive.org/details/springer_10.1007-3-540-69685-7/page/n47

van Wachem, P. B., Vreriks, C. M., Beugeling, T., Feijen, J., Bantjes, A., Detmers, J. P., and van Aken, W. G. (1987). The influence of protein adsorption on interactions of cultured human endothelial cells with polymers," *Journal of Biomedical Materials Research*, 21(6), 701–718.

Vasilets, V. N., Hermel, G., König, U., Werner, C., Müller, M., Simon, F., and Jacobasch, H. J. (1997). Microwave CO_2 plasma-initiated vapour phase graft polymerization of acrylic acid onto polytetrafluoroethylene for immobilization of human thrombomodulin. *Biomaterials*, 18(17), 1139–1145.

Wei, Q., Becherer, T., Angioletti-Uberti, S., Dzubiella, J., Wischke, C., Neffe A. T., and Haag, R. (2014). Protein interactions with polymer coatings and biomaterials. *Angewandte Chemie International Edition*, 53(31), 8004–8031.

Zhang, J., Xu, X., Liu Y., Liu, C., Chen X., Li, C., and Zhang, X. (2012). Design of an 'active defense' system as drug carriers for cancer therapy. *Advanced Functional Materials*, 22(8), 1704–1710.

Zhou T., Zhu, Y., Li, X., Liu, X., Yeung, K. W. K., Wu, S., and Chu, P. K. (2016). Surface functionalization of biomaterials by radical polymerization. *Progress in Materials Science*, 83, 191–235.

Zhu, J. (2010). Bioactive modification of poly(ethylene glycol) hydrogels for tissue engineering. *Biomaterials*, 31(17), 4639–4656.

8 Nano-approach

8.1 BACKGROUND

Nanotechnology can be considered as a novel branch of science that deals with the applications of different materials and devices at the nanolevel. Because of the capabilities for improving the properties of well-established products and to develop materials with completely new functions and characteristics, this technology is always under continuous expansion. It can control the physical and chemical properties of a material at the molecular stage. Nanotechnology changes the direction of our thinking. It breaks the boundaries between chemistry, biology, and physics. The dismissal of these barriers will give birth to unexpected challenges and unique methods for scientists and engineers working in this area. The continuous attempt at the nano-approach in the research area of biomedical science has created a hybrid branch called nanobiotechnology. This novel branch deals with the development of atomic scale devices which mimic biological structures at the atomic level to understand or improve the structural properties of biological systems more precisely.

The concept behind the expansion of biotechnology is to utilize the fundamental knowledge of biology for the manipulation of genetic, cellular, and molecular activities that can help to develop the existing services and products, especially in the area of medical science (Sahoo and Labhasetwar, 2003). The unique combination of biotechnology, nanotechnology, and information technology has the ability to solve many unidentified questions that restrict the growth of biomedical research. This newly developed technology is widely expected to cause revolutionary change and significant expansion in different medical applications, not only in gene therapy and drug delivery but also in biosensors and molecular imaging. Targeted specific drug delivery and applicable methods for quick treatment of different diseases are the primary research areas where the nano-approach would perform a vital role (Sahoo and Labhasetwar, 2003; Rosen et al., 2011).

In the future, the most common implementation of nano-approach will probably be in the field of pharmaceutical research. The involvement of the nano-approach in this field includes the application of unique characteristics of nanoparticles as drugs or drug carriers. Significant research is also in progress with special attention to the development of new techniques for drug targeting, controlled release, and recovery of drugs with low bioavailability (Roy et al., 2003; Brigger et al., 2002; Crommelin et al., 2003). Polymer capsules can be engineered at the nanolevel to break down and release drugs in a controlled manner. Nanocapsules can be developed by utilizing monomers directly or by the nanodeposition of predesigned polymers. Nanobiotechnology has also helped with the synthesis of biochips and has a significant role to play in producing drugs with no issues of biocompatibility or

by Yaser Dahman and Bhavik Vyas

biocomplexity. Other applications involve the development of biofluids, nanofiltration, in-vitro fertilization of livestock, and the traceability of genetically reformed food (Na and Bae, 2002).

Growing attention to the health care applications of nanotechnology has given a birth to a new discipline called nanomedicine. Nanomedicine is a novel research area with the capability to design, direct, and manipulate the growing future of various medical treatments. One of the major obstacles that restricts the expansion of nanobiotechnology is the development of nanotherapies which need to target the diseased organs and tissues without damaging the healthy cells. This approach offers tremendous help in preventing the unexpected side effects of existing treatments. Nanomedicine has a capability to fill the gaps in nanotherapies by regenerating the tissues and organs that are badly damaged during the treatment of the original diseases (Mirza and Siddiqui, 2014; Patrick and Bertrand, 2011).

Nanotechnology and nanoscience are a critical part in many areas of our regular life, and they have a significant impact on medical science and health care. The area of medical research is highly complex, and therefore all the advantages to health care from nanobiotechnology will take time before they reach the commercialization stage, but other advantages will quickly come. The tools for medical practice and research will be more powerful and less expensive. The research and development of diagnostic techniques will become much more efficient, enabling quicker response, and the capability of curing new diseases. Many novel types of treatment methods can be identified, while medical expenses will decrease, diseases will be treated more carefully, and the advantages will be experienced by more and more people all around the world.

8.2 BIOMATERIALS

Any natural or synthetic materials which can be used in a biological system are defined as biomaterials. In the literature, we can find a number of definitions for the term “biomaterial”. It can also be defined more scientifically as any material implemented in medical devices with the primary objective of interacting with biological systems (Ratner et al., 2004). According to the definition, it is clear that the use of biomaterials is broadly in the area of medical science, i.e. in devices such as vascular or orthopedic implants. In the dental and medical areas, biomaterials are widely used to evaluate, treat, and replace the damaged organs and tissues (Ratner et al., 2004; O'Brien, 2011). Since there is a strong interaction between a biological system and a material, we can expect a variety of responses from a biological system to a biomaterial up to a certain level. In conjunction with this, the term “biocompatibility” has been described as follows: biocompatibility is the capability of a material to act with a reasonable host response in a targeted application (Ratner et al., 2004; Yoruç and Şener, 2012). Moreover, the term biocompatibility is not only restricted to the response of a biological system to a material, but it also includes the characteristics of a material’s ability to remain stable without any notable degradation or corrosion in a human body (Black, 2005).

The development of biomaterials over the past century can be categorized into four generations according to their growth with time. These generations may vary depending on the complexity and ability of biomaterials to react with the biological environment of the human body. The first generation of biomaterials involved the application of off the shelf commodity materials like polyethylene or steel. Approximately around the middle of the past century, production of the second generation of biomaterials for the biological applications began, although the majority of these types of materials were synthesized to serve other areas such as aeronautics. This second generation of biomaterials such as titanium and its alloys still exist and cover the major portion of the commercial market. For these inert materials, the body had to change to accept the material, which in most cases did not succeed. The main reason for that is that the in-depth medical and biological interaction between the body and the material was an enigma. The majority of development in biomaterials was on the basis of empirical data (Migonney, 2014).

During the early 1990s, a revolutionary change took place in the area of biomaterials. The human body should no longer have to adapt to the biomaterials, but the materials expanded in properties and features directed towards adjustment of the biomaterials to the body. This opened the door for the third generation of biomaterials, often called the "active" materials. They have nearly similar composition and features to those found in an actual biological system. Bioglasses are the most common example of these biomaterials, which cover a notable part of today's commercial market. The fourth generation includes extensive changes in synthesis techniques, with enhanced characteristics of developed biomaterials. This generation of materials has a minor market penetration as most of them are still in the growth stage and accreditation and testing procedures are slow as well as money-consuming. Tissue engineered systems and materials belong to the category of the fourth generation of biomaterials. These materials are often called smart biomaterials, i.e. materials that can change their characteristics based on stimuli or may sometimes have sensor functions to control their interactions with biological processes. In addition, this class of biomaterials may use the fundamental principles of biological design, often called biomimetics, in terms of the structure they exhibit and the materials they contain. In the last 15 years, polymeric biomaterials have gained special attention because of their higher biostability and successful performance in different clinical applications (Ratner et al., 2004; Migonney, 2014).

Nowadays, the implementation of polymers is important in medical surgery and in the field of pharmacology for drug development and controlled drug release. Many polymeric compounds with well-known names are easily found in the range of established biomaterials. They have the capabilities to compete with inorganic biomaterials (ceramics, alloys, metals, etc.) and modified biopolymers. However, their performances are sometimes far away from the ideal, but they are nevertheless significant in many aspects (Vert, 2007).

8.3 CHARACTERISTIC ENHANCEMENT OF BIOMATERIALS

The synthesis of novel materials always leads to technological improvements and develops interesting solutions to older difficulties. Nanomaterials and nanotechnologies are the existing key areas with great possibilities and challenges. Nanomaterials, often called nanostructured materials, are single phase or multi-phase materials with average sizes less than 100 nm. The most promising feature of nanotechnology is its capability to develop nanostructured materials that will have novel characteristics on the bulk materials. The major characteristics of nanomaterials are neither the free molecules and atoms nor the bulk materials with similar structure, but rather the many relevant phenomena taking place at the nanoscale, primarily due to the following unique changes in its structure:

- Enlarged surface area of grains
- Enlarged volume of grain boundaries
- Creation of discrete electronic energy level

The surface area of nanoparticles increases with a decrease in size. This is primarily because nanoparticles of 1 nm size have 100% of their atoms on their surfaces, while nanoparticles with a size of 3 nm have only 50% of their atoms on their surfaces (Figure 8.1) (Klabunde et al., 1996). Actually, the boundary region between the grains starts increasing with rapid decrements in the grain size. Therefore, nanoparticles which are of smaller grain size have a larger magnitude of grain boundary area. In regular microsized materials, major impurities favourably diffuse to the region of the grain boundary and may result in intergranular fracture. Since nanoparticles have

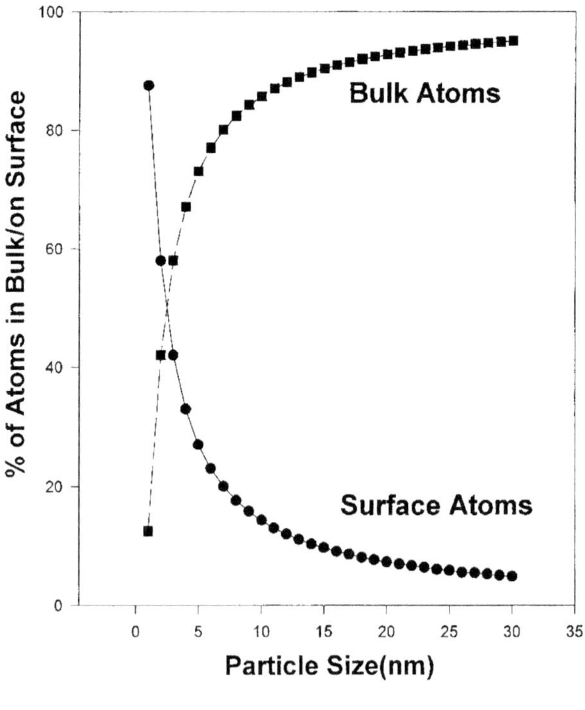

FIGURE 8.1 Percentage of surface and bulk atoms as a function of particle size (Klabunde et al., 1996).

a large number of specific grain boundaries to diffuse, there is a strong possibility of homogenization of impurities and thereby fewer chances of failure.

According to the famous Hall–Petch equation, the strength increases with decreasing grain size.

$$\sigma_y = \sigma_0 + kd^{-1/2}$$

Here,

y represents the yield strength
o represents lattice yield stress
k is constant
d is the grain size

A material's hardness also has a similar dependency on grain size. Therefore, mechanical properties like the hardness and strength of polycrystalline nanoparticles increase drastically with decreasing grain size. The increment in the strength increases due to the piling up of the main dislocations at the grain boundaries. The increment in the surface area of grains causes rapid growth in the dislocation density at the surface, which in general increases the strength of nanoparticles. On the other hand, when the grain size is smaller than the dislocation spacing, the previously mentioned relation is no longer usable and the strength starts decreasing with decreases in the grain size (Figure 8.2) (Nieh and Wadsworth, 1991).

The presence of a large number of surface atoms in a nanocrystal delivers many active reaction sites, consequently increasing its catalytic and chemical reactivity. The thermal properties of nanoparticles also change because of the enhancement in diffusivity at the grain boundaries. Decreases in the thermal conductivity and melting point have been observed in nanostructured ceramics and metals (Ramakrishna et al., 2010).

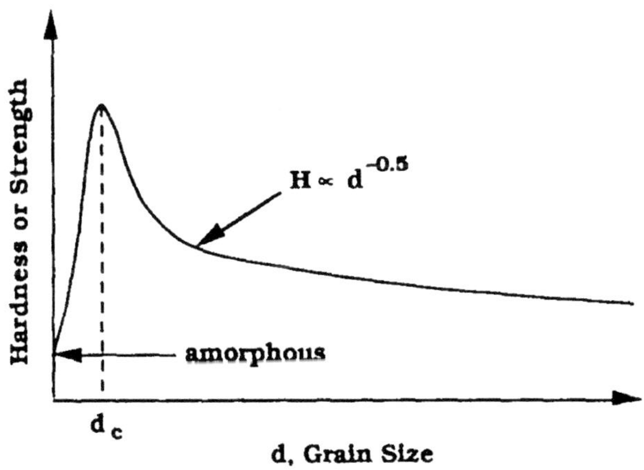

FIGURE 8.2 Hardness or strength vs grain size (Nieh and Wadsworth, 1991).

8.4 APPLICATIONS OF NANO-APPROACH FOR ANALYSIS AND TREATMENT

8.4.1 In-Vitro-Based Analysis and Treatments

The objectives of in-vitro-based treatment of cells are manifold. It is quite necessary to protect the supply of blood for transfusion reasons, to inspect the level in the blood of different drugs supplied to the patient, and to provide important information for the treatment of diseases. The ultimate goal of any treatment procedure is an early non-invasive and accurate identification of disease markers through the technique of routine screening, enabling selection of the proper treatment regimen. Different nano-approach-based platforms are established to allow for the efficient determination of a broad variety of disease markers by non-invasive methods (Maurya et al., 2007; Truskett and Watts, 2006; Lee et al., 2003; Bruckbauer et al., 2003).

Fluorescent dyes can be considered an important class of nanoimaging tools primarily used for the visualization of molecules and cells. A major disadvantage of fluorescent dye is its photo-instability, with a tendency to degrade within 1 minute. The bleaching of these dyes limits their application range. Inorganic quantum dots offer significantly higher photostability. However, sulfides and selenides commonly used for that reason are cytotoxic and therefore are specifically used for inspection of biological samples outside of the human body. Due to their high photoluminescence quantum efficiency, high biocompatibility, and stability against photobleaching, silicon-based quantum dots are considered the best candidates for taking the place of fluorescent dyes in biological assays. It is possible to synthesize the Si nanocrystals by electron beam lithography or wet chemistry and reactive ion etching. This forms Si nanopillars that are later oxidized to develop luminescent silicon cores (Valenta et al., 2002). Their size is so small that the removal or addition of even a single atom can change their optical appearance considerably. Other specific characteristics of quantum dots (QD) are composition, size, tunable emission, narrow emission spectra, and wide large absorption spectra (Figure 8.3) (Fu et al., 2005).

The highly improved photostability and brilliance of QD make them suitable systems for the identification of low abundance antigens (Zheng et al., 2006; De Rosa et al., 2001). According to the different test results, it has been found that QD are not completely stable in-vivo due to the degradation effects that result in a quenching of the fluorescence. Gao et al. (2004, 2007) revealed that a hydrophobic exterior on QD protects them from this outcome and modifies them for different medical applications.

A rapid development in different surface imaging techniques (scanning probe microscopy, atomic force microscopy, etc.) has helped tremendously to improve the concept of structural analysis of different surfaces at the nanolevel. Dynamic force microscopy techniques are currently best for the analysis of soft systems such as biological cell assemblies, and it also provides significant help in the tracking of proteins and the surface imaging of macromolecules in liquids (Schirmeisen et al., 2005). For example, cytoskeletal structures, particularly stress fibers, can be efficiently imaged with the atomic force microscopy (AFM) (Riethmuller et al., 2007; Shahin et al., 2005). The detailed visualization of different cells is also possible without affecting their surface using AFM. The focal adhesion plaques imaging

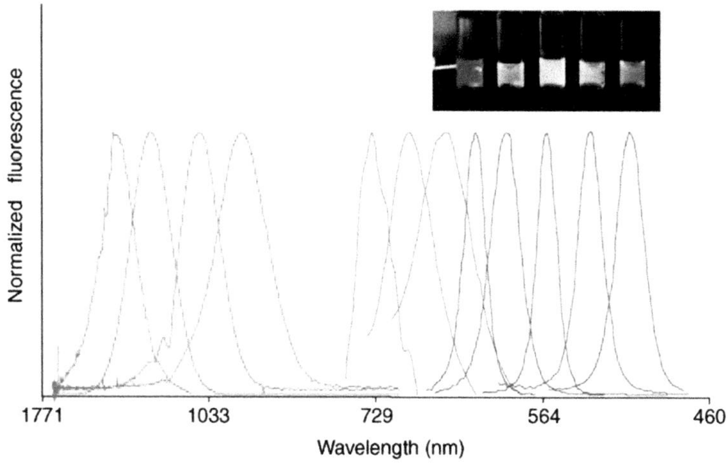

FIGURE 8.3 Size and material-dependent emission spectra of different surfactant-coated semiconductor nanocrystals in a variety of (Fu at al., 2005).

technique is commonly used for high resolution imaging of membrane transport (Gorelik et al., 2004; Franz and Müller 2005; Kueng et al., 2005).

An innovative experimental procedure combining quantum-dot-labeled antibodies with AFM can be used as surface markers to identify the cystic fibrosis transmembrane conductance regulator (CFTR) primarily responsible for the development of cystic fibrosis (CF). By comparing erythrocyte plasma membranes adopted from healthy donors and CF patients, it was found that erythrocytes clearly reflect the CFTR status of the subject organism and that detailed quantification of CFTR in a blood sample can be helpful in the analysis of CFTR related diseases (Lange et al., 2006).

Promising advancements in AFM-based technology have helped significantly in the development of its application for in-vivo imaging. As Imer et al. (2006) proposed, AFM could be utilized as a minimally invasive tool in the analysis of rheumatoid arthritis to exhibit disease related expansions in specific cartilage. It has also been suggested to use AFM as an appropriate tool to study the surface morphology of cell in intact native human stratum corneum. The stratum corneum consists of corneocytes structured within the whole stratum corneum layer like bricks in a wall. Aging of the skin or skin related diseases have been shown to alter the composition of the corneocyte morphology. AFM-based technology has also been proposed as a suitable methodology for the nanoscale study of native stratum corneum in terms of surface morphology. The continuous expansion in the features of the AFM technique may open a new door for highly sensitive representation of cellular membranes by recording images and simultaneously measuring different physical properties such as porosity, stiffness, and elasticity (Gorzelanny et al., 2006).

A technique, established by Rothenhäusler and Knoll (1988), based on surface plasmon resonance is used to explain the interactions of different biomolecules in a substance without any label. Surface plasmon resonance observes changes in refractive index occurring in the vicinity of a subjective surface. The effect starts to occur when a source of light is focused at a particular angle on the metal/glass interface of a metallic film to stimulate the surface plasmons, i.e. the oscillation of available free electrons starts to propagate along the surface of the film. Later, when immobilized biomolecules are bound by the ligands, the interfacial optical condition gets altered and affects the propagation of plasmons on the free surface. The binding of subjected biomolecules is measured by the observed changes in the refractive index. Surface plasma resonance imaging offers a strong possibility of measuring the magnitude of a force responsible for binding the interacting biomolecules together. In general, the kinetic study of the majority of the biomolecules like protein–lipid, protein–protein, protein–drug, and protein–nucleic acid is accessible by surface plasmon resonance methods. The method is also applied to observe the effects of plasma exchange during blood circulation. The method can be described as time-saving and accurate for measurements of anti-A/B IgG titers that can be standardized easily, and it can also be used for the detailed analysis of blood required during transplantations (Campbell and Kim, 2007).

Optical laser methods have experienced a rapid advancement towards nanoscopic medicine (Peters, 2006). Seisenberger and his group have shown that a specific confocal laser optical technique for the tracing of a single virus permits the direct analysis of the different pathways through which the viruses can enter into the living cells using the fluorescence labeled particles called adeno-associated viruses (Seisenberger et al., 2001). Digital holography can be used to retrieve specific information related to the unidentified questions of biomedical science. This technique helps to perform marker-free quantitative study in the subcellular and cellular range (Marquet et al., 2005). Digital holography can also be applied to characterize the fast movement of cells and the results can be used to predict the metastatic characteristics of a tumour. Meyer and his coworkers developed a pioneering tool in the area of digital imaging. They successfully reduced the focal spot area by nearly an order of magnitude below the diffraction limit of optical path using stimulated emission depletion (Figure 8.4). This development opens a totally new pathway for high-resolution optical imaging in the area of nanomedical science (Meyer et al., 2008).

8.4.2 In-Vivo-Based Analysis and Treatments

The development of nanotechnology and the requirement for personalized medicine give an impactful drive towards the evolution of point-of-care diagnosis with higher specificity, reliability, and sensitivity. In-vivo analysis creates data in a very quick form from the patient and permits the follow up of disease therapy and further development. Advancement in in-vivo-based analysis and treatment is highly dependent on the implantable devices and the imaging of different molecules. Nanotechnology offers a unique perspective for the creation of a new generation of medical imaging methods and biosensors with high precision of recognition.

Microscopic systems are unable to observe the electronic and optical effects originating from specific spatial confinements. Developments in this area include semiconductor nanoclusters, nanopowders, and quantum dots (Kinge et al., 2008). Many of these particles can be successfully used as markers within the human body for

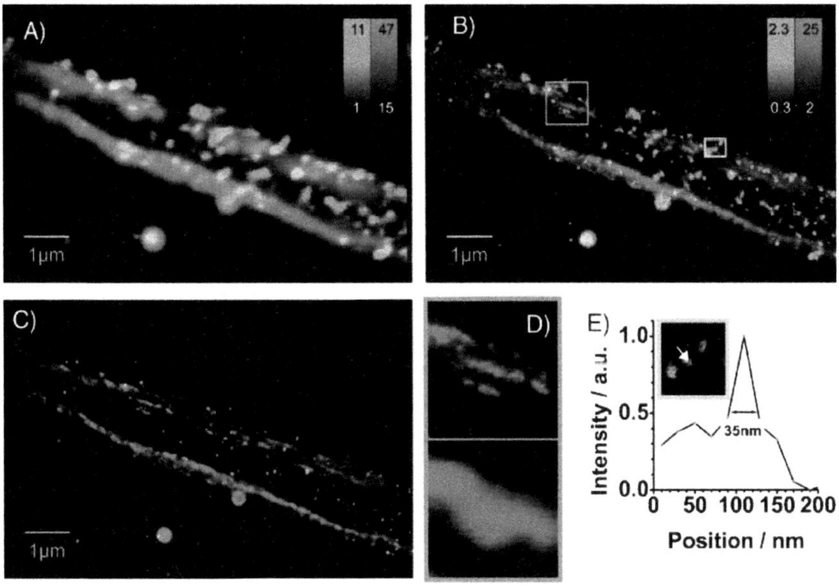

FIGURE 8.4 (a) Confocal, (b) stimulated emission depletion, (c) Richardson–Lucy deconvolved stimulated emission depletion images of neurofilaments (green: light subunits, red: α-internexin). (d) In contrast to the confocal image, stimulated emission depletion reveals three well-separated α-internexin strands of the axon. (e) Structures of the light subunits exhibit a full width at half-maximum (FWHM)<40 nm. Note the different organization of the light subunits and α-internexin (Meyer et al., 2008).

molecular imaging. Incorporation of these particles can significantly help to provide a dramatic increase the sensitivity and resolution for earlier identification of diseases (Nam et al., 2003). As a result, cheaper clinical measures can be used for therapy. It is well known that nanoparticles modified by surface functionalization show vectorial character. They have the capability to recognize the complementary groups present on the cell surfaces that are suggestive of different diseases. For example, superparamagnetic iron oxide nanoparticles functionalized by a phosphorothioate-modified oligodeoxynucleotide were specifically synthesized to detect neurodegenerative diseases by magnetic resonance (Liu et al., 2008).

A well-established application of ultrasmall superparamagnetic iron oxide labeled body cells in coordination with MR imaging is the complete follow up of immune cells (macrophages/ monocytes) during the growth of an inflammation. This result-oriented technique can be used for the analysis of cardiovascular diseases. Moreover, the constraint of the blood barrier can be easily surpassed by iron oxide nanoparticles using macrophages as carriers, opening doors for the wide investigation of neurodegenerative brain diseases (Liu et al., 2007; Chang et al., 2008; Kim et al., 2008). Other specific applications for targeted imaging are superparamagnetic iron oxide nanoparticles which are used for the tracking of stem-cells, of liposomes for targeting atheroma components, and of perfluorocarbon nanoparticles for analysis of angiogenesis (Barrett et al., 2007; Maluf et al., 2005).

8.5 NANO-APPROACH-BASED NOVEL DRUG DELIVERY SYSTEMS

After the impactful involvement of nanotechnology in the area of medical science, significant advancements have been possible in conventional drug delivery systems. A variety of inorganic/organic nanobiomaterials and devices have been used as drug delivery vehicles to develop efficient therapeutic techniques. In the last two decades, the Food and Drug Administration (FDA) has accepted more than two dozen nanotechnology-based products for medical use, and many more are currently in clinical trials. Of these products, most consist of a non-targeted delivery system (e.g. polymers and liposomes) and a drug (Wagner et al., 2006).

8.5.1 New Therapeutic Delivery Systems

RNA interference and gene therapy, the two most notable therapies, hold great promise as treatment methods for different diseases (Whitehead et al., 2009). Nevertheless, the detailed systematic administration of newly developed therapeutics is affected by several constraints such as kidney filtration, uptake by the reticuloendothelial system, limited intracellular entry, and enzymatic degradation. Therefore, cationic nonviral nanocarriers have been commonly applied to encapsulate and make newly developed therapeutics more effective. However, it is unfortunate that a substantial number of these kinds of materials can generate several obstacles associated with toxicity, serum instability, and inflammatory or immune responses, making them incompatible for clinical uses. Subsequently, efficient and safe delivery materials are becoming the bottleneck for the extensive use of RNA interference and gene therapy.

Recent developments are now paving the way for the successful medical application of these new varieties of therapeutics. For example, high throughput screening platform methodologies have been explained and the proof of concept has been positively demonstrated by identifying lipid-like materials and cationic polymers for RNA interference and gene delivery (Anderson et al., 2004). On the other hand, rational design approaches also represent tremendous potential for developing successful delivery materials (e.g. cyclodextrin containing polymer and DLin–KC2–DMA). However, it is more feasible to use previously FDA accepted materials for the implementation of new therapeutic delivery, as they have specific advantages such as sustained release and safety. For sustained gene silencing, poly(lactic-co-glycolic acid) is one such FDA approved material and can be used to synthesize nanoparticles loaded with siRNA (Woodrow et al., 2009). Beyond assuring the efficiency and safety of various delivery materials, more research is still required to screen other factors, including the targeting ligands, physicochemical properties of nanocarriers, and formulation methods – all of which can significantly affect the delivery of new therapeutics.

8.5.2 Targeted Delivery Systems

It is widely believed that ligand-modified nanoparticles have tremendous potential to improve therapeutic efficiency and lower the number of additional side effects as compared to the regular therapeutics in active targeting. The capability to actively target particular cells instead of tissues also permits ligand-modified nanocarriers to outperform the very first generation, nanotargeted nanoassemblies. While the requirement of a targeted delivery strongly depends on a number of factors (e.g. drugs, diseases, and delivery vehicles), a myriad of significant advantages have been demonstrated (Bartlett et al., 2007).

In cancer treatment, the presence of different targeting ligands can impressively improve the retention as well as the cellular uptake of nanoparticles, even though the accumulation of a tumour is mostly evaluated by the physicochemical characteristics of nanoparticles. This can lead to increased therapeutic activity and higher intracellular drug concentration, which is crucial specifically for bioactive macromolecules (e.g. siRNA and DNA) that need intracellular delivery for bioactivity (Bartlett et al., 2007). In the case of immunological tissue targeting or endothelial targeting for cardiovascular diseases, localization of nanoparticles is usually monitored by ligand-receptor interactions instead of enhanced permeability and retention (EPR) effect (Farokhzad and Langer, 2009). Similarly, for the transcytosis of nanoscale drugs across endothelial barriers, tight epithelial ligand-mediated targeting is of primary importance (Gabathuler 2010). Moreover, targeted delivery has also been used in some cases to combat multidrug resistance. Regardless of the targeted nanoparticles, nanoparticle assemblies are now in clinical trials. The clinical translation of targeted drug delivery is not as straightforward as we assume. One possible constraint stems from the high complexity in the manufacturing process of viable targeted nanoparticles. Targeted nanoparticle synthesis generally needs different steps such as assembly of biomaterials, ligand insertion, purification, etc. This could create quality related issues. The development of a single step synthesis procedure of targeted nanoparticles via self-assembling pre-functionalized nanobiomaterials offers a scalable and simple manufacturing strategy (Gu et al., 2008). Some other factors for consideration that could affect the cell targeting include ligand surface density as well as spacer type, and the length that divides nanoparticles and ligand molecules. It is expected that with gradual developments in ligand screening and nanoparticle optimization, targeted drug delivery systems will probably become a backbone in the upcoming generation of drug therapy (Jung et al., 2008).

8.5.3 Co-Delivery Systems

When compared to single drug therapy, combination therapy has presented several benefits and has been proven as more efficient in many ways (Greco and Vicent, 2009). However, because of the different pharmacokinetic profiles of individual drugs, the preoptimized drug ratio will definitely vary after a conventional administration of drug 'cocktails' – a result that could end up with inefficient therapeutic results. To this end, polymer- and/or lipid-based nanoscale delivery systems previously constructed for individual drug delivery have been used to facilitate the co-delivery systems. It is also possible to tune the relative dosage of different drugs in single particle levels for some specific drug combinations. Other than that, we require the development of a new class of drug delivery vehicles with preferred functionalities that enables combined encapsulations of hydrophilic and hydrophobic drugs, temporary controlled release, and active targeting (Sengupta et al., 2005). Such special properties are especially important for the co-delivery of nucleic acid and drugs, which needs intracellular delivery to show bioactivity. For example, in the case of the co-delivery of RNAi therapy and chemotherapy to treat cancers, the ideal nanoparticles would be required to first release siRNA to lower the expression of MDR transporters, followed by a controlled release of drugs used for chemotherapy.

Another excellent development in co-delivery is the capability of combining therapeutic agents and targeted imaging with the same nanoparticle, permitting us to deliver therapeutics and analyse the drug delivery sites simultaneously (Debbage and Jaschke, 2008). This technique is novel in concept and holds important promise for making a huge medical impact within a few decades. It can possibly offer us significant information on intracellular targets and help to ensure that therapeutic drugs are effectively reaching their sites for the efficient treatment of different diseases. Research in this area is primarily focused on the development of such multifunctional nanoscale systems for the implementation of smart drug delivery systems (Bagalkot et al., 2007). To advance the formulation, calculate the clinical doses, and assure the safety of the drug delivery system, it is necessary that we completely understand the pharmacokinetics of a given delivery system (Sanhai et al., 2008). The radionuclide labelling technique is commonly used to get in situ quantitative information, but the instability of radio emitters prevents them from conjugating with nanobiomaterials. With the help of nanoimaging tools like quantum dots, magnetic nanoparticles, and carbon nanotubes, improved imaging modalities may become accessible for the detailed evaluation of nanotherapeutics in the body (Dubertret et al., 2002; De la Zerda et al., 2008).

8.5.4 MEM/NEM Devices for an Efficient Drug Delivery

Parallel to the advancement of specific delivery systems, the area of micro-/nanoelectromechanical (MEM/NEM) device-based delivery has also seen significant improvements over the past two decades. Particularly, implantable microchips containing nanosized tanks have been constructed to deliver the necessary drugs for long times in a predesigned manner, and microsized needles are being checked for use in transdermal delivery. The addition of nanocharacteristics (e.g. nanoparticles, nanochannels, nanopores) in nanofabricated systems is improving drug delivery and immunoisolation methods (Ainslie and Desai, 2008). Moreover, these novel devices can be further advanced to deliver therapeutics, co-deliver multiple agents, and achieve targeted delivery. Extensive efforts are being put into developing smart devices that could sense when and which drug dosage is required and then automatically deliver it from the reservoir tanks (Staples et al., 2006). To achieve this, one criterion that must be met is the stable and continuous monitoring of physical as well as biochemical conditions in-situ. The recent growth of nano-approach-based sensors (e.g. nanotubes and nanowire) may provide new paths to address this concern (Patolsky et al., 2006).

Along with drug delivery, various micro- and nanofabricated devices have presented great potential for the successful development of nanocarriers with highly controlled physical and biochemical characteristics (e.g. shape, size, and surface chemistry). It is possible to replicate polymeric particles with variable shapes at submicron scale using perfluoropolyether moulds, developed by conventional top-down approaches (Gratton et al., 2008). As compared to bulk synthesis, different microfluidic devices have been considered for the nanoprecipitation synthesis of homogeneous PLGA–PEG nanoparticles (Karnik et al., 2008).

Beyond the number of challenges and opportunities, the following are also important for the expansion of next generation of delivery systems:

- Comprehensive physical and biochemical characterization of nanosystems (which will need sophisticated methods)
- Limiting the uptake of undesired non-target organs (e.g. spleen and liver)
- Improvement of programmable and stimuli-triggered drug release systems
- Advancing the ways by which we can deeply understand the fundamental principles of different diseases and its microenvironment
- Biological constraints that restrict drug delivery
- Endosomal trafficking pathways

With continual expansion in these areas, we can assume that the field of drug delivery systems will benefit tremendously from the presence of finely engineered nanobiomaterials and devices.

8.6 NANOSTRUCTURING AND NANOCOATING OF SURFACES

In parallel to the advancement of nanobiomaterials, knowledge of the nanostructuring of biomaterial surfaces develops rapidly. Nanostructuring of surfaces provides a massive platform for improving fouling control methodologies for different biomaterials intended to act as implants for medical applications. In addition to fouling control, nanostructuring of surfaces also controls other characteristics like friction, wettability, porosity, roughness, chemical and physical reactivity, conductivity, etc. Especially in the field of artificial organs, there is a rising need for smart surfaces which have the capability of achieving high biocompatibility (Zeifang et al., 2008).

Another potential implementation of nanotechnology exists in the possibility of mimicking a variety of self-organized systems found in nature. Complex structures and substructures like cells, folded proteins, and molecular motors represent a specific type of self-organized nanoscale machine that presently cannot be prepared synthetically. Nature, nevertheless making use of dynamic molecular systems, has shown that self-organized molecular systems are highly successful. This idea will be partially transferred into synthetic assemblies in the future, and their execution may direct us to new developments that cannot be gained by traditional large-scale synthesis processes. Nanostructuring can be done chemically, physically, or by self-assembly. Perhaps the most common naturally occurring example is the upper surface structure of the leaf of a lotus flower. The specifically structured surface with a bimodal size distribution in the submicron regime leads to a self-cleaning ability as an outcome of surface chemistry and structure of the material. Another very good example is cell behaviour manipulation by making changes in the surface structure while maintaining almost the same chemical composition (Graeter et al., 2007). By

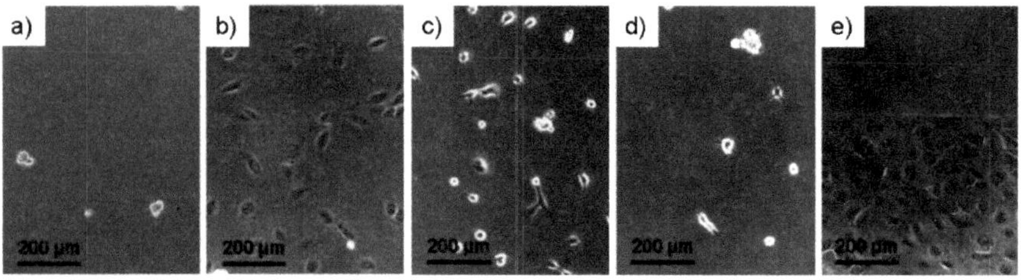

FIGURE 8.5 Phase contrast optical micrographs of 3T3 fibroblasts on PEGDA 700 hydrogels. (a) Cells on a non-RGD-functionalized gold nanoparticle pattern. (b–d) Cells on cyclo(-RGDfK-)-functionalized gold particles; cyclo(-RGDfK-) patches are separated by varying distances (b) 40 nm, (c) 80 nm, and (d) 100 nm, after 24h in culture. (e) Dense cell layer on a PEG support after 14 days in culture. The bottom part of the sample was patterned with cyclo(-RGDfK-) peptide-functionalized gold nanoparticles spaced 40 nm apart (Graeter et al., 2007).

changing the distance between the functionalized gold nanoparticles attached tightly to the upper surface, attachment behaviour and growth of fibroblasts were found, as is visible in Figure 8.5 (Graeter et al., 2007).

The detailed understanding of the fundamental principles of these effects opens up technical pathways for the generation of surfaces with non-fouling characteristics, and also tremendously helps to develop surfaces with optimized structures for the growth of a particular cell. Modifying the outer surface of titanium alloys through nanostructuring using calcium phosphate in the form of a hydroxyapatite coating results in advanced mechanical properties (Sato et al., 2008). It was also found that by altering the surface into a nanotubular structure, the artificial joints were more suitably incorporated without inducing chronic inflammation (Popat et al., 2007).

The implementation of nanostructured surfaces with well-defined characteristics also opens the path to diagnostic approaches. It has been observed that cell diagnostics are failing frequently due to unfavourable interactions between the cells and the device. This specifically refers to the case when immune cells are investigated. Receptors present on the surface of leucocytes react with artificial structures, which creates unwanted differentiation or activation of cells. In this case, it is necessary that surfaces do not induce any type of activation after contact. This is crucial for all microfluidic devices used for proteomics and biochips. The advancement in the development in the microelectromechanical assemblies provides nanostructuring of surfaces with a new level of significance (Zuruzi et al., 2005).

8.7 TOXICITY AND BIOCOMPATIBILITY OF NANOBIOMATERIALS

The creation of nanoscale materials may be a primary concern for discussion in nanobiotechnology with respect to toxicology. Nanomaterials possess a highly reactive nature because of large surface to volume ratio. This leads to the strong possibility of the presence of latent risk associated with all newly developed nanosystems. The structure of existing rules and regulations for new pharmaceuticals and chemicals seems to be appropriate to treat this type of materials. The synthesis procedures of nanomaterials have to be examined carefully in conjunction with their probable side effects within the human body. It has been observed that scientific and public awareness of nanobiotechnological labs is growing and there is always an increasing intensity of detailed discussions on the societal and ethical issues with different research organizations.

The medical application of nanotechnology needs a large number of regulations to ensure the proper use of drugs and medical devices originating from the implementation of the nano-approach (Rickerby 2007). The potential of molecular diagnostics by nanotechnology also requires a focus from the political corner. This involves an important questioning of the development of life quality in the cases of different diseases, cost effective analysis, treatment of patients, the novel methods of an artificial extension of natural senses, etc., which might be available or affordable to only a limited number of people. Due to possible biocompatibility and potential toxicity, the implementation of nanomaterials creates several unexpected problems in a human body. This may include immunoreaction, inflammation, or cancer. Fundamental mechanisms of all of these side effects are not yet completely studied, but they may be due to an enhanced hydrophobic reaction to biological matter or an increased generation of active radicals from the surface catalysis (Emerich and Thanos, 2006).

Nanomaterials which have entered into an organism, if not removed or recovered, accumulate in tissues and cells, developing a strong and unexpected potential for causing disease on a long-term basis. It has been found that nanoparticles can easily enter a human body through different ports. Involuntary or accidental contact during synthesis or application is the most common mode of entry, and usually happens via the lung (Nemmar et al., 2001). Nanoparticles used specifically for oral drug delivery have a tendency to accumulate in the liver and excessive immune response may turn into permanent damage in that area. This type of accumulation effect in cells is found in the case of pulmonary fibrosis caused by silicosis and asbestosis, diseases that usually originate from breathing in quartz or silica (Chong et al., 2006). It has also been found that a high concentration of nanomaterials may result in a cell's transformation to a tumorous state, which is primarily responsible for causing cancer. Thorough research by X-ray microanalysis via an energy-dispersive detector and scanning electron microscopy on the renal and hepatic tissues affected by granulomatosis represented a clear correlation of the presence of non-biodegradable, inert, and exogenous nanoparticles with different kinds of diseases that conventional histopathology could not usually account for (Figure 8.6) (Gatti and Rivasi, 2002).

Nanomaterials used for the delivery of drugs are widely exposed to different biomolecules in the lung or to the endothelial barrier. This type of contact causes the uptake of nanomaterials via endocytosis (Hoet et al., 2004). These examples show that the effects of nanomaterials on human health could be possibly two-edged, similar to traditional drug exposure though based on totally different schemes. The majority of the investigated systems so far show relatively few short-term risks. However, since all new methodologies may possess hidden risks, it is necessary to develop a systematic risk assessment procedure in parallel with technological expansion to keep hazardous potential as low as possible.

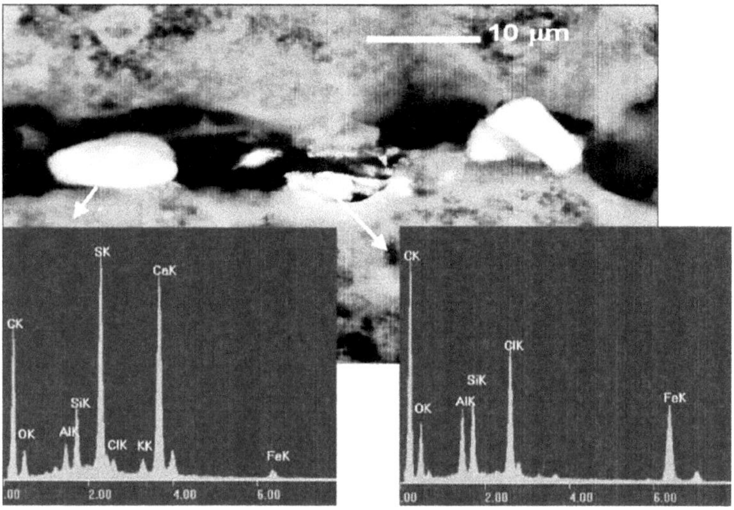

FIGURE 8.6 Scanning electron microscopic (SEM) microphotograph of a granulomatous liver section with two small particles and a cluster of nanodebris in between. The energy dispersive system (EDS) spectra reveal that the debris has different compositions, and probably have different origins (Gatti and Rivasi, 2002).

8.8 CONCLUSION

The implementation area of the nano-approach for effective diagnosis and treatment of diseases is currently very broad. Therefore, besides creativity and visionary power, practical application of nanotechnology in the field of biomedical science requires a systematic development to achieve real progress and allow us to reach new edges. In this chapter, I tried to provide an overview of some interesting developments using the nano-approach in the area of medical research. Since the area of medical science is currently under rapid expansion, this detailed study cannot cover all aspects of the nano-approach in detail. My objective was primarily to explain the role of the nano-approach in the successful utilization of recent clinical advancements. Nanotechnology has already offered a significant impact on medical applications, which is definitely expected to be explored during the next decades. Situated on the intersection of many fundamental branches, applications of the nano-approach completely rely on:

- Detailed knowledge of pathophysiology and disease biology to enable effective targeting and therapy
- An understanding of the chemical formulations to offer necessary modifications to the nanovector surface as well as to allow conjugation of the contrast/drug agent
- An awareness of the physical characteristics of the nanosystems to be able to magnificently engineer and manipulate material for the development of novel drug delivery systems

Most of the research and the majority of nanomedical applications are currently targeting the diagnostics and treatment of cancer. In order to effectively understand malignancies, molecular alterations should be identified as early as possible. Therefore, it is necessary that highly sensitive methods have to find their way for quick and efficient diagnostics. The concept of the nano-approach can play a fundamental role in achieving this goal. The growing attention to this area will be a major backbone in the establishment of personalized medicine.

REFERENCES

Ainslie, K. M. and Desai, T. A. (2008). Microfabricated implants for applications in therapeutic delivery, tissue engineering, and biosensing. *Lab on a Chip*, 8(11), 1864–1878.

Alivisatos, A. P., Gu, W., and Larabell C. (2005). Quantum dots as cellular probes. *The Annual Review of Biomedical Engineering*, 7, 55–76.

Anderson, D. G., Peng, W., Akinc, A., Hossain, N., Kohn, A., Padera, R., and Sawicki, J.A. (2004). A polymer library approach to suicide gene therapy for cancer. *Proceedings of the National Academy of Sciences of the United States of America*, 101(45), 16028–16033.

Bagalkot, V., Zhang, L., Levy-Nissenbaum, E., Jon, S., Kantoff, P. W., Langer, R., and Farokhzad, O. C. (2007). Quantum dot–aptamer conjugates for synchronous cancer imaging therapy, and sensing of drug delivery based on bi-fluorescence resonance energy transfer. *Nano Letters*, 7(10),3065–3370.

Barrett, T. , Brechbiel, M. , Bernardo, M. , and Choyke, P. L. (2007). MRI of tumor angiogenesis. *Journal of Magnetic Resonance Imaging*, 26(2), 235–249.

Bartlett, D. W., Su, H., Hildebrandt, I. J., Weber, W. A., and Davis, M. E. (2007). Impact of tumor-specific targeting on the biodistribution and efficacy of siRNA nanoparticles measured by multimodality in vivo imaging. *Proceedings of the National Academy of Sciences of the United States of America*, 104(39), 15549–15554.

Black, J. (2005). *Biological Performance of Materials: Fundamentals of Biocompatibility* (4th ed.). CRC Press.

Brigger, I., Dubernet, C., and Couvreur, P. (2002). Nanoparticles in cancer therapy and diagnosis. *Advanced Drug Delivery Reviews*, 54(5), 631–651.

Bruckbauer, A., Zhou, D., Ying, L., Korchev, Y. E., Abell, C., and Klenerman, D. (2003). Multicomponent submicron features of biomolecules created by voltage controlled deposition from a nanopipet. *Journal of the American Chemical Society*, 125(32), 9834–9839.

Campbell, C. T., and Kim, G. (2007). SPR microscopy and its applications to high-throughput analyses of biomolecular binding events and their kinetics. *Biomaterials*, 28(15), 2380–2392.

Chang, H. H., Moura, J. M. F., Wu, Y. L., and Ho, C. (2008). Automatic detection of regional heart rejection in USPIO-enhanced MRI. *IEEE Transactions on Medical Imaging*, 27(8), 1095–1106.

Chong, S., Lee, K. S., Chung, M. J., Han, J., Kwon, O. J., and Kim, T. S. (2006). Pneumoconiosis: comparison of imaging and pathologic findings. *Radiographics*, 26(1), 59–77.

Couvreur, P., Barratt, G., Fattal, E., Legrand, P., and Vauthier, C. (2002). Nanocapsuletechnology: a review. *Critical Reviews in Therapeutic Drug Carrier Systems*, 19(2), 99–134.

Crommelin, D. J., Storm, G., Jiskoot, W., Stenekes, R., Mastrobattista, E. and Hennink, W. E. (2003). Nanotechnological approaches for the delivery of macromolecules. *Journal of Controlled Release*, 87(1–3), 81–88.

De la Zerda, A., Zavaleta, C., Keren, S., Vaithilingam, S., Bodapati, S., Liu, Z., and Gambhir, S. S (2008). Carbon nanotubes as photoacoustic molecular imaging agents in living mice. *Nature Nanotechnology*, 3(9), 557–562.

De Rosa, S. C., Herzenberg, L. A., Herzenberg, L. A., and Roederer, M. (2001). 11-color, 13-parameter flow cytometry: identification of human naive T cells by phenotype, function, and T-cell receptor diversity. *Nature Medicine*, 2, 245–248.

Debbage, P., and Jaschke, W. (2008). Molecular imaging with nanoparticles: giant roles for dwarf actors. *Histochemistry and Cell Biology*, 130(5), 845–875.

Dubertret, B., Skourides, P., Norris, D. J., Noireaux, V., Brivanlou, A. H., and Libchaber, A. (2002). In vivo imaging of quantum dots encapsulated in phospholipid micelles. *Science*, 298(5599), 1759–1762.

Emerich, D. F., and Thanos C. G. (2006). The pinpoint promise of nanoparticle-based drug delivery and molecular diagnosis. *Biomolecular Engineering*, 23(4), 171–184.

Farokhzad, O. C., and Langer, R. (2009). Impact of nanotechnology on drug delivery. *ACS Nano*, 3(1), 16–20.

Franz, C. M., and Müller, D. J. (2005). Analyzing focal adhesion structure by atomic force microscopy. *Journal of Cell Science*, 118(22), 5315–5323.

Fu, A., Gu, W., Larabell, C., and Alivisatos, A. P. (2005). Semiconductor nanocrystals for biological imaging. *Current Opinion in Neurobiology*, 15(5), 568–575.

Gabathuler, R. (2010). Approaches to transport therapeutic drugs across the blood-brain barrier to treat brain diseases. *Neurobiology of Disease*, 37(1), 48–57.

Gao, X., Cui, Y., Levenson, R. M., Chung, L.W., and Nie, S. (2004). In vivo cancer targeting and imaging with semiconductor quantum dots. *Nature Biotechnology*, 22, 969–976.

Gao, X., Chung, L.W., and Nie, S. (2007). Quantum dots for in vivo molecular and cellular imaging. *Methods in Molecular Biology*, 374, 135–145.

Gatti, A. M., and Rivasi, F. (2002). Biocompatibility of micro- and nanoparticles. Part I: in liver and kidney. *Biomaterials*, 23(11), 2381–2387.

Gorelik, J., Zhang, Y., Shevchuk, A. I., Frolenkov, G. I., Sánchez, D., Lab, M. J., and Korchev, Y. E. (2004). The use of scanning ion conductance microscopy to image A6 cells. *Molecular and Cellular Endocrinology*, 217(1–2), 101–108.

Gorzelanny, C., Goerge, T., Schnaeker, E. M., Thomas, K., Luger, T. A., and Schneider, S.W. (2006). Atomic force microscopy as an innovative tool for nanoanalysis of native stratum corneum. *Experimental Dermatology*, 15(5), 387–391.

Graeter, S. V., Huang, J., Perschmann, N., Lopez-Garcıa, M., Kessler, H., Ding, J., and Spatz, J.P. (2007). Mimicking cellular environments by nanostructured soft interfaces. *Nano Letters*, 7(5), 1413–1418.

Gratton, S.E., Williams, S. S., Napier, M. E., Pohlhaus, P. D., Zhou, Z., Wiles, K. B., and Desimone, J. M. (2008). The pursuit of a scalable nanofabrication platform for use in material and life science applications. *Accounts of Chemical Research*, 41(12), 1685–1695.

Greco, F., and Vicent, M. J. (2009). Combination therapy: opportunities and challenges for polymer-drug conjugates as anticancer nanomedicines. *Advanced Drug Delivery Reviews*, 61(13), 1203–1213.

Gu, F., Zhang, L., Teply, B. A., Mann, N., Wang, A., Radovic-Moreno, A. F., and Farokhzad, O. C. (2008). Precise engineering of targeted nanoparticles by using self-assembled biointegrated block copolymers. *Proceedings of the National Academy of Sciences of the United States of America*, 105(7), 2586–2591.

Hoet, P.H., Brüske-Hohlfeld, I., and Salata, O. V. (2004). Nanoparticles–known and unknown health risks. *Journal of Nanobiotechnology*, 2(12), 1–15.

Imer, R., Akiyama, T., De Rooij, N. F., Stolz, M., Aebi, U., Friederich, N. F., Staufer, U. (2006). Development of atomic force microscope for arthroscopic knee cartilage inspection. *Japanese Journal of Applied Physics, Part 1: Regular Papers and Short Notes and Review Papers*, 45(3B), 2319–2323. doi:10.1143/JJAP.45.2319

Jung, H., Yang, T., Lasagna, M. D., Shi, J., Reinhart, G. D., and Cremer P.S. (2008). Impact of hapten presentation on antibody binding at lipid membrane interfaces. *Biophysical Journal*, 94(8), 3094–3103.

Karnik, R., Gu, F., Basto, P., Cannizzaro, C., Dean, L., Kyei-Manu, W., … and Farokhzad, O. C. (2008). Microfluidic platform for controlled synthesis of polymeric nanoparticles. *Nano Letters*, 8(9), 2906–2912.

Kim, J. Kim, D. I., Lee, S. K., Kim, D. J., Lee J. E., and Ahn, S. K. (2008). Imaging of the inflammatory response in reperfusion injury after transient cerebral ischemia in rats: correlation of superparamagnetic iron oxide-enhanced magnetic resonance imaging with histopathology. *Acta Radiologica*, 49(5), 580–588.

Kinge, S., Crego-Calama, M., and Reinhoudt, D. N. (2008). Self-assembling nanoparticles at surfaces and interfaces. *Chemphyschem*, 9(1), 20–42.

Klabunde, K. J., Stark, J., Koper, O., Mohs, C., Park, D. G., Decker, S., and Zhang, D. (1996). Nanocrystals as stoichiometric reagents with unique surface chemistry. *The Journal of Physical Chemistry*, 100(30), 12142–12153.

Kueng, A., Kranz, C., Lugstein, A., Bertagnolli, E., and Mizaikoff, B. (2005). AFM-tip-integrated amperometric microbiosensors: high-resolution imaging of membrane transport. *Angewandte Chemie (International Ed. in English)*, 44(22), 3419–3422.

Lane, R., Craig, B., and Babcock, W. (2002). Materials engineering with nature's building blocks. *The AMPTIAC Newsletter*, 6(1), 31–36. Retrieved from https://web.archive.org/web/20180501191308/https://www.dsiac.org/sites/default/files/journals/AMPQ6_1.pdf

Lange, T., Jungmann, P., Haberle, J., Falk, S., Duebbers, A., Bruns, R., … and Schillers, H. (2006). Reduced number of CFTR molecules in erythrocyte plasma membrane of cystic fibrosis patients. 23(4), 317–323.

Lee, K. B., Lim, J. H., and Mirkin, C. A. (2003). Protein nanostructures formed via directwrite dip-pen nanolithography. *Journal of the American Chemical Society*, 125(19), 5588–5589.

Liu C. H., Huang, S., Cui, J., Kim, Y. R., Farrar, C. T., Moskowitz, M. A., … and Liu, P. K. (2007). MR contrast probes that trace gene transcripts for cerebral ischemia in live animals. *The FASEB Journal*, 21(11), 3004–3015.

Liu, C. H. You, Z., Ren, J., Kim, Y. R., Eikermann-Haerter, K., and Liu, P. K. (2008). Noninvasive delivery of gene targeting probes to live brains for transcription MRI. *The FASEB Journal*, 22(4), 1193–1203.

Maluf, D., Cotterell, A., Clark, B., Stravitz, T., Kauffman, H. M., and Fisher, R. A. (2005). Hepatic angiosarcoma and liver transplantation: case report and literature review. *Transplantation Proceedings*, 37(5), 2195–2199.

Marquet, P., Rappaz, B., Magistretti, P. J., Cuche, E., Emery, Y., Colomb, T., and Depeursinge, C. (2005). Digital holographic microscopy: a noninvasive contrast imaging technique allowing quantitative visualization of living cells with subwavelength axial accuracy. *Optics Letters*, 30(5), 468–470.

Maurya, D. K, Ng,W. Y., Mahabadi, K. A., Liang, Y. N., and Rodríguez, I. (2007). Fabrication of lab-on chip platforms by hot embossing and photo patterning. *Biotechnology Journal*, 2(11), 1381–1388.

Meyer, L., Wildanger, D., Medda, R., Punge, A., Rizzoli, S. O., Donnert, G., and Hell, S. W. (2008). Dual-color STED microscopy at 30-nm focal-plane resolution. *Small*, 4(8), 1095–1100.

Migonney, V. (Ed.) (2014). History of Biomaterials. In *Biomaterials*. John Wiley & Sons.

Mirza, A. Z., and Siddiqui, F. A. (2014). Nanomedicine and drug delivery: a mini review. *International Nano Letters*, 4(94), 1–7.

Na, K., and Bae, Y. H. (2002). Self-assembled hydrogel nanoparticles responsive to tumour extracellular pH from pullulan derivative/sulphonamide conjugate: characterization, aggregation and adriamycin release in vitro. *Pharmaceutical Research*, 19(5), 681–683.

Nam, J. M., Thaxton, C. S., and Mirkin, C. A. (2003). Nanoparticle-based bio-bar codes for the ultrasensitive detection of proteins. *Science*, 301(5641), 1884–1886.

Nemmar, A., Vanbilloen, H., Hoylaerts, M. F., Hoet, P. H., Verbruggen, A., and Nemery, B. (2001). Passage of intratracheally instilled ultrafine particles from the lung into the systemic circulation in hamster. *American Journal of Respiratory and Critical Care Medicine*, 164(9), 1665–1668.

Nieh, T. G., and Wadsworth, J. (1991). Hall–Petch relation in nanocrystalline solids. *Scripta Metallurgica et Materiala*, 25(4), 955–958.

O'Brien, F. J. (2011). Biomaterials & scaffolds for tissue engineering. *Materials Today*, 14 (3), 88–95.

Patolsky, F., Zheng, G., and Lieber, C. M. (2006). Nanowire sensors for medicine and the life sciences. *Nanomedicine*, 1(1), 51–65.

Patrick, B., and Bertrand, L. (2011). Nanomedicine, nanotechnology in medicine. *Comptes Rendus Physique*, 12(7), 620–636.

Peters, R. (2006). Nanoscopic medicine: the next frontier. *Small*, 2(4), 452–456.

Popat, K. C., Leoni, L., Grimes, C. A., and Desai, T. A. (2007). Influence of engineered titania nanotubular surfaces on bone cells. *Biomaterials*, 28(21), 3188–3197.

Ramakrishna, S., Ramalingam, M., Kumar, T. S. S., and Soboyejo, W. O. (2010). *Biomaterials: A Nano Approach*, Chapter 5, CRC Press .

Ratner, B., Hoffman, A., Schoen, F., and Lemons, J. (2004). *Biomaterials Science: An Introduction to Materials in Medicine* (2nd ed.). Academic Press.

Rickerby, D. G. (2007). Nanotechnological medical devices and nanopharmaceuticals: the European regulatory framework and research needs. *Journal for Nanoscience and Nanotechnology*, 7(12), 4618–4625.

Riethmuller, C., Schäffer, T. E., Kienberger, F., Stracke, W., and Oberleithner, H. (2007). Vacuolar structures can be identified by AFM elasticity mapping. *Ultramicroscopy*, 107(10–11), 895–901.

Rosen, J. E., Yoffe, S., Meerasa, A., Verma, M., and Gu, F. X. (2011). Nanotechnology and diagnostic imaging: new advances in contrast agent technology. *Journal of Nanomedicine & Nanotechnology*, 2, 1–12.

Rothenhäusler, B., and Knoll, W. (1988). Surface–Plasmon Microscopy, 332, 615–617.

Roy, I., Ohulchanskyy, T. Y., Pudavar, H. E., Bergey, E. J., Oseroff, A. R., Morgan, J., Prasad, P. N, et al. (2003). Ceramic-based nanoparticles entrapping water-insoluble photosensitizing anticancer drugs: a novel drug-carrier system for photodynamic therapy. *Journal of American Chemical Society*, 125(26), 7860–7865.

Sahoo, K. S., and Labhasetwar, V. (2003). Nanotech approaches to drug delivery and imaging. *Drug Discovery Today*, 8(24), 1112–1120.

Sanhai, W.R., Sakamoto, J. H., Canady, R., and Ferrari, M. (2008). Seven challenges for nanomedicine. *Nature Nanotechnology*, 3(5), 242–244.

Sato, M., Aslani, A., Sambito, M. A., Kalkhoran, N. M., Slamovich, E. B., and Webster, T. J. (2008). Nanocrystalline hydroxyapatite/titania coatings on titanium improves osteoblast adhesion. *Journal of Biomedical Materials Research Part A*, 84(1), 265–272.

Schirmeisen, A., Holscher, H., Anczykowski, B., Weiner, D., Schafer, M. M., and Fuchs, H. (2005). Dynamic force spectroscopy using the constant-excitation and constant-amplitude modes. *Nanotechnology*, 16(3), S13–S17.

Seisenberger, G., Ried, M. U., Endress, T., Büning, H., Hallek, M.,and Bräuchle, C. (2001). Real-time single-molecule imaging of the infection pathway of an adeno-associated virus. *Science*, 294(5548), 1929–1932.

Sengupta, S., Eavarone, D., Capila, I., Zhao, G., Watson, N., Kiziltepe, T., and Sasisekharan, R. (2005). Temporal targeting of tumour cells and neovasculature with a nanoscale delivery system. *Nature*, 436(7050), 568–572.

Shahin, V., Albermann, L., Schillers, H., Kastrup, L., Schafer, C., Ludwig, Y., and Oberleithner, H. (2005). Steroids dilate nuclear pores imaged with atomic force microscopy. *Journal of Cellular Physiology*, 202(2), 591–601.

Staples, M., Daniel, K., Cima, M. J., and Langer, R. (2006). Application of micro- and nano-electromechanical devices to drug delivery. *Pharmaceutical Research*, 23(5), 847–863.

Truskett, V. N., and Watts, M. P. (2006). Trends in imprint lithography for biological applications. *Trends in Biotechnology*, 24(7), 312–317.

Valenta, J., Juhasz, R., and Linnros, J. (2002). Photoluminescence spectroscopy of single silicon quantum dots. *Applied Physics Letters*, 80, 1070–1072.

Vert, M. (2007). Polymeric biomaterials: strategies of the past vs. strategies of the future. *Progress in Polymer Science*, 32(8–9), 755–761.

Wagner, V., Dullaart, A., Bock, A. K., and Zweck, A. (2006). The emerging nanomedicine landscape. *Nature Biotechnology*, 24(10), 1211–1217.

Whitehead, K. A., Langer, R., and Anderson, D. G. (2009). Knocking down barriers: advances in siRNA delivery. *Nature Reviews Drug Discovery*, 8(2), 129–138.

Woodrow, K. A., Cu, Y., Booth, C. J., Saucier-Sawyer, J. K., Wood, M. J., and Saltzman, W. M. (2009). Intravaginal gene silencing using biodegradable polymer nanoparticles densely loaded with small-interfering RNA. *Nature Materials*, 8(6), 526–533.

Yoruç, A. B. H., and Sener, B. C. (2012). Biomaterials. In S. Kara (Ed.), *A Roadmap of Biomedical Engineers and Milestones* (pp. 67–114). doi:10.5772/48057

Zeifang, F., Grunze, M., Delling, G., Lorenz, H., Heisel, C., Tosounidis, G., and Holstein, J.H. (2008). Improved osseointegration of PTFEP-coated titanium implants. *Medical Science Monitor*, 14(2), 35–40.

Zheng, J., Ghazani, A. A., Song, Q., Mardyani, S., Chan, W. C., and Wang, C. (2006). Cellular imaging and surface marker labeling of hematopoietic cells using quantum dot bioconjugates. *International Journal of Laboratory Hematology*, 12(2), 94–98.

Zuruzi, A. S., Ward, M. S., and MacDonald, N. C. (2005). Fabrication and characterization of patterned micrometre scale interpenetrating Au-TiO_2 network nanocomposites. *Nanotechnology*, 16(8), 1029–1034.

9 Applications in Nanomedicine and Drug Delivery Systems

9.1 INTRODUCTION

Nanomedicine is the design and development of therapeutics and diagnostic tools, distinguished by the nanoscopic scale of its delivery vehicles and diagnostic agents (Sanna, 2014). The nanomedical field is swiftly gaining popularity and recognition by developing methods of administering treatment with high efficiency and unprecedented safety. In the human body, nanosized particulates behave very differently from larger particles. Based on this finding, plenty of studies within the discipline have been done by researchers and have advanced the current standards of drug delivery by relating them to biodistribution, intracellular uptake, and dosing efficacy by utilizing nanoparticles (NPs), which are able to encapsulate therapeutics and target them to specific sites or diseased cells (Farokhzad and Langer, 2009). These successful tactics have proved that drug delivery via useful NPs is a progressive technique for disease remedy. Currently, nanosized drug delivery systems (DDS) carry out various features in the biomedical arena and promote nanomedicine as having an extensive variety of promising packages, together with the development of medicine. A variety of nanosized devices are present in the market and have been authorized by the Food and Drug Administration (FDA) for clinical use (Sanna, 2014). For instance, liposomes and several polymer-based NPs, which include nanospheres, nanomicelles, nanocapsules, and nanofibers, are specially used for biodegradable tissue engineering scaffolds. There is also nanoimaging using quantum dots (QDs), which includes single-molecule imaging as well as whole-body imaging at a very high resolution. Other advances in molecular imaging, such as magnetic resonance imaging using superparamagnetic iron oxide and microcomputed tomography, have also improved significantly. Other applications of nanomedicine are delivery systems using various polyplexes that are composed of nanocarrier molecules and plasmid DNA or nucleic acid drugs such as antisenses and siRNA. The last decade has witnessed the use of many new biotechnological approaches to the treatment of cancer. For example, the 2006 Nobel Prize in Physiology or Medicine brought a renewed focus on gene silencing and the therapeutic opportunities offered by precise regulation of gene expression using siRNA and miRNA technologies (Whitehead et al., 2009).

Polymeric materials have also played a crucial role in the research and development of nanomedicines. The polymeric materials that are useful in this field can be

by Yaser Dahman and Mohamed Khattab

classified by their chemical nature into natural polymers, biodegradable synthetic polymers, water-soluble synthetic polymers, polyelectrolytes, and polyion complexes. Nevertheless, the higher-order structures and morphology of the materials used in the form of nanowires, particles, vesicles, capsules, shells, gels, cages, skeletons, and film membranes can also be considered a key factor in nanomedicine.

9.1.1 Delivery Nanoplatforms

According to the National Nanotechnology Initiative (NNI), nanoparticles are defined as structures of sizes ranging from 1 to 100 nm in at least one dimension. However, the prefix "nano" is commonly used to express particles that are up to several hundred nanometers in size. Their nanoscopic size facilitates greater intracellular uptake than larger molecules. NPs have different abilities to encapsulate therapeutics and release them in a controlled manner to specifically target diseased cells. The encapsulation process improves the solubility and stability of drug compounds (Chrastina et al., 2011). The method of conjugating the drug to the nanocarrier and the strategy of its targeting is highly significant for targeted therapy. A drug may be encapsulated into the nanocarrier or adsorbed onto or covalently attached to its surface. Covalent linking has the advantage over other means of drug attachment as it enables more control over the number of drug molecules connected to the nanocarrier and consequently ensures precise control over the amount of the therapeutic compound delivered.

Additional advantages of NPs are their modifiable external shell, biodegradability, and low cytotoxicity (Davis et al., 2008). Controlled DDS have several advantages compared to the traditional methods of drug release. The drug is transported directly to the place of action, and hence, its influence on vital tissues and undesirable side effects are minimized. The accumulation of therapeutic compounds in the target site also increases and consequently the required doses of drugs are lessened. Cell-specific targeting can be achieved by attaching drugs to specially designed carriers. Various nanostructures that have been tested and currently used as DDS are presented in (Figure 9.1).

Nevertheless, engineered therapeutic NPs offer numerous clinical advantages. Surface modification of NPs with polyethylene glycol (PEG) protects them from clearance by the mononuclear phagocytic system (MPS), markedly increasing both blood circulation times and drug uptake by target cells. Functionalization of the NP's surface with multivalent targeting moieties not only improves drug efficacy but also simultaneously reduces dosage and provides a novel method for optimizing drug pharmacokinetics (Bamrungsap et al., 2012), see Figure 9.2.

The undesirable effects of nanoparticles strongly depend on their hydrodynamic size, shape, number, surface chemistry, route of administration, the reaction of the immune system (especially the manner of uptake by macrophages and granulocytes), and the residence time in the bloodstream. Due to many factors which may affect the toxicity of nanoparticles, their estimation is rather challenging and, thus, toxicological studies of each new DDS formulation are needed. However, based on their sizes, one can make some generalizations; the smaller particles with a greater surface area are more reactive and consequently more toxic (Ai et al., 2011). It is accepted that

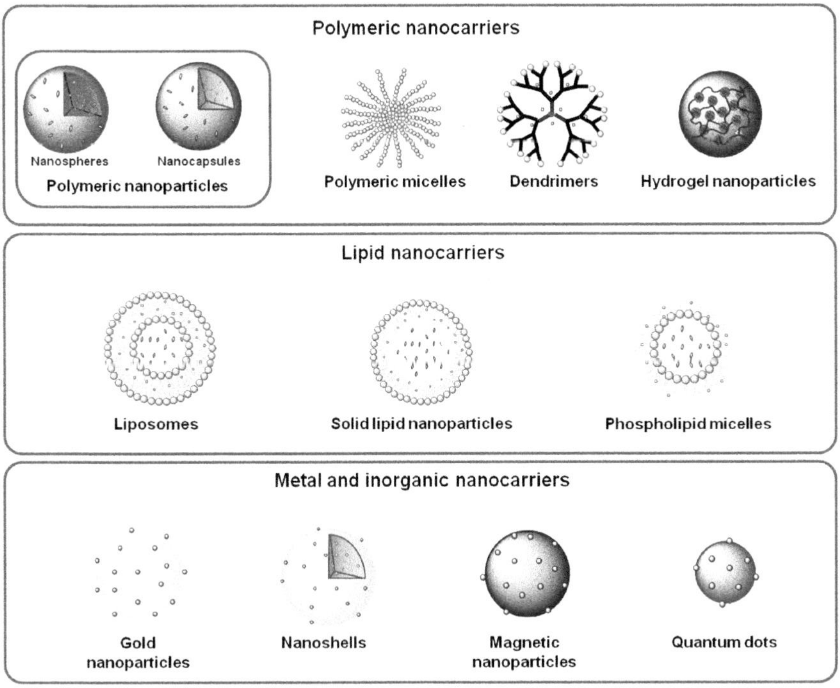

FIGURE 9.1 Examples of nanotechnology-based drug delivery platforms; polymeric, lipid and metal and inorganic nanocarriers. Adapted from (Conniot et al., 2014).

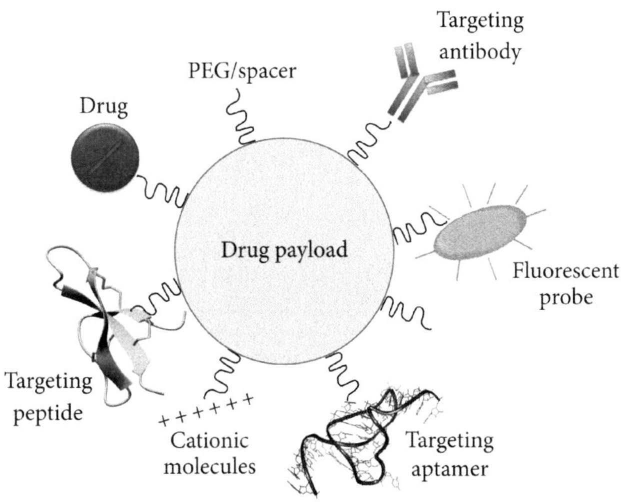

FIGURE 9.2 Graphical representation of surface-modified NPs with drugs (incorporated within the core or conjugated to the surface of NPs), targeting molecules (antibodies, peptides, aptamers, and cationic molecules), PEG for stealthiness, and with the fluorescent probe as a tracer (Masserini, 2013).

nanoparticles with a hydrodynamic diameter of 10–100 nm have optimal pharmacokinetic properties for in-vivo applications. Smaller nanoparticles are subject to tissue extravasations and renal clearance, whereas larger nanoparticles are quickly opsonized and removed from the bloodstream via macrophages of the reticuloendothelial system (Cole et al., 2011).

9.1.2 Methods of Nanoparticle Preparation

NPs require a very specialized formulation approach. The most common methods employ spray drying, salting out, nanoprecipitation, and self-assembly by the emulsion-based process. The latter includes oil-in-water (O/W) single emulsion and water-in-oil-in-water (W/O/W) double emulsification (Rao and Geckeler, 2011). The most recent development in the synthesis of NPs involves the discipline of microfluidics, which is capable of manipulating nanoscale volumes in microscale fluidic channels (Capretto et al., 2013). Microfluidic reactors offer precise control and manipulation of the fluids used to create NPs which promotes higher NP yield and uniform size. Other significant advantages are that microfluidics devices are tunable and can use three-dimensional hydrodynamic focusing to create NPs of different sizes and target ligand densities with various polymers, while simultaneously encapsulating drugs, which is not readily feasible with conventional approaches and can, in turn, produce various NPs (Figure 9.3). However, the challenges associated with the high costs of glass/silicon fabrication and large-scale production for clinical use still need to be addressed.

9.1.3 "Stealth" Modifications of NPs

Modification with PEG is currently the standard coating for NPs. Surface functionalization of NP with PEG was shown to dramatically reduce protein adsorption through hydrophilicity and steric repulsion effects, extending the circulation time

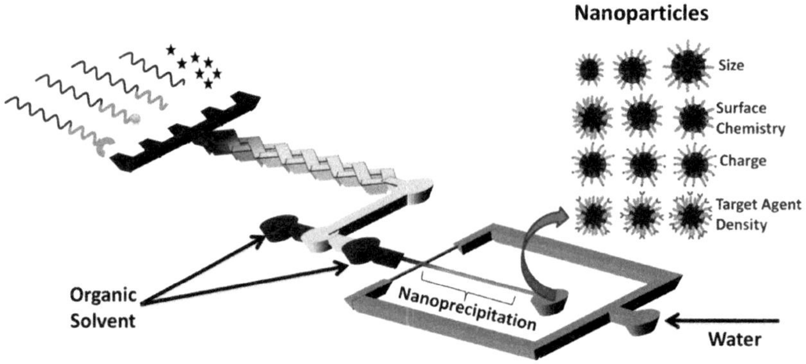

FIGURE 9.3 Schematic of a system in which programmable mixing of polymer precursors allows for the synthesis of NPs with a broad range of sizes, surface chemistry, charge, and targeting agent densities (Valencia et al., 2013).

of NPs in the blood. This allowed the "stealth" NP carriers to persist in the bloodstream long enough to reach or recognize their therapeutic site of action (Peracchia et al., 1999). Examples of stealth nanocarriers include PEGylated liposomal doxorubicin (Doxil®) and the PLA-PEG micelle form of paclitaxel (Genexol-PM®). In addition to PEGylation, new biomaterials and delivery strategies have been developed to prolong the circulation time of NPs, such as zwitterionic polymer-based NPs which are resistant to non-specific protein adsorption due to electrostatically induced hydration (Zhu, 2014). Another concept prevalent in scientific innovation is the coating of poly(D,L-lactic-co-glycolic acid) (PLGA) NPs with natural peptide erythrocyte membranes derived from RBC, (see Figure 9.4). Cellular membranes provide a robust native functionality to NPs. In comparative studies with PEG-coated NPs, RBC membrane coated NPs exhibited 39.6 hr half-life compared with 15.8 hr for PEG NPs.

9.1.4 Passive and Active Targeting of Nanomedicine

Passive targeting is based on both the minute size of drug carriers as well as the leaky neovasculature of the tumour. With the longer blood circulation effect brought about by "stealth" modification (e.g. PEGylation), increased accumulation of NPs is possible through the enhanced permeability and retention (EPR) effect (Torchilin, 2011). The EPR effect refers to the fact that tumours retain more polymeric NPs, proteins, liposomes, and micelles than other tissues (Bertrand et al., 2014). Most tumours have a permeable vasculature created through stimulation by the vascular

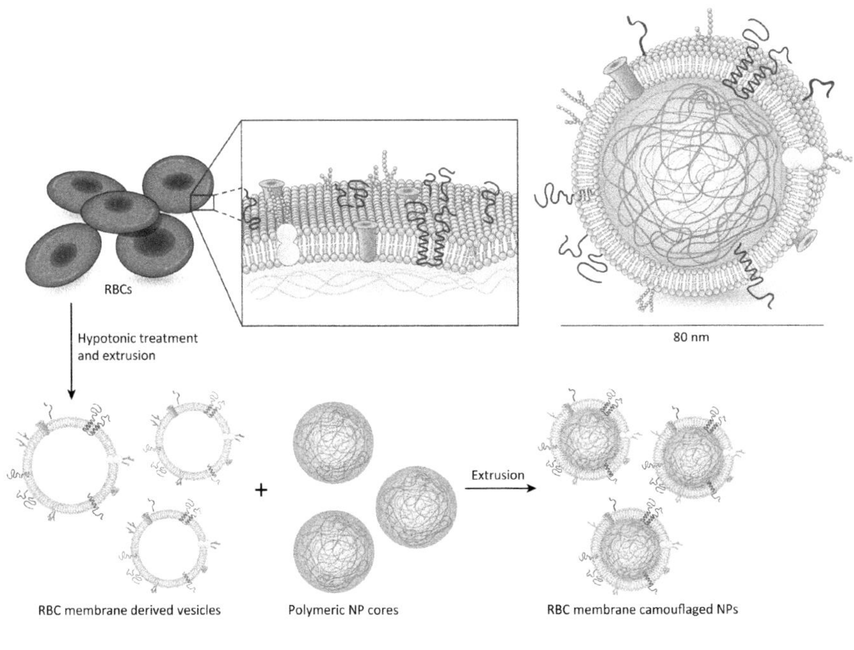

FIGURE 9.4 Red blood cell membrane-coated PLGA NPs (Xu et al., 2015).

endothelial growth factor (VEGF). Tight junctions in normal vasculature prevent particles larger than 2 nm from crossing between endothelial cells (Figure 9.5). However, the tight junctions and membranes of tumour vasculatures are disordered, allowing entities from 10–500 nm in size to extravasate and accumulate within the tumour interstitium. An example of passive targeting nanomedicine is Genexol-PM®, which was approved in Korea in 2007. This polymeric micelle DDS formulation includes poly(D, L-lactide), which allows for the controlled release of therapeutics. Active targeting is being explored as a method to achieve spatial localization by intentionally targeting NPs to actively diseased sites while eliminating off-target adverse effects in healthy tissue. The polyvalent decoration of the NP's surface with a ligand can facilitate binding to a biomarker located in targeted cells. The ligands used to modify NPs include antibodies, engineered antibody fragments, proteins, peptides, small molecules, and aptamers (Figure 9.2). Ligand-receptor interaction can be utilized to concentrate the nanomedicine at the diseased tissue in-vivo, producing a preferred distribution profile. The ligand density can be further tuned in the formulation process to optimize cell avidity to the therapeutic material. In combinatorial delivery, two or more therapeutic agents inhibit different or identical disease pathways for a synergistic effect.

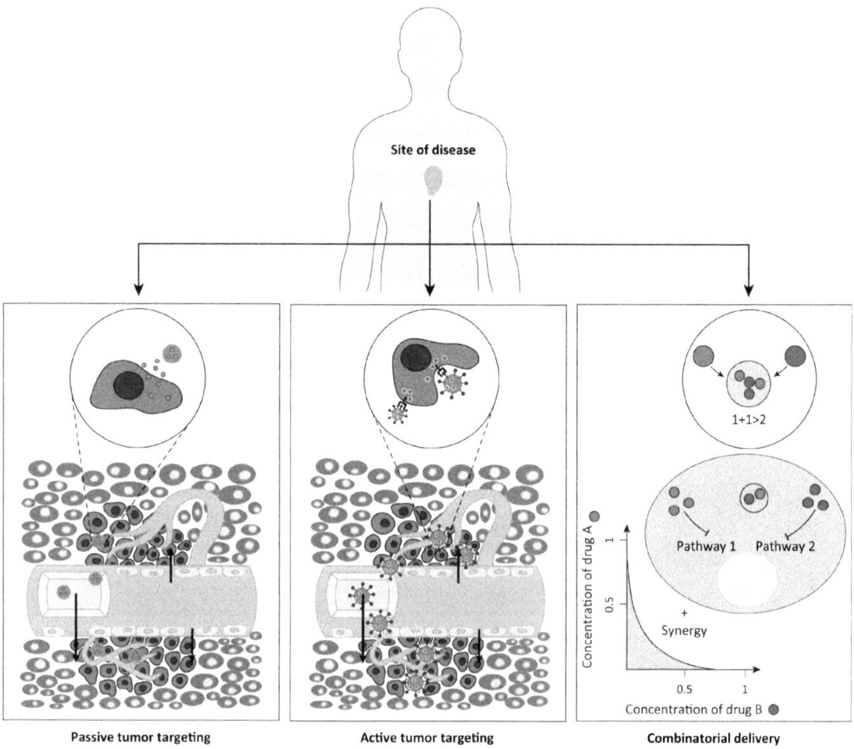

FIGURE 9.5 Demonstration of passive and active targeting, and combinatorial delivery (Xu et al., 2015).

9.2 NANOCARRIERS

9.2.1 Polymeric Nanocarriers

9.2.1.1 Polymeric Nanoparticles (PNPs)

Polymeric nanoparticles (PNPs) are solid nanocarriers that have a diameter ranging from 10 to 100 nm. Two different types of polymeric NPs are usually considered: nanospheres and nanocapsules (Figure 9.1). Nanospheres consist of a polymeric matrix in which the drug is homogenously dispersed, whereas nanocapsules are vesicular systems formed by a polymer wall that surrounds a core containing the drug. They can be either made of natural or synthetic polymers and may or may not be biodegradable. Natural polymers used to obtain PNPs include albumin (Martínez et al. 2011), DNA (Mao et al., 2001), chitosan (Rejinold et al., 2011), and gelatin (Saraog et al., 2010). Synthetic polymers from the ester family such as poly(lactic acid) (PLA), poly(cyanoacrylates) (PACA), poly(acrylic acid), poly(anhydrides), poly(amides), poly(ortho esters), poly(ethylene glycol), poly(vinyl alcohol) (PVA), and others like poly(iso-butyl cyanoacrylate) (PIBCA), poly(ethylene oxide) (PEO), poly(ε-caprolactone) (PCL), and polyurethane are suitable for drug delivery due to their biodegradability. In fact, PLGA is another synthetic polymer that has been extensively used in DDS (Vij et al., 2010) and other medical applications such as suture materials and bone fixation nails and screws (Lim et al., 2009). Upon biodegradation, it forms compatible moieties of lactic acid and glycolic acid, which are further removed by the citric acid cycle. High entrapment efficiencies (>80%) can be achieved with PLGA when sonication is used during the emulsification steps. PLGA exhibits an inherently good loading because of its high molecular weight and high hydrophilicity. Recently, poly(β-amino ester) (PbAE) has emerged in the spotlight because it demonstrates a pH sensitive release, in which at acid pH in the vicinity of the tumour mass it rapidly releases its payload. This polymeric material has shown to be less toxic than other cationic polymers such as poly(ethyleneimine) and poly(L-lysine) (PLL) (Shenoy et al., 2005). Biodegradable polymeric nanocarrier drug conjugates used for drug delivery (Figure 9.6) are stable in blood, non-toxic, and non-thrombogenic. They are also non-immunogenic, non-proinflammatory, and neither activates neutrophils nor affects the reticuloendothelial system (des Rieux et al., 2006). To reduce immunological interactions and intermolecular interactions between the surface chemical groups of PNPs, e.g. hydrogen bonding, hydrophobic interactions, and van der Waals forces, PNPs are usually coated with nonionic surfactants (Torchilin, 2008).

Drugs can be conjugated on a PNP's surface after a polymerization reaction (Luo et al., 2010) or can be encapsulated in the PNP's structure in-situ during a polymerization reaction (Mora-Huertas et al., 2010). Moreover, drugs can be released from PNPs by different mechanisms including desorption, diffusion, or nanoparticle erosion in the target tissue (Torchilin, 2008). Examples of drug–polymeric nanocarrier conjugates currently used as DDS are shown in Table 9.1.

9.2.1.2 Dendrimer Nanocarrier

Dendritic architecture is one of the most famous structures observed throughout all biological systems. In contrast to linear polymers, dendrimers are tree-like branched

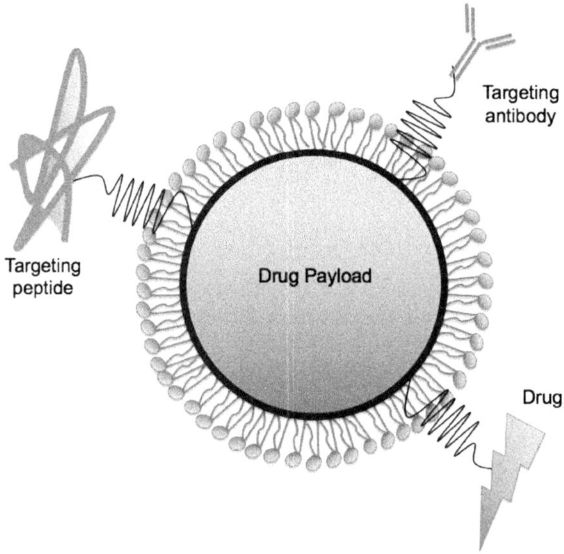

FIGURE 9.6 Drug-biodegradable polymeric nanocarrier conjugates (Trementozzi, 2013).

synthetic polymer macromolecular nanoparticles, and their structures are characterized by the following elements: core, dendrons, and surface-active groups. The core is a single atom or molecule (only if it has at least two identical functional groups) which dendrons are attached to. The dendrons (dendrimer arms) are molecules of monomers linked with the core, forming layers and building successive generations (their growth is spatially limited). They have a remarkably well-defined control over size (comparable in size to proteins) with narrow polydispersity (Kaminskas et al., 2011). They also have a large surface functionality, providing a broad range of applications such as drug and gene delivery (Lim et al., 2002), biological adhesives (Velazquez et al., 2004), and imaging agents (e.g. MRI) (Menjoge et al., 2010). Thus, they can be used for oral, transdermal, ocular, and intravenous deliveries (Sugisaki et al., 2008). Moreover, dendrimers can easily cross cell barriers by both paracellular and transcellular pathways (Menjoge et al., 2010). Dendrimers can be structurally modified. This modification can be made to the core or dendrons, giving polyfunctional capacity to the dendritic structure through different synthetic routes such as self-assembling synthesis, lego chemistry, and click chemistry (Medina et al., 2011). Dendrimer synthesis can start from the core element (i.e. divergent polymerization) or the peripheral branching units (i.e. convergent polymerization), resulting in a structure with a hydrophilic surface and a hydrophobic central core, (Figure 9.7). There are a few ways for connecting biologically active compounds to dendrimers. The drug may be encapsulated in the internal structure of dendrimers (Caminade et al., 2005), chemically attached, or physically adsorbed onto the dendrimer surface (Menjoge et al., 2010). The choice of the immobilization method depends on the drug's properties. Dendrimers either incorporated in the core and branches or conjugated to the terminal groups can transport molecules.

TABLE 9.1
Anticancer Nanomedicines Based on PNPs and Micelles in Clinical Trials and on the Market

Nanomedicine Type	Drug	Product Name/ Company	Indication	Phase
Polymeric conjugate	Camptothecin	CRLX101 (cyclodextrin adamantine)/Cerulean	Renal cancer Small cell lung cancer Ovarian cancer	Phase II
	Asparaginase	OncasparTM (PEG)/ Baxalta	Acute lymphoblastic leukaemia	Approved
	Paclitaxel	OpaxioTM (Polyglycerol adipate)/CTI Biopharrna	Ovarian cancer Non-small lung cancer (women)	Phase III
	Irinotecan	NKTR102 (PEG)/Nektar	Metastatic breast cancer	Phase III
	Camptothecin	CRLX101 (nanoparticle)/ Cerulean	Renal cell carcinoma (3rd/4th line) Ovarian cancer (2nd/3rd line)	Phase II
	Diaminocyclohex-ane (DACH) Platinum	AP 5346 (Hydroxypropyl methacrylate)/ ProLindac™	Ovarian cancer	Phase II
	Docetaxel	DEpTM (G5 PEG-Polylysinel/StarPharma	Advanced cancers Solid tumours	Phase I
Polymeric nanoparticles	Docetaxel + Prostate-Specific Membrane Antigen (PSMA)	BIND-014 (AccurinTM)/ BIND Therapeutics	Cholangiocarcinoma Cervical cancer Bladder cancer Head and neck cancer	PhaseII
	AZD2811 (AZD1152 hydroxyquinazoline pyrazol anilide; Aurora-B Kinase Inhibitor)	AZD2811 (AccurtnTM) nanoparticle/ AstraZeneca	Advanced solid tumours	Phase I
Polymeric micelles	Paclitaxel	Genexol-PMTM/ Samyang Biopharmaceuticals	Breast cancer Ovarian cancer	Approved
	Paclitaxel	NK105/NanoCarrier™	Stomach cancer Breast cancer	Phase III
	Paclitaxel	NC -4016/NanoCarrier™	Solid tumours	Phase I
	Paclitaxel	Nanoxel™/Samyang Biopharmaceuticals	Advanced breast cancer	Phase I
	DACH-platin	NC-6004 Nanoplatin™/ NanoCarrierTM	Pancreatic cancer Head and neck cancer Bladder cancer	Phase III

Source: Reproduced from Hare et al. (2017).

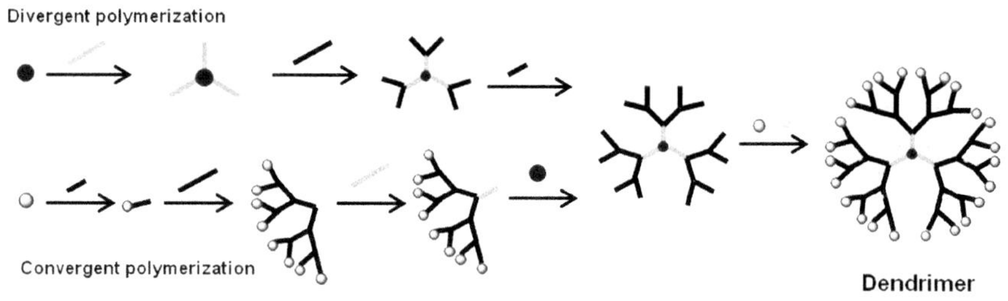

FIGURE 9.7 Divergent and convergent polymerization techniques for dendrimers hyper-branching (Conniot et al., 2014).

In general, dendrimers are terminated with amine surface groups (G1, G2, G) or carboxylate groups (G1.5, G2.5, G.) (Kitchens et al., 2007). Dendrimer size, molecular weight, and the number of surface functional groups can be modulated through increases in generation number (1 nm per generation) (Menjoge et al., 2010). For instance, poly(amido amide) (PAMAM) is a frequently used dendrimer in biomedical applications. Both the structure and drug or gene distribution inside these molecules have intensively investigated. PAMAM dendrimers grow through generations G = 1–10 and their size increases from 1.1 to 12.4 nm (Figure 9.8). The size of PAMAM generations is comparable to the size of selected proteins, e.g. dendrimer G3 with insulin and G7 with hemerythrin (Tomalia, 2005).

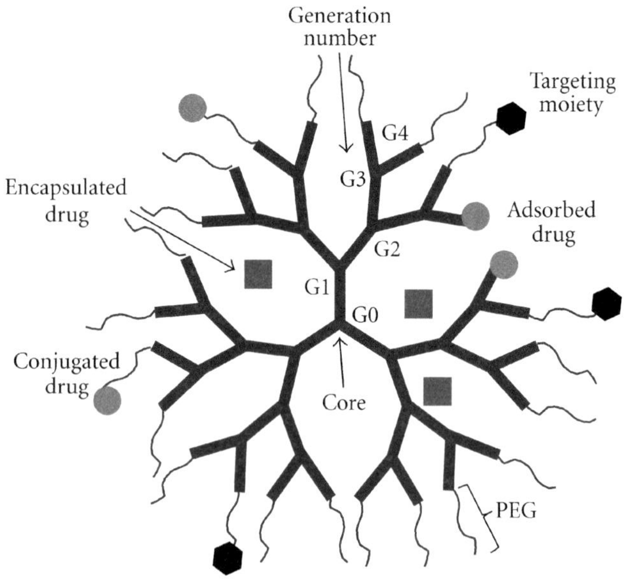

FIGURE 9.8 Combination DDS based on dendrimers: drugs adsorbed to the surface (ionic interaction) (black hexagons), encapsulated within hydrophobic cavities inside branching clefts (blue squares), or direct conjugated to the surface functional groups (red circles). (Lee and Nan, 2012).

The biocompatibility and physicochemical properties of dendrimers are determined by surface functional groups (Caminade et al., 2005). Selection of a core, type of a monomer, and surface functional groups determine the usability of dendrimers in medical applications. Dendrimer cytotoxicity depends on the core material and is strongly influenced by the nature of the dendrimer's polyvalence (i.e. the number of active groups on a dendrimers surface). For example, changing the surface amine groups into hydroxyl groups may result in lower levels of cytotoxicity.

Polymer micelles have emerged as a vital system for targeted drug delivery. Self-assembly of amphiphilic block or graft copolymers in aqueous media in which the most hydrophobic parts of their chains form a hydrophobic core and the most hydrophilic regions of the polymer chain are displayed outward. The previous assembly leads to nanoparticles with sizes ranging from 10 to 100 nm with hydrophobic cores for drug encapsulation, as well as hydrophilic shells for stabilization and specific targeting (Nasongkla, 2004). Using prostate cancer as a model, Farokhzad et al. (2006) demonstrated the potential utility of polymer micelle–aptamer bioconjugates that encapsulate docetaxel for therapeutic applications. The structural integrity of the hydrophilic corona of polymer micelles is selectively crosslinked, leading to static entities that are not easily disrupted. Smart core-shell nanoparticles have also been produced, e.g. thermoresponsive micelles which are biodegradable nanoparticles made from poly(D, L-lactide)-graft-poly(N-isopropyl acrylamide-co-methacrylic acid) (PLA-g-P(NIPAm-co-MAA)) to yield a hydrophilic outer shell and a hydrophobic inner core that exhibit a phase transition temperature above 37°C, rendering them suitable for biomedical applications (Lo, 2005). Another example is the pH-sensitive carrier micelles that self-assemble from pH-responsive poly(ethylene glycol) methyl ether-b-(poly lactic acid-co-poly(β-aminoesters)) (MPEG-b-(PLA-co-PAE) (Zhang et al., 2014) (Figure 9.9).

Nanoscale hydrogels, or "nanogels", are hydrophilic, three-dimensional cross-linked polymer networks that swell in the presence of water. They can be designed to respond to many physiological stimuli, including ionic strength, pH, and temperature. Nanogels offer a straightforward synthesis and relatively high drug loading capacity (Vinogradov, 2002). Nanogel particles combine the properties of gels with the properties of colloids: high surface to volume ratio, microheterogeneous structure, and small size (Figure 9.10). Hybrid polymerized nanogels that incorporate both physical and chemical crosslinking structures have been synthesized (Morimoto et al., 2005).

In addition to their high biocompatibility and tunable properties, hydrogel matrices have advantages for use in drug delivery because of their ability to prevent payload aggregation (Morimoto et al., 2005). Perhaps the greatest advantage of nanohydrogels as drug carriers is the possibility of their synthesis in the absence of drugs, which decreases the potential of drug inactivation (Shin et al., 2001). Both charged and hydrophobic drug molecules can be subsequently incorporated into hydrogel networks via self-assembly processes based on non-covalent interactions (Vinogradov et al., 2005).

Polymeric nanogels display extended stability, sustained release, low cytotoxicity, and protection from enzymatic degradation (Vinogradov et al., 2005). However, their widespread use in nanomedicine has been limited by the lack of a general methodology for hydrogel surface coating (De Geest et al., 2006). While hydrogels

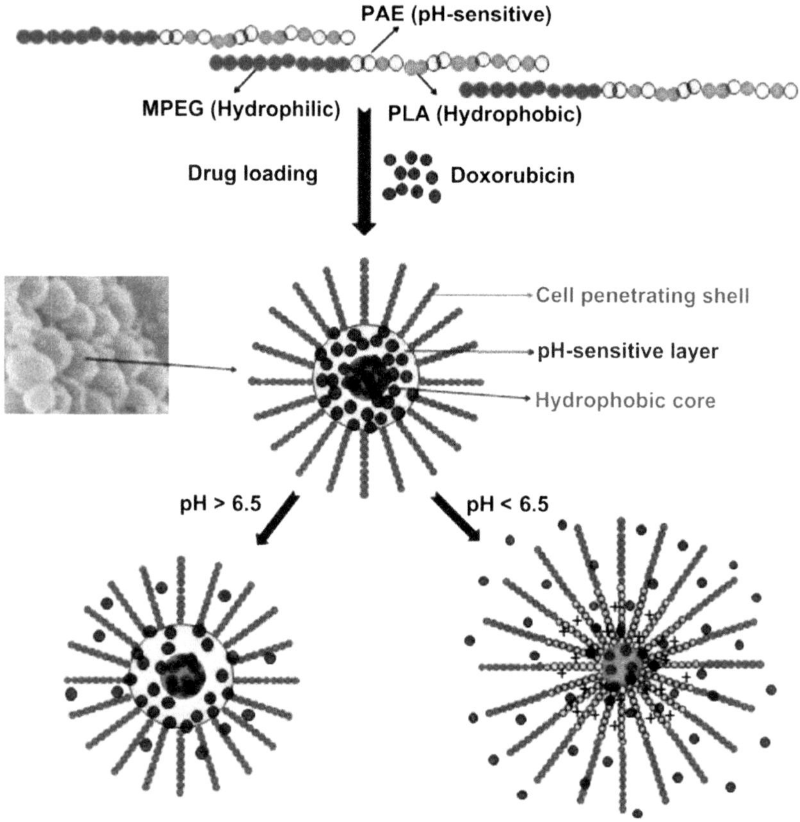

FIGURE 9.9 Polymeric micelles formed by self-assembly of amphiphilic block copolymers in aqueous medium and pH-dependent drug release from MPEG-b-(PLA-co-PAE) micelles (Zhang et al., 2014).

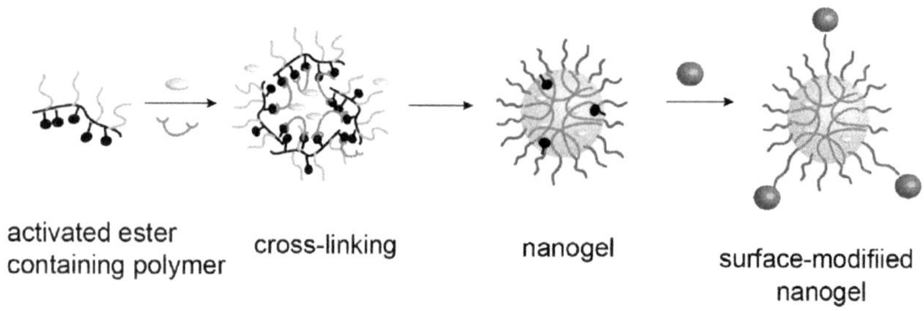

FIGURE 9.10 Showing surface modified nanogel assembly (Zhuang et al., 2012).

can be coated with lipids to preserve colloidal stability, these procedures are limited by small coating efficiency (Kraft and Moore, 2001).

It was recently shown that well-defined, stable, crosslinked, amphiphilic hydrogel nanoparticles can be prepared by inverse emulsion photo-polymerization (Missirlis et al., 2005). The size of the colloidal material, which loaded with hydrophobic drugs, can be tuned by varying polymerization conditions. Thus, the hydrophobic core solubilizes lipophilic molecules, and the hydrophilic corona prevents aggregation, protein adsorption, and immunogenic response (Missirlis et al., 2005). This platform, agreeable to many amphiphilic precursors, may give rise to a new class of colloidal nanoscale DDS.

9.2.2 Inorganic Nanoparticles

9.2.2.1 Silica Nanomaterials (Xerogels and Mesoporous Silica Nanoparticles (MSNs))

Among inorganic nanoparticles, silica materials are the carriers, which most often are chosen for biological purposes. Silica materials used in controlled DDS are classified as xerogels (Czarnobaj, 2008) and mesoporous silica nanoparticles (MSNs) (Wei et al., 2010). They exhibit several advantages as carrier systems, including biocompatibility, highly porous framework, and ease regarding functionalization (Amato, 2010). Silica xerogels possess an amorphous structure with high porosity and enormous surface area. The shape and pore dimensions depend on synthesis parameters, such as the ratio of reagents, temperature, the concentration of the catalyst, and the pressure of drying, which allow the altering of the properties of xerogels used in controlled drug release (Echeverría et al., 2010). The sol-gel technique is most frequently used to form silica xerogels loaded with drugs. Phenytoin, doxorubicin, cisplatin, metronidazole, nifedipine, diclofenac, and heparin are examples of drugs which have been loaded into xerogels using this technique.

In comparison with xerogels, the MSNs possess a more homogenous structure, lower polydispersity, and higher surface area for the adsorption of therapeutic or diagnostic agents (Di Pasqua et al., 2009). The mechanism of drug loading into mesoporous silica material is chemical or physical adsorption (Di Pasqua et al., 2009), see (Figure 9.11). By these processes, diverse types of drugs including anticancer drugs (He et al., 2011), antibiotics (Li et al., 2010), and heart disease drugs (Popovici et al., 2011) have been embedded into MSNs. The drug release is usually controlled by diffusion (Li et al., 2010). The structure of MSNs enables the incorporation of both small and large molecules, adsorption of DNA, and gene transfer. This gives the possibility of using these nanomaterials in a combined therapy.

Some data indicate that silica nanoparticles (SNPs) are biocompatible and have considerable potential for diagnostic and therapeutic applications. However, recent studies have revealed in-vitro and in-vivo toxicity and particular hazards of using nanosilica. SNPs have an impact on the generation of oxidative stress in cells via the formation of reactive oxygen species (Liu and Sun, 2010), elevated production of malondialdehyde (Lin et al., 2006), decreasing glutathione level (Yu et al., 2009), and the induction of antioxidant enzymes (Park and Park, 2009). All the later events are responsible for lipid peroxidation and cell membrane damage (Lin et al., 2006).

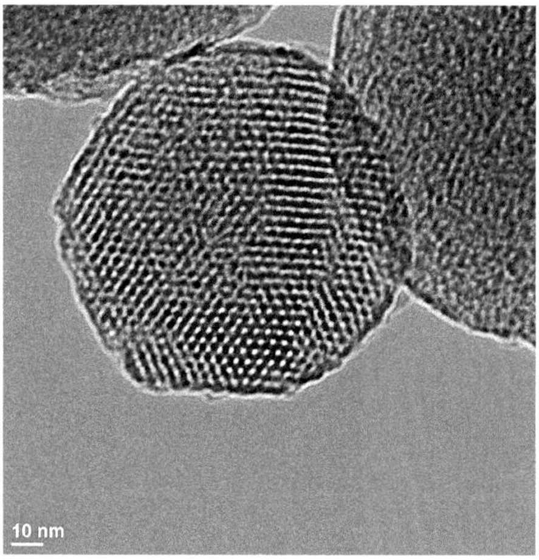

FIGURE 9.11 TEM image of a mesoporous silica nanoparticle (Lin, 2008).

9.2.2.2 Carbon Nanomaterials

Carbon nanocarriers used in DDS are differentiated into carbon nanotubes (CNTs) and nanohorns (CNH). CNTs are characterized by a unique architecture formed by the rolling of single-walled carbon nanotubes (SWCNTs) or multi-walled carbon nanotubes (MWCNTs) layers of graphite with an enormous surface area and an excellent electronic and thermal conductivity (Beg et al., 2011). The biocompatibility of nanotubes may be improved by chemical modification of their surface. Such adjustments can be implemented by covalent anchoring of PAMAM dendrimers (Zhang et al., 2010), amphiphilic diblock copolymers (Di Crescenzo et al., 2011), or PEG layers (Bhirde et al., 2010) onto the CNTs' surfaces, or dispersion within a hyaluronic acid matrix (Shin et al., 2011). Owing to the high mechanical strength of SWCNTs, they are used as supports for improving the properties of other carriers, e.g. polymeric or non-polymeric composites (Shin et al., 2011).

There are three ways of drug immobilization on carbon nanocarriers. This can be achieved by encapsulation of a drug in the carbon nanotube (Arsawang et al., 2011), chemical adsorption on the surface or in the spaces between the nanotubes (by electrostatic, hydrophobic and p–p interactions) (Chen et al., 2011), and attachment of active agents to functionalized carbon nanotubes (f-CNTs). Of these, encapsulation has the advantage over the two other methods since the drug is protected from degradation during transportation to the cells and is released only under specific conditions (Perry et al., 2011). Examples of drugs that were attached to CNTs are listed in Table 9.2. The drug release from CNTs can be either electrically or chemically controlled. To prevent the unwanted release of the drug, the open ends of CNTs were sealed with polypyrrole (PPy) films (Luo et al., 2011).

On the other hand, nanohorns are single-walled nanotubes that exhibit similar properties to nanotubes. However, their formation process does not require a metal catalyst. Therefore, they can be easily prepared with little cost and high purity (Shiba et al., 2006). The immobilization of drugs may rely on adsorption on the nanohorns'

TABLE 9.2
Applications of Liposomes

Applications	Indication	Drug Formulation	Company	Status	Reference
		Liposomes in Clinical Trials			
ThermoDox™	Primary hepatocellular carcinoma, Refractory chest wall breast cancer	liposomal doxorubicin	Celsion	Phase III	Hare et al. (2017)
MBP-426	Gastrointestinal adenocarcinoma	Oxaliplatin	Mebiopharm	Phase II	
ThermoDox	Breast cancer, primary liver cancer	Thermosensitive liposomal doxorubicin	Celsion	Phase II/III	
LEP-ETU	Breast cancer	Paclitaxel	Insys	Phase II	

walls (Murakami et al., 2008) or the nanoprecipitation of drugs with nanohorns (Ajima et al., 2008). The incorporation of cisplatin into nanohorns showed more efficient nanoprecipitation, which has a 3-fold increase over adsorption in the number of molecules entrapped in nanohorns (Ajima et al., 2008).

The toxic potential of CNTs results from their unique geometric structure (length to diameter ratio) the toxicity of the sole graphite material (Jia et al., 2005), and residual metal impurities left by the catalysts used in their preparation. In addition, the presence of amorphous carbon contributes to increased levels of reactive oxygen species (ROS) that induce oxidative stress in cells (Dobrovolskaia and McNeil, 2007). Unmodified MWCNTs cause a pro-inflammatory response in keratinocyte cell lines and lead to the formation of cytoplasmic vacuoles (Monteiro-Riviere et al., 2005).

9.2.2.3 Magnetic Nanoparticles (MNPs)

Magnetic drug targeting is characterized by the conjugation of a magnetic material which can accumulate in target tissue areas under the action of an external magnetic field (Arvizo et al., 2011). Depending on their magnetic properties, MNPs involve pure metals, such as cobalt, nickel, manganese, and iron, in addition to their alloys and oxides. However, for biomedical applications, the choice of MNPs is significantly reduced due to the lack of knowledge about the adverse effects of these nanomaterials on the human body. Iron oxides (magnetite and maghemite) occur naturally in the human heart, spleen, and liver (Grassi-Schultheiss et al., 1997), which indicate their biocompatibility and non-toxicity at a physiological concentration. Due to the desirable features of iron oxide NPs, they are the only type of MNPs that approved for clinical use by FDA. However, MNPs alone are not suited for use as drug vehicles because of limitations in the controlled release. Therefore, for improved therapeutic activity, MNPs can be chemically modified by coating their cores with various types of shell, such as golden (Tamer et al., 2010), polymeric (Chomoucka et al., 2010), silane (Chang et al., 2008), or dendrimeric (Pan et al., 2005), see (Figure 9.12).

MNPs have been examined for multitasking treatment, where they simultaneously act as diagnostic biosensors and drug carriers. Connecting a drug with an MNP can be achieved by covalent binding (Yallapu et al., 2011), electrostatic interactions (Figuerola et al., 2010), adsorption (Yallapu et al., 2011), or encapsulation processes (Wu et al., 2010). The targeting of drug–MNP conjugates to diseased tissues is dependent on their size and surface chemistry and can be carried out by a passive or active mechanism. The influence of MNPs on cellular function and toxicity has been investigated and reported. Depending on the method of their administration, MNPs may interact with extracellular components, cell membranes of macrophages, endothelial cells, skin epithelium, and respiratory or gastrointestinal tracts. After inhalation, MNPs accumulate in the brain, which indicates their ability to cross the blood-brain barrier (BBB) (Mahmoudi, 2010). MNPs coated with long-chain polymers are less cell-toxic than MNPs coated with shorter polymer chains of the same type (Hafeli et al., 2009).

However, in most cases where magnetic nanocarriers are used, difficulties in achieving their objectives appeared. This is most likely associated with inappropriate features of magnetic nanoparticles, for instance, they tend to aggregate into larger clusters, losing the specific properties connected to their nanoscale dimensions and making physical handling difficult. In turn, the magnetic force of an external magnet may not be strong enough to overcome the force of blood flow and to accumulate magnetic drugs only at the target site (Neuberger et al., 2005). Therefore, the design of magnetic DDS requires the consideration of many factors, e.g. magnetic properties, size of particles, the strength of the magnetic field, drug loading capacity, the place of accessibility of target tissue, and the rate of blood flow (Cao et al., 2011).

9.2.2.4 Metal Nanoparticles and Quantum Dots

Inorganic nanoparticles emerged a few years ago as drug and gene delivery systems, imaging agents, and diagnostic biosensors (Mody et al., 2010). As mentioned earlier, the

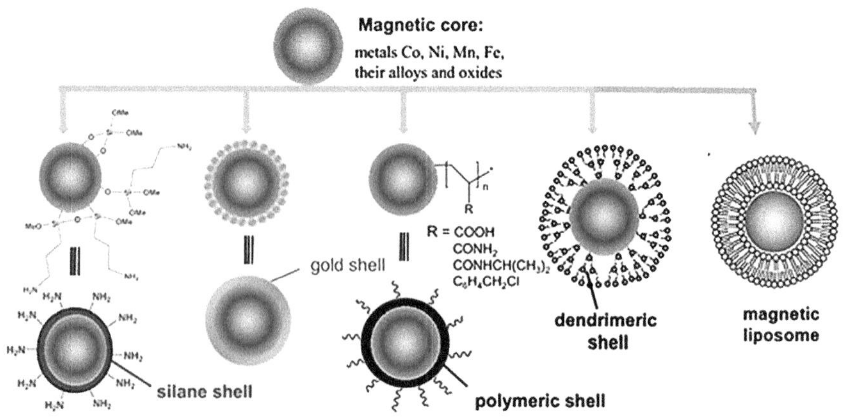

FIGURE 9.12 Magnetic nanoparticles with various shells (Wilczewska et al., 2012).

conjugation of a magnetic material due to a magnetic field is how magnetic drug targeting systems are classified (Arias, 2008). However, magnetic particles alone are not suitable for drug delivery because of limitations in the controlled release. However, a mixed composition of a magnetic nucleus and a polymeric shell could take advantage of the two components (Arias, 2008). Furthermore, titanium dioxide and zinc oxide demonstrate the potential of nanoparticles to improve therapeutic/prevention performance, being particularly useful as sunscreen agents (Jain and Jain, 2010). Due to the presence of surface plasmon, gold, copper, and silver nanoparticles strongly absorb light in the visible spectrum, making it possible to study their size-dependent light absorption through surface plasmon resonance (SPR). Gold nanoparticles (GNPs) and nanorods have many unique properties, which have been explored for potential applications in biomolecular detection (Kim et al., 2006). Regarding biocompatibility and non-cytotoxicity, GNPs, as approved by the FDA, have distinct advantages over other metallic nanoparticles and could also be utilized as a favourable carrier for delivery of drugs (Dhar et al., 2008). These nanoparticles have shown a selective transportation of drugs to cancer cell nucleus especially when conjugated with amino acid (Selvakannan et al., 2004) and proteins (Niemeyer, 2003), arginine-glycine-aspartic acid peptide (RGD) and PEG (Chen, 2008). When reaching the tumour cells, they can induce hyperthermia using non-invasive radiofrequency. These functionalized nanoparticles are promising candidates for drug delivery as biomarkers of drug-resistant cancer cells (Hwu, 2009).

QDs are small-sized semiconductor nanocrystals (1–10 nm) composed of an inorganic elemental core (e.g. Cd and Se) surrounded by a metallic shell (ZnS), see (Figure 9.13). They are widely used as drug carriers or simply as fluorescent labels for other drug carriers (Michalet, 2005). These nanoparticles are of extreme importance for diagnosis. Fluorescent-labeled QDs are more stable in the body and increase the detection sensitivity and photostability. In the same platforms, a high number of probe molecules can be incorporated, in opposite to a single conventional molecule. QDs can have better properties than organic fluorescence probes. These include high resistance to degradation and photobleaching, high quantum yields, high molar extinction coefficients, continuous absorption spectra covering from UV to near-infrared, long fluorescence lifetimes (>10 ns), narrow emission spectra, and very long effective Stokes shifts (Cai, 2007). QDs have been used for numerous applications, from cell tracking to mapping to simultaneously identifying several ligands, using multiple colours and intensities to detect different structures (Ballou, 2007).

9.2.3 Vesicular Systems

9.2.3.1 Liposomes, Transferosomes, Ethosomes, Niosomes, Virosomes, Cochleate, and Cubosomes

Liposomes are spherical vesicles composed of phospholipid and steroid (e.g. cholesterol) bilayers or other surfactants. They are formed spontaneously when certain lipids are dispersed in aqueous media and sonicated. According to their size and lamellarity, they divide into small unilamellar vesicles (SUV, from 20 nm to 100 nm), large unilamellar vesicles (LUV, from 100 to 500 nm) and multilamellar vesicles (MVL, exceeding 500 nm) (Korting, 2010) (Figure 9.14). These systems have an

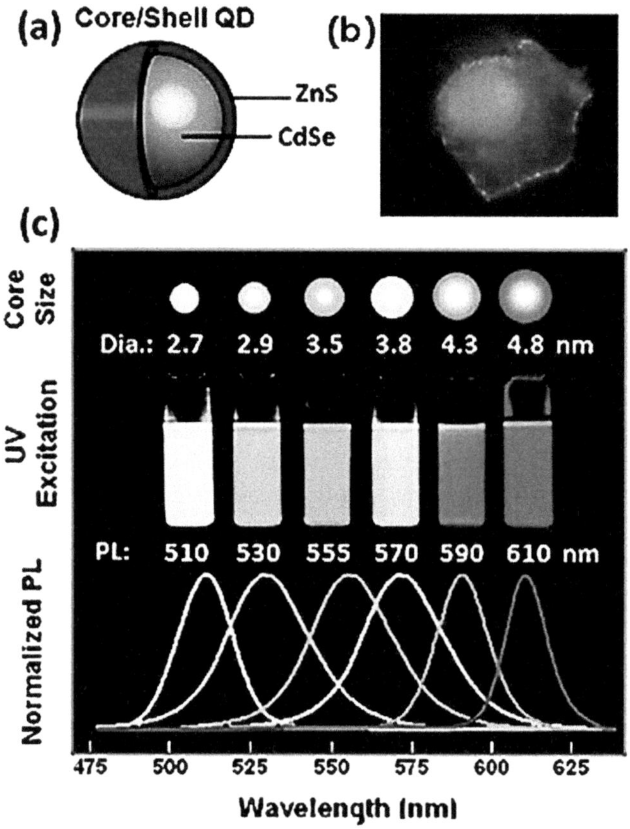

FIGURE 9.13 (a) Schematic drawing representing of CdSe–ZnS core-shell QD, (b) cell imaging using QDs and (c) Illustrative size and photoluminescence spectrum showing a progressive color change of CdSe–ZnS QDs with increasing diameter (Algar et al., 2011).

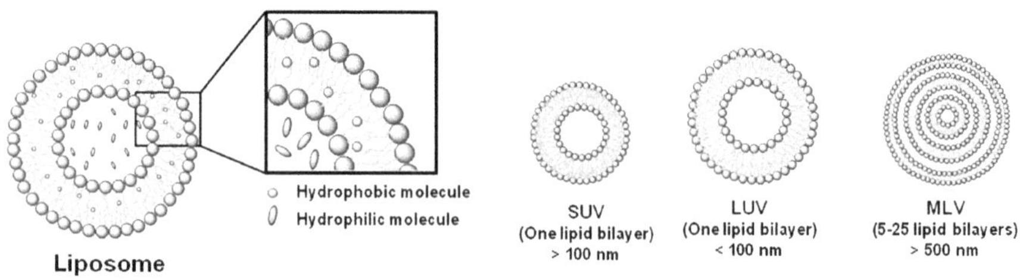

FIGURE 9.14 Liposomes self-assembled bilayer structure and their classification according to size and lamellarity (Conniot et al., 2014).

ability to encapsulate both lipophilic drugs within their membrane, and hydrophilic drugs inside or outside the aqueous core and the membrane of these carriers can be altered and tuned (Lattin, 2011).

Due to their biocompatibility, biodegradability, and ability to cross lipid bilayers and cell membranes, liposomes have been proposed as delivery platforms for vaccines, anticancer drugs, and gene therapy, see Table 9.2. Liposomes have been reported to increase drug solubility, improve drug pharmacokinetic properties such as the therapeutic index, rapid metabolism, and cause a reduction of harmful side effects and an increase of in-vitro and in-vivo anticancer activity (Dos Santos Giuberti, 2011). A drug is incorporated into liposomes by the encapsulation process. The release of a drug from the liposomes depends on the liposome's composition, pH, osmotic gradient, and the surrounding environment (Dos Santos Giuberti, 2011). A prolonged residence time will also increase the duration of their action but decreases their number. Interactions of liposomes with cells can be due to adsorption, fusion, endocytosis, and lipid transfer.

However, weak chemical and physical protection of sensitive drugs, aggregation into large particles, hydrolysis with the formation of oxidation products, difficulties in industrial-scale production, and stability problems during storage have also been reported (Korting, 2010). As a result, unusual types of modified liposomes, such as ethosomes and transferosomes, which overcome limitations commonly found in liposomes have increased flexibility due to the addition of ethanol and surfactants, respectively (Figure 9.15). Niosomes are non-ionic surfactant vesicles made up from polyoxyethylene alkyl ethers, polyoxyethylene alkyl esters or saccharose diesters (Korting, 2010). These systems are specially designed for skin delivery (ethanol is a known permeability enhancer) due to their facilitated fusion and malleability. The other types of liposomes are classified as virosomes, which are liposomes that carry viral proteins on their surface. This strategy has been proposed for immunization (Cusi, 2006). Furthermore, cochleates are stable particles (moreso than other lipidic structures) derived from liposomes composed mainly of charged phosphatidylserine in the presence of divalent counter ion such as Ca^{++} which forms a continuous large lipid bilayer sheet with no internal aqueous space (Sesana, 2011). Cochleate delivery has shown potential use for amphotericin B, factor VIII delivery, proteins, peptides, and DNA (Sesana, 2011). Finally, there are cubosomes, which are similar to cochleates, but are considered as novel lipid delivery systems due to their continuous multilayers of lipid bilayer structure. They have self-assembly cubic-like appearance, are biocompatible, and show bioadhesive properties ideal for oral administration (Wu, 2011).

9.2.3.2 Nanoparticles Based on Solid Lipids (SLN)

Solid lipid nanoparticles (SLN), nanostructured lipid carriers (NLC), and lipid drug conjugates (LDC) are types of nanocarrier systems made of a solid lipid matrix at room and body temperature, such as glycerol behenate, glycerol palmitostearate, lecithin, triglycerides, and tristearin glyceride (Fricker, 2010). Contrary to liposomes, SLNs have shown more stability for an extended period, protect labile compounds from chemical degradation and can be processed up to large-scale production. However, they still present problems related to their loading efficiency due

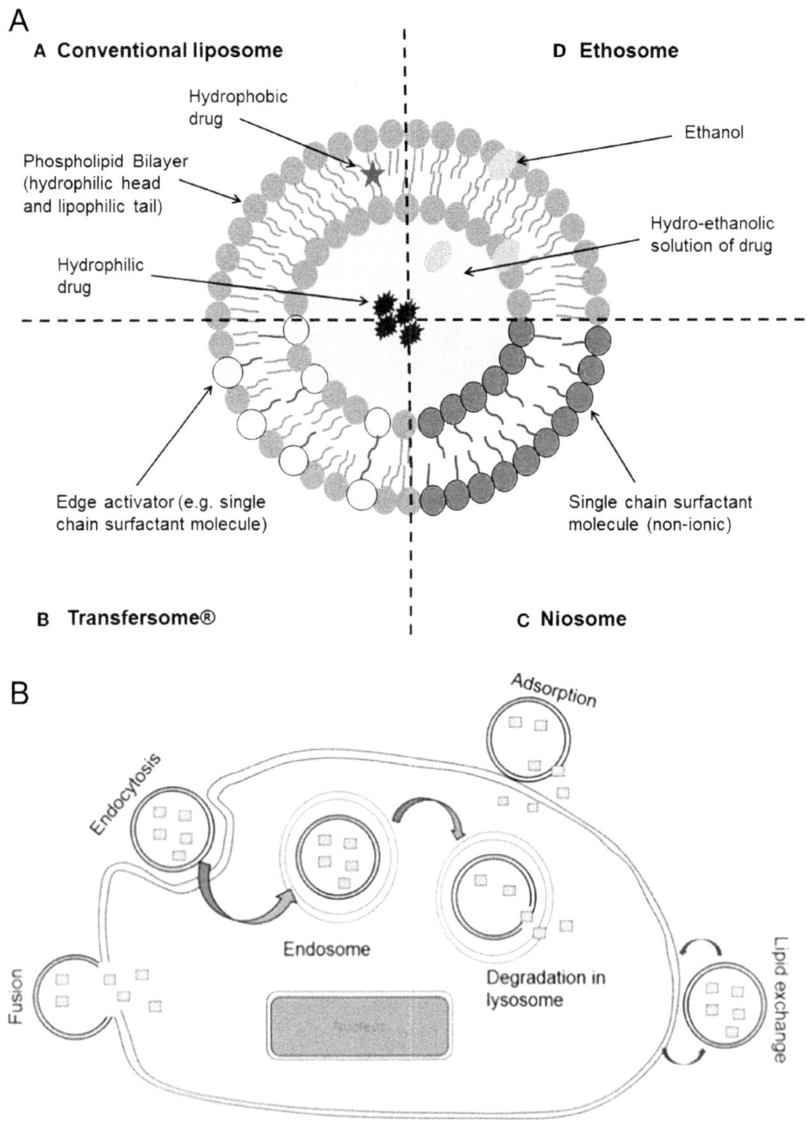

FIGURE 9.15 Schematic representation of the different types of lipid-based vesicular delivery systems. (a) Conventional liposomes, (b) transfersomes, (c) niosomes, (d) ethosomes, and (e) liposome joins cell and releases drug (Hua, 2013; Chen et al., 2018).

to the formation of a lipid crystal matrix and possible changes in the physical state of the lipids (Fricker, 2010). To overcome those limitations, novel structures produced by mixing solid lipids with liquid lipids at room temperature (semi-liquid formulations) named NLCs were produced (Korting, 2010). These systems show high encapsulation efficiency and loading capacity due to the formation of less ordered lipid matrix, and they show long-term stability with a controlled release and without burst effect. These colloidal carriers have emerged as a potential alternative to other recent colloidal systems like PNPs (Fricker, 2010). LDCs were developed to expand the applicability of lipid-based carriers to lipophobic drug molecules. These insoluble drug–lipid conjugates can be prepared by salt formation or by covalent linking, followed by homogenization (Müller and Kick, 2004).

9.2.4 Cyclodextrins

Cyclodextrins are cyclic oligosaccharides containing at least 6 D-(+)-glucopyranose units attached by α-1,4-linkage. Three types of cyclodextrins are found in nature, named α (6 units), β (7 units) and γ-cyclodextrins (8 units), (Figure 9.16). β-Cyclodextrin as a nanocontainer is ideal for drug delivery due to the cavity size, efficiency of drug complexation and loading, availability, and relatively low cost (Karande and Mitragotri, 2009). They can prevent drug degradation and improve drug stability and solubility, resulting in a higher bioavailability (Wu and Senter, 2005). An example of cyclodextrin application in a drug delivery system is 2-hydroxypropyl-beta-cyclodextrin (HPβCD), which is a powerful solubilizer with a hydrophilic outside and a hydrophobic inside (Manosroi, 2005). For absorption in the GI tract, the complexes must make contact with the surface, thus promoting dissociation and drug permeation across the membrane (Karande and Mitragotri, 2009). Moreover, cyclodextrins can work synergistically as penetration enhancers to improve their absorption across the skin.

9.2.5 Nanoemulsions

Nanoemulsions are isotropic mixtures of oil/water stabilized by surfactants in combination with co-surfactants (Karande and Mitragotri, 2009). They show high solubilization, dissolution properties, thermodynamic stability, and stabilizers that prevent particle agglomeration and drug leakage. Thus, they have improved permeation enhancement ideal for transdermal delivery as they act in synergy (Karande and Mitragotri, 2009). Nanoemulsions may work by enhanced disruption of skin–lipid structure or by improving the stability of the drug in the formulation.

9.2.6 Immunoconjugates

Antibody–drug conjugates or immunoconjugates are recombinant antibodies covalently bound through a linker to a drug. The idea behind this technology is to target potent drugs to the specific site by using the specificity of monoclonal antibodies (mAb), thus avoiding non-targeted organ toxicity (Wu and Senter, 2005). These immunoconjugates can be utilized across a broad spectrum of diseases by selecting the appropriate molecular domains. However, initial works showed some limitations such as short half-lives, immunogenicity, or even lack of efficient interaction (Nelson, 2010). To avoid this limitation, strategies such as PEGylation, conjugation with proteins such as albumin, or the use of fully human mAbs has been envisioned (Nelson, 2010). Mylotarg (gemtuzumab ozogamicin) was the first approved immunoconjugate and is utilized for the treatment of acute myeloid leukemia (Wu and

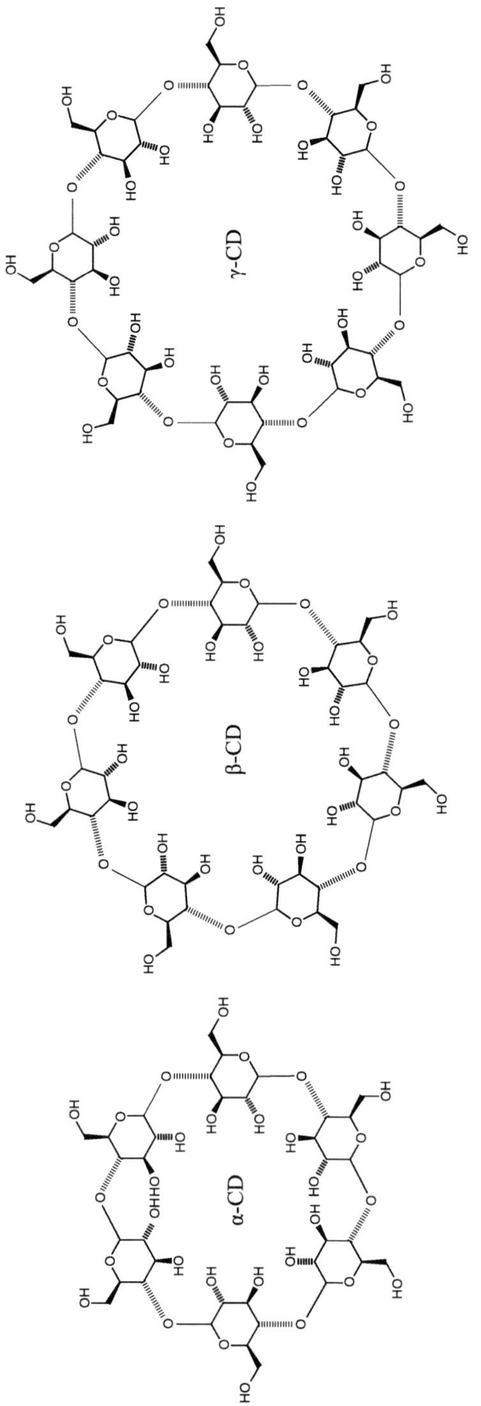

FIGURE 9.16 Three of the most common types of cyclodextrins (CDs): α-, β-, and γ-CDs (with 6, 7 or 8 glucose units respectively) and formation of the inclusion complex between drug molecules and CD (Skowron, 2006).

Senter, 2005). Several other immunoconjugates are in phase 3 clinical trials, such as Naptumomab estafenatox, for the treatment of the advanced renal disease, or Brentuximab vedotin, for the treatment of Hodgkin lymphoma (Reichert, 2011). Nevertheless, immuno-nanoparticles and immuno-liposomes are new strategies that have been developed to use antibodies attached on nanoparticles and liposomes, respectively (Olivier, 2002). These systems can be applied to encapsulate multiple drugs to protect them from the environment and exert a controlled release. Moreover, they are utilized for treating hard-to-target tissues, such as the BBB, by targeting transferrin, insulin, or glutathione receptors, triggering their activation and consequent internalization (Olivier, 2005).

9.2.7 Viruses

Viruses are considered one of the potential vehicles for drug and gene therapies due to their natural ability to infect specific cells and transport genomic material to the nucleus. Using a recombinant virus can improve transfection efficiency by enhancing drug delivery and avoiding degradation by lysosomes (Eliyahu, 2005). Various viruses have been tested, and the most commonly used are retroviruses, lentiviruses, and adenoviruses (Taratula, 2009). However, the use of viruses raises concerns related to their safety because of the risks of insertion mistakes, the activation of proto-oncogenes, viral replication, and strong immune responses (Blau, 1995). Moreover, retroviruses have size loading limitations as they can only infect dividing cells. Therefore, they are mostly used for ex-vivo delivery. Lentiviruses, on the other hand, can deliver a gene into non-dividing cells as well as adenoviruses (the virus remains extrachromosomal which reduces the chances of disrupting the cellular genome) (Blau, 1995). These systems are most likely to be applied in cytotoxic gene therapy (Thaci, 2011). In contrast to these, the nonviral vectors such as virosomes and nanoparticles have rapidly increased due to their weak immune response and ease of synthesis (Cusi, 2006).

9.2.8 Nucleic Acids

Gene therapy and RNA interference (RNAi) are two therapies of substantial interest: the former (gain-of-function) eliminates the need for subsequent drug administration, while the latter (loss-of-function) confers highly specific gene silencing. Transcriptional targeting (i.e. incorporation of genetic regulatory elements and transductional targeting (i.e. conjugation of targeting ligands) can control gene expression.

As well as being much safer and less expensive than viruses, non-viral vectors can be synthesized easily and reproduced in large quantities and afford much greater nucleic acid-carrying capacity (Dincer, 2005). To achieve non-viral delivery of nucleic acids, polycationic lipids or polycationic polymers to form nanocomplexes typically condense DNA and RNA.

To generate protein expression, DNA payloads must first enter the cell, escape the endosome, dissociate from their carrier, cross the nuclear membrane, and finally be transcribed by host machinery (Luo, 2000). Extracellular interactions between the complexes and anionic glycosaminoglycans (GAGs) located on the cell surface inhibit gene transfer by nanoscale DDS (Ruponen, 2003). If the target cell takes up the compound, it should promote endosomolysis via a pH-responsive functionality in the carrier backbone, such as a "proton sponge" (Remy, 1998).

Poly(ethylene imine) (PEI) is the most widely characterized polymeric vector for nucleic acid delivery. Linear polymers with a molecular mass of 22 kDa overcome the nuclear barrier most effectively resulting in the highest transfection rates (Brunner, 2002); however, in vivo, the use of PEI is limited by its relatively high toxicity. Alternative classes of endosomolytic polymers have been synthesized, including amphoteric poly(amido amines) (Wightman et al., 2001), poly(acrylic acids) (Ferruti et al., 2002), poly(lysine imidazoles) (Stayton, 2000), and poly(β-amino esters) (Putnam, 2001). Other materials include PLGA (Akinc, 2003), poly(lysine) (Kumar, 2004), protamine (Lee, 2002), chitosan (Zhang, 2003), calcium phosphate (Ravi Kumar, 2004), cationized gelatin (Fu, 2005), and carbonate apatite (Kushibiki, 2005).

9.3 RECENT ADVANCES IN NANO DRUG DELIVERY SYSTEMS

9.3.1 Anisotropic Nanoparticles

Biocompatible polymeric NPs have been investigated for their use as delivery vehicles for therapeutic and diagnostic agents. Although polymeric nanoconstructs have traditionally been fabricated as isotropic spheres, anisotropic, nonspherical NPs have gained interest in the biomaterials community owing to their unique interactions with biological systems. PNPs with different forms of anisotropy have been manufactured using a variety of novel methods in recent years (Meyer, 2016) see (Figure 9.17). In addition, they have enhanced physical, chemical, and biological properties compared with spherical nanoparticles, including increased targeting avidity and decreased nonspecific in-vivo clearance. With these desirable properties, anisotropic nanoparticles have been successfully utilized in many biomedical settings and have performed superiorly to analogous spherical NPs.

Self-assembled anisotropic or Janus particles were designed to load anticancer drugs for lung cancer treatment by inhalation. The particles were synthesized using binary mixtures of biodegradable and biocompatible materials (Figure 9.18). The particles did not demonstrate cyto- and genotoxic effects. Janus particles were internalized by cancer cells and accumulated both in the cytoplasm and the nuclei (Garbuzenko et al., 2014).

9.3.2 Drug-Free Macromolecular Therapeutics

Conventional polymeric nanomedicines utilize polymers as delivery vehicles to modify the drugs' biodistribution and enhance their efficacy. Recent designs of nanomedicines have added another function to trigger and improve the therapeutic effects of drugs through natural biological responses. Drug-free macromolecular therapeutics is a new paradigm that avoids the use of low molecular weight drugs by employing biomimetic strategies to stimulate or control specific cellular activities (Te-Wei Chu, 2015). The basic idea is to induce apoptosis by the crosslinking of cell-surface

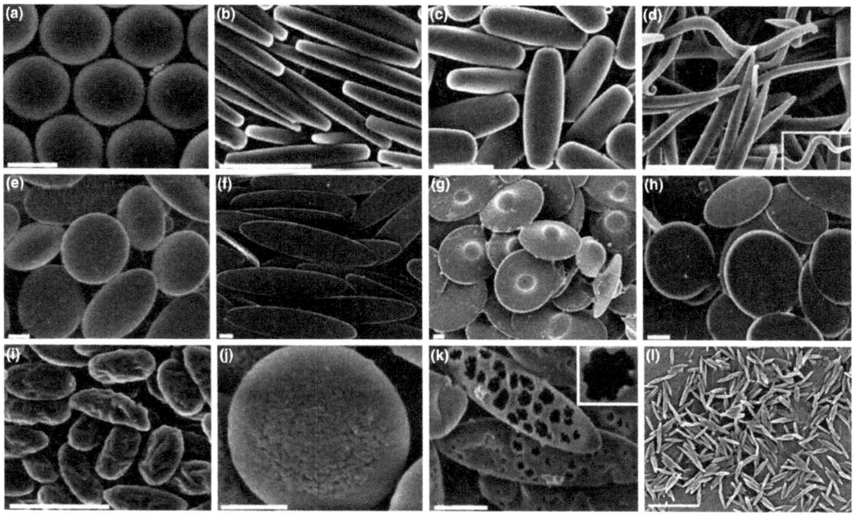

FIGURE 9.17 A wide repertoire of particle shapes can be produced with the thin film stretching method. (a) Spherical, (b) rectangular disk, (c) prolate ellipsoidal, (d) worm-like, (e) oblate ellipsoidal, (f) prolate ellipsoidal disk, (g) UFO-like, (h) flattened circular disk, (i) wrinkled prolate ellipsoidal, (j) wrinkled oblate ellipsoidal, and (k) porous prolate ellipsoidal particles. Scale bars are 2 µm. Adapted from (Meyer and Green, 2016).

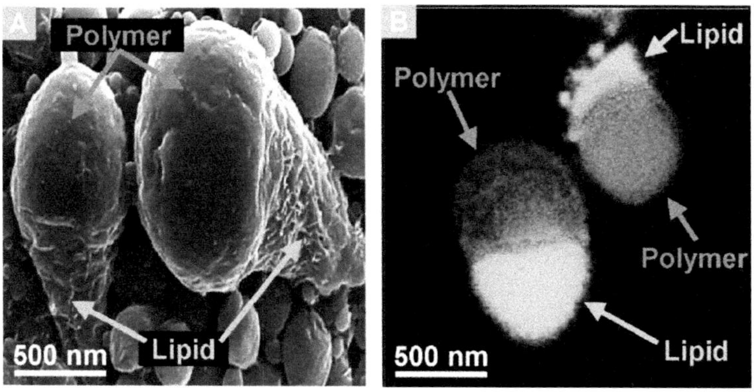

FIGURE 9.18 Anisotropic biodegradable biphasic polymer/lipid Janus nanoparticles. (a) Scanning electron and (b) fluorescence. Polymer/lipid combinations yielded "ice cream cone" shaped particles. Polymeric phase of nanoparticles was labeled with FITC (green fluorescence); and lipid phase was labeled with DiR (red fluorescence). Adapted from (Kuzmov and Minko, 2015).

internalized receptors mediated by the biorecognition of high-fidelity natural binding motifs (see Figure 9.19). The incorporation of native biorecognition motifs as grafts attached to synthetic polymer chains results in hybrid macromolecules that have the potential for self-assembly. For example, two N-(2-hydroxypropyl) methacrylamide (HPMA) copolymers containing grafts of complementary coiled-coil

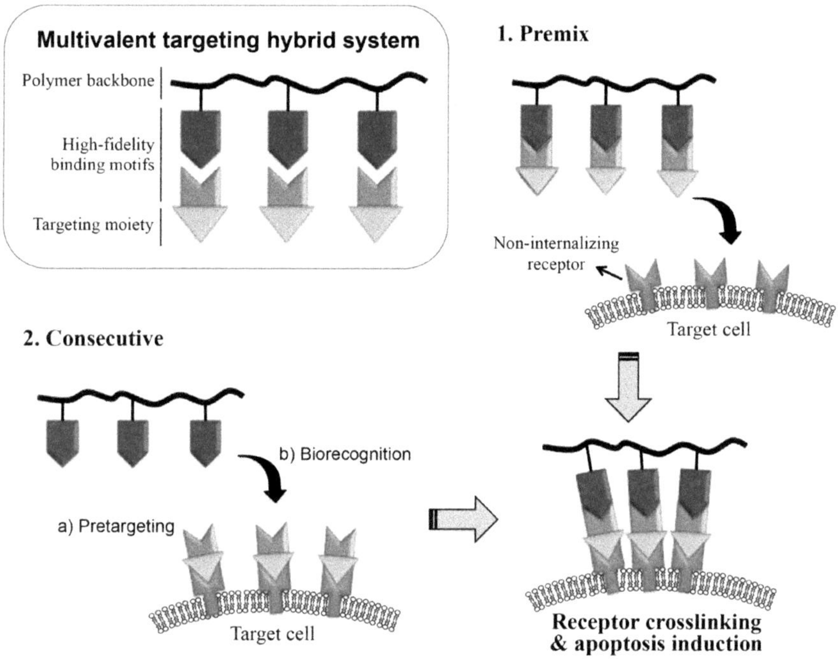

FIGURE 9.19 Drug-free macromolecular therapeutics for apoptosis induction. Crosslinking of cell surface non-internalizing receptors is mediated by the biorecognition of natural binding motifs. (Adapted from Chu and Kopeček, 2015.)

forming peptides (Yang, 2006) or HPMA copolymers grafted with peptide nucleic acids self-assemble into three-dimensional hydrogels (Chu, 2015).

9.3.3 Nanoparticle-Based Combination Therapy

Combined therapies by simultaneously encapsulated drugs are systems that have the potential to deliver more than one drug at once. This method is very useful, especially for complex diseases like cancers that involve multiple pathways and whose progression is marked by many successive mutations in cell lines. Therefore, the inhibition of a pathway by a single drug might not be sufficient to bring about tumour recession. In combination chemotherapy, the synergistic effect of two or more drugs targeting different disease pathways raises the chances of eliminating the cancer, see (Figure 9.5). For instance, PLGA nanoparticles were simultaneously loaded with vincristine sulfate and verapamil hydrochloride to deliver the effective chemotherapeutic agents while inhibiting the P–gP efflux system. This system allows for overcoming of the tumour's lack of sensitivity and increases the therapeutic index (Song, 2008). Another strategy is to have two different release rates of the two drugs to improve treatments. For example, paclitaxel and a C6-ceramide were encapsulated in a controlled blend polymer of PLGA-PbAE to overcome the cancer drug resistance mechanisms effectively (van Vlerken, 2008). Table 9.3 shows different combination DDS based on liposomes, dendrimers, PNPs, and micelles.

TABLE 9.3
Dug Delivery Systems Based on Liposomes, Dendrimers, Polymeric Nanoparticles, and Micelles

Formulation	Therapeutics	Indication	Status	Targeting
Liposome				
PEG-Liposome	Topotecan + Vincristine	Brain cancer	In-vivo	Passive
Liposome	Cytarabine + Daunorubicin	Acute myeloid leukemia	Phase II	Passive
Liposome	Irinotecan + Floxuridine	Colorectal cancer	Phase II	Passive
Mixture of two Liposomes	Irinotecan + Cisplatin	Small-cell lung cancer	In-vivo	Passive
Cationic, anionic PEG-Liposome	siRNA + Doxorubicin	MDR-breast cancer	In-vivo	Passive
Transferrin- (Tf) conjugated PEG-Liposome	Doxorubicin + Verapamil	MDR-leukemia	In-vitro	Active (Tf receptor)
Polymeric nanoparticles				
PLGA	Vincristine + Verapamil	Hepatocellular carcinoma	In-vitro	Passive
PACA	Doxorubicin + Cyclosporine A	Various cancers	In-vitro	Passive
Dendrimer				
GS PAMAM dendrimer	Antisense-miRNA21 + 5-FU	Glioblastoma	In-vitro	Active; miRNA
Dendritic PEG	Paclitaxel + alendronate	Cancer bone metastases	In-vivo	Active; Bone metastasis
Folate-GS poly-propyleneimine dendrimer with ethylenediamine core	Methotrexate + all-trans-retinoic acid	Leukemia	In-vitro	Active; folate receptor
Polymer-polymer micellar nanoparticles				
PEG-PLGA	Lonidamine + Paclitaxel	MDR breast cancer	In-vivo	Active; EGFR
Methoxy PEG- PLGA	Doxorubicin + Paclitaxel	Various cancers	In-vitro	Passive
PDMAEMA-PCL-PDMAEMA	Paclitaxel + siRNA	Prostate cancer	In-vitro	Active; VEGF
Polymer-Lipid micellar nanoparticles				
PEG- DSPE/PLGA	Combretastatin + Doxorubicin	Lung carcinoma	In-vivo	Passive
PEG-PLA and PEG- DSPE/TPGS	Paclitaxel + 17-AAG (HSP90 inhibitor)	Ovarian cancer	In-vitro	Active; HSP90

Source: Jun (2004).

9.4 DESIGNING NANOMATERIALS AS DRUG CARRIERS

Advances in nanomedicine arise through the improvement of novel nanocarriers and technology for drug delivery. An ideal nanocarrier should fit the subsequent profile: (1) biodegradable and biocompatible; (2) effectively homing the majority of therapeutics to the target site; (3) designed with optimal biophysicochemical properties for superior drug loading; (4) sustained drug release during infrequent administration times; (5) ideal circulation half-life; and (6) amenable to scale-up for cost effectiveness and commercialization. Incorporating and refining the previous qualities in one nanocarrier is very challenging and needs cutting-edge knowledge and interdisciplinary technologies from different fields of science, such as medicine, chemistry, engineering, and physics.

9.5 OBSTACLES AND CURRENT LIMITATIONS

Besides the complications that are facing the experimental design, fabrication, and scaling-up of NPs, there are other existing challenges confronting their regulation and approval for clinical use. Compliance with quality-control guidelines such as good laboratory quality practice (GLP), good manufacturing practice (GMP), as well as passing the three phases of FDA investigations must be done before bringing a new nanoformulation to market. Moreover, there are other types of potential problems posed by nanomedicine, such as regulatory issues due to classification difficulties and lack of scientific expertise. Environmental risks have been raised by the National Science Foundation and the Environmental Protection Agency concerning their potential impact on the environment and their adverse effects due to their small size, which can cause respiratory disorders and affect the health of individuals. There are also social issues due to high-cost production and concerns that patients from low-income groups would be deprived of these novel therapeutics. Ethical matters have been raised by bioethical researchers who believe that nanomedicine could be manipulated to harm the human body rather than healing it, besides their uses for terrorism purposes.

9.6 CONCLUSION

In conventional oral or intravenous drug delivery, the medicine is distributed indiscriminately throughout the body, with arbitrary concentrations reaching both the disease site and healthy tissue. Nanoparticle-based DDS offer revolutionary opportunities to develop highly efficient targeted therapeutics with improved bioavailability, biodistribution, blood circulation time, pharmacokinetics, and safety profiles. Nevertheless, nanoparticles are crucial for maintaining synergistic drug ratios in combinational therapy and offer the first possibility of delivering therapeutic agents such as nucleic acids and unstable proteins.

However, there is still much to be learned in this emerging field of nanomedicine. We have yet to develop a carrier that can efficiently deliver a payload intracellularly and intratumorally with clinically validated results. Extending circulation time closer to the time scale of red blood cells and the retention of NPs at the disease site rather than the reticuloendothelial organs remain significant challenges.

REFERENCES

Ai, J., Biazar, E., Montazeri, M., Majdi, A., Aminifard, S., Safari, M., and Akbari, H. R. (2011). Nanotoxicology and nanoparticle safety in biomedical designs. *International Journal of Nanomedicine*, 6, 1117–1127.

Ajima, K., Murakami, T., Mizoguchi, Y., Tsuchida, K., Ichihashi, T., Iijima, S., and Yudasaka, M. (2008). Enhancement of in vivo anticancer effects of cisplatin by incorporation inside singlewall carbon nanohorns. *ACS Nano*, 2(10), 2057–2064.

Akinc, A., Anderson, D. G., Lynn, D. M., and Langer, R. (2003). Synthesis of poly(β-amino ester)s optimized for highly effective gene delivery. *Bioconjugate Chemistry*, 14(5), 979–988. doi:10.1021/bc034067y

Algar, W. R., Susumu, K., Delehanty, J. B., and Medintz, I. L. (2011). Semiconductor quantum dots in bioanalysis: Crossing the valley of death. *Analytical Chemistry*, 83(23), 8826–8837.

Amato, G. (2010). Silica encapsulated efficient and stable si quantum dots with high biocompatibility. *Nanoscale Research Letters*, 5(7), 1156–1160.

Apriyani, M. G., Foe, K., and Manosroi, A. (2005). Enhancement of the release of azelaic acid through the synthetic membranes by inclusion complex formation with hydroxypropyl-beta-cyclodextrin. *International Journal of Pharmaceutics*, 293(1–2), 235–240.

Arias, J. L. (2008). Novel strategies to improve the anticancer action of 5-fluorouracil by using drug delivery systems. *Molecules*, 13(10), 2340–2369.

Arsawang, U., Saengsawang, O., Rungrotmongkol, T., Sornmee, P., Wittayanarakul, K., Remsungnen, T., and Hannongbua, S. (2011). How do carbon nanotubes serve as carriers for gemcitabine transport in a drug delivery system? *Journal of Molecular Graphics and Modelling*, 29(2), 591–596.

Arvizo, R. R., Miranda, O. R., Moyano, D. F., Wal-den, C. A., and Giri, K. (2011). Modulating pharmacokinetics, tumor uptake and biodistribution by engineered nanoparticles. *PLoS One*, 6(9), e24374.

Ballou, B., Lagerholm, B. C., Ernst, L. A., Bruchez, M. P., and Waggoner, A. S.(2004). Noninvasive imaging of quantum dots in mice. *Bioconjugate Chemistry*, 15(1), 79–86.

Bamrungsap, S., Zhao, Z., Chen, T., Wang, L., Li, C. Fu, T., and Tan, W. (2012). A focus on nanoparticles as a drug delivery system. *Nanomedicine*, 7(8), 1253–1271.

Beg, S., Rizwan, M., Sheikh, A. M., Hasnain, M. S., Anwer, K., and Kohli, K. (2011). Advancement in carbon nanotubes: basics, biomedical applications and toxicity. *Journal of Pharmacy and Pharmacology*, 63(2), 141–163.

Bertrand, N., Wu, J., Xu, X., Kamaly, N., and Farokhzad, O. C.(2014). Cancer nanotechnology: The impact of passive and active targeting in the era of modern cancer biology. *Advanced Drug Delivery Reviews*, 66, 2–25.

Bhirde, A. A., Patel, S., Sousa, A. A., Patel, V, Molinolo, A. A., Ji, Y., and Rusling, J. F. (2010). Distribution and clearance of PEG-single-walled carbon nanotube cancer drug delivery vehicles in mice. *Nanomedicine*, 5(10), 1535–1546.

Blau, H. M., and Springer, M. L. (1995). Gene therapy—a novel form of drug delivery. *The New England Journal of Medicine*, 333(18), 1204–1207.

Brunner, S., Furtbauer, E., Sauer, T., Kursa, M., and Wagner, E. (2002). Overcoming the nuclear: cell cycle independent nonviral gene transfer with linear polyethylenimine or electroporation, 5(1), 80–86.

Cai, W., Chen, K., Li, Z. B., Gambhir, S. S., and Chen, X. (2007). Dual-function probe for PET and near-infrared fluorescence imaging of tumor vasculature. *Journal of Nuclear Medicine*, 48(11), 1862–1870.

Caminade, A. M., Laurent R., and Majoral J. P.(2005). Characterization of dendrimers. *Advanced Drug Delivery Reviews*, 57(15), 2130–2146.

Cao, Q., Han, X., and Li, L.(2011). Enhancement of the efficiency of magnetic targeting for drug delivery: Development and evaluation of magnet system. *Journal of Magnetism and Magnetic Materials*, 323(15), 1919–1924.

Capretto, L., Carugo, D., Mazzitelli, S., Nastruzzi, C., and Zhang, X. (2013). Microfluidic and lab-on-a-chip preparation routes for organic nanoparticles and vesicular systems for nanomedicine applications. *Advanced Drug Delivery Reviews.* 65(11–12): 1496–1532.

Chang, J. H., Kang, K. H., Choi, J., and Jeong, Y. K. (2008). High efficiency protein separation with organosilane assembled silica coated magnetic nanoparticles. *Superlattices and Microstructures*, 44(4–5), 442–448.

Chen, H., Jin, Y., Sun, L., Li, X., Nan, K., Liu, H., and Wang, B. (2018). Recent developments in ophthalmic drug delivery systems for therapy of both anterior and posterior segment diseases. *Colloid and Interface Science Communications*, 24, 54–61. doi:10.1016/j.colcom.2018.03.008

Chen, Z, Pierre, D., He, H., Tan, S., Pham-Huy, C., Hong, H., and Huang, J. (2011). Adsorption behavior of epirubicin hydrochloride on carboxylated carbon nanotubes. *International Journal of Pharmaceutics*, 28(405), 153–161.

Chen, P. C., Mwakwari, S. C., and Oyelere, A. K. (2008). Gold nanoparticles: From nanomedicine to nanosensing. *Nanotechnology, Science and Applications*, 1, 45–66.

Chomoucka, J., Drbohlavova, J., Huska, D., Adam, V., Kizek, R., and Hubalem, J. (2010). Magnetic nanoparticles and targeted drug delivering. *Pharmaceutical Research*, 62(2), 144–149.

Chrastina, A., Massey, K. A., and Schnitzer, J. E. (2011). Overcoming in vivo barriers to targeted nanodelivery. *Wiley Interdisciplinary Reviews: Nanomedicine and Nanobiote chnology*, 3(4), 421–437.

Chu, T. W., and Kopeček, J. (2015). Drug-free macromolecular therapeutics: a new paradigm in polymeric nanomedicines. *Biomaterial Science*, 3(7), 908–922.

Chu, T. W., Feng, J., Yang, J., and Kopeček, J. (2015). Hybrid polymeric hydrogels via peptide nucleic acid (PNA)/DNA complexation. *Journal of Controlled Release*, 220(Pt. B), 608–616.

Cole, A. J., Yang, V. C., and David, A. E. (2011). Cancer theranostics: the rise of targeted magnetic nanoparticles. *Trends in Biotechnology*, 29(7), 323–332.

Conniot, J., Silva, J. M., Fernandes, J. G., Silva, L. C., Gaspar, R., Brocchini, S. and Barata, T. S. (2014). Cancer immunotherapy: nanodelivery approaches for immune cell targeting and tracking. *Frontiers in Chemistry*, 26 November 2014.

Cusi, M. G. (2006). Applications of influenza virosomes as a delivery system. *Human Vaccine*, 2(1), 1–7.

Czarnobaj, K. (2008). Preparation and characterization of silica xerogels as carriers for drugs. *Drug Delivery*, 15(8), 485–492.

Davis, M. E., Chen, Z. G., and Shin, D. M. (2008). Nanoparticle therapeutics: an emerging treatment modality for cancer. *Nature Reviews Drug Discovery*, 7(9), 771–782.

De Geest B. G., Stubbe, B. G., Jonas, A. M., Van Thienen, T., Hinrichs, W. L., Demeester, J., and De Smedt, S. C. (2006). Self-exploding lipid-coated microgels. *Biomacromolecules.* 7(1), 373–379.

des Rieux, A., Fievez, V., Garinot, M., Schneider, Y. J., and Préat V. (2006). Nanoparticles as potential oral delivery systems of proteins and vaccines: a mechanistic approach. *Journal Control Release*, 116(1), 1–27.

Dhar, S., Reddy, E. M., Shiras, A., Pokharkar, V., and Prasad, B. L. V.(2008). Natural gum reduced/stabilized gold nanoparticles for drug delivery formulations. *Chemistry European Journal*, 14(33), 10244–10250.

Di Crescenzo, A., Velluto, D., Hubbell, J. A., and Fontana, A..(2011)Biocompatible dispersions of carbon nanotubes: a potential tool for intracellular transport of anticancer drugs. *Nanoscale*, 3(3), 925–928.

Di Pasqua, A. J., Wallner, S., Kerwood, D. J., and Dabrowiak, J. C. (2009). Adsorption of the Pt(II) anticancer drug carboplatin by mesoporous silica. *Chemistry & Biodiversity*, 6(9), 1343–1349.

Dincer, S., Turk, M., and Piskin, E. (2005). Intelligent polymers as nonviral vectors. *Gene Therapy*, 12(1), S139–145.

Dobrovolskaia, M. A., and McNeil, S. E. (2007). Immunological properties of engineered nanomaterials. *Nature Nanotechnology*, 2(8), 469–478.

Dos Santos Giuberti, C., De Oliveira Reis, E. C., Ribeiro Rocha, T. G., Leite, E. A., Lacerda, R. G., Ramaldes, G. A. and De Oliveira, M. C. (2011). Study of the pilot production process of long-circulating and pH-sensitive liposomes containing cisplatin. *Journal of Liposome Research*, 21(1), 60–69.

Echeverría, J. C., Estella, J., Barbería, V, Musgo, J., and Garrido, J. J. (2010). Synthesis and characterization of ultramicroporous silica xerogels. *Journal of Non-Crystalline Solids*, 356(6–8), 378–382.

Eliyahu, H., Barenholz, Y., and Domb, A. J. (2005). Polymers for DNA Delivery. *Molecules*, 10(1), 34–64.

Farokhzad, O. C., and Langer, R. (2009). Impact of nanotechnology on drug delivery. *ACS Nano*, 3(1), 16–20.

Farokhzad, O. C., Cheng, J. J., Teply, B. A., Sherifi, I., Jon, S., Kantoff, P. W., and Langer, R. (2006). Targeted nanoparticle-aptamer bioconjugates for cancer chemotherapy in vivo. *Proceedings of the National Academy of Sciences of the United States of America*, 103(16), 6315–6320.

Ferruti, P., Marchisio, M. A., and Duncan, R. (2002). Poly(amido-amine)s: biomedical applications. *Macromolecular Rapid Communications*, 23(5–6), 332–355.

Fighting Cancer with Heat Triggered "Liposome Grenades". (n.d.). Retrieved from Key Compounding Pharmacy website: https://www.keycompounding.com/fighting-cancer-with-heat-triggered-liposome-grenades/.

Figuerola, A., Di Corato R., Manna L., and Pellegrino, T. (2010). From iron oxide nanoparticles towards advanced iron-based inorganic materials designed for biomedical applications. *Pharmacological Research*, 62(2), 126–143.

Fricker, G., Kromp, T., Wendel, A., Blume, A., and Zirkel, J. (2010). Phospholipids and lipid-based formulations in oral drug delivery. *Pharmaceutical Research*, 27(8), 1469–1486.

Fu, H., Hu, Y., McNelis, T., and Hollinger, J. O. (2005). A calcium phosphate-based gene delivery system. *Journal of Biomedical Materials Research*, 74, (1)40–48.

Garbuzenko, O. B., Winkler, J., Tomassone, M. S., and Minko, T. (2014). Biodegradable janus nanoparticles for local pulmonary delivery of hydrophilic and hydrophobic molecules to the lungs. *Langmuir: The ACS Journal of Surfaces and Colloids*, 30(43), 12941–12949.

Grassi-Schultheiss, P. P., Heller, F., and Dobson, J. (1997). Analysis of magnetic material in the human heart, spleen and liver. *Biometals*, 10(4), 351–355.

Hafeli, U. O., Riffle, J. S., Harris-Shekhawat, L., Carmichael-Baranauskas, A., Mark, F., Dailey, J. P., and Bardenstein, D. (2009). Cell uptake and in vitro toxicity of magnetic nanoparticles suitable for drug delivery. *Molecular Pharmacology*, 6(5), 1417–1428.

Hare, J. J., Lammers, T., Ashford, M. B., Puri, S., Storm, G., and Barry, S. T. (2017). Challenges and strategies in anti-cancer nanomedicine development: An industry perspective. *Advanced Drug Delivery Reviews*, 108, 25–38.

He, Q., Gao, Y., Zhang, L., Zhang, Z., Gao, F., Ji, X., and Shi, J. (2011). A pH-responsive mesoporous silica nanoparticles-based multi-drug delivery system for overcoming multi-drug resistance. *Biomaterials*, 32(30), 7711–7720.

Hu, C. M., Zhang L., , Aryal S. , Cheung C., Fang, R. H., and Zhang, L. (2011). Erythrocyte membrane-camouflaged polymeric nanoparticles as a biomimetic delivery platform. *Proceedings of the National Academy of Sciences of the United States of America*, 108(27), 10980–10985.

Hua, S. (2015). Lipid-based nano-delivery systems for skin delivery of drugs and bioactives. *Frontiers in Pharmacology*, 6, 219.

Hwu, J. R., Lin, Y. S., Josephrajan, T., Hsu, M. H., Cheng, F. Y., Yeh, C. S., and Shieh, D. B. (2009). Targeted paclitaxel by conjugation to iron oxide and gold nanoparticles. *Journal of the American Chemical Society*, 131(1), 66–68.

Jain, S. K., and Jain, N. K. (2010). Multiparticulate carriers for sun-screening agents. *International Journal of Cosmetic Science*, 32(2), 89–98.

Jia, G., Wang, H., Yan, L., Wang, X., Pei, R., Yan, T., and Guo, X. (2005). Cytotoxicity of carbon nanomaterials: single-wall nanotube, multi-wall nanotube, and fullerene. *Environmental Science & Technology*, 39(5), 1378–1383.

Kaminskas, L. M., Boyd, B. J., and Porter, C. J. (2011). Dendrimer pharmacokinetics: The effect of size, structure and surface characteristics on ADME properties. *Nano Medicine*, 6(6), 1063–1084.

Karande, P., and Mitragotri, S. (2009). Enhancement of trans-dermal drug delivery via synergistic action of chemicals. *Biochimica et Biophysica Acta*, 1788(11), 2362–2373.

Kharisov, B. I., Dias, H. V. R., Kharissova, O. V., Vázquez, A., Peña, Y. and Gómez, I. (2014). Solubilization, dispersion and stabilization of magnetic nanoparticles in water and non-aqueous solvents: recent trends. *RSC Advances*, 4(85), 45354–45381.

Kim, C. K., Kalluru, R. R., Singh, J. P., Fortner, A., Griffin, J., Darbha, G. K., and Ray, P. C. (2006). Gold-nanoparticle-based miniaturized laser-induced fluorescence probe for specific DNA hybridization detection: Studies on size-dependent optical properties. *Nanotechnology*, 17(13), 3085.

Kitchens, K. M., Foraker, A. B., Kolhatkar, R. B., Swaan, P. W., and Ghandehari, H. (2007). Endocytosis and interaction of poly(amidoamine) dendrimers with Caco-2 cells. *Pharmaceutical Research*, 24(11), 2138–2145.

Korting, H. C., and Schafer-Korting, M. (2010). Carriers in the topical treatment of skin disease. *Handbook of Experimental Pharmacology*, 197, 435–468.

Kraft, M. L., and Moore, J. S. (2001). Surfactant-induced lysis of lipid-modified microgels. *Journal of the American Chemical Society*, 123(51), 12921–12922.

Kumar, M. N., Mohapatra, S. S., Kong, X., Jena, P. K., Bakowsky, U., and Lehr, C. M. (2004). Cationic poly(lactide-co-glycolide) nanoparticles as efficient in vivo gene transfection agents. *Journal for Nanoscience and Nanotechnology*, 4(8), 990–994.

Kushibiki, T., and Tabata, Y. J. (2005). Preparation of poly(ethylene glycol)-introduced cationized gelatin as a non-viral gene carrier. *Journal of Biomaterials Science, Polymer Edition*, 16(11), 1447–1461.

Kuzmov, A., and Minko, T. (2015). Nanotechnology approaches for inhalation treatment of lung diseases. *Journal of Controlled Release*, 219, 500–518.

Lattin, J. R., Belnap, D. M., and Pitt, W. G. (2011). Formation of eliposomes as a drug delivery vehicle. *Colloids and Surfaces B: Biointerfaces*, 89, 93–100.

Lee, H., Jeong, J. H., and Park, T. G. (2002). PEG grafted polylysine with fusogenic peptide for gene delivery: high transfection efficiency with low cytotoxicity. *Journal of Controlled Release*, 79(1–3), 283–291.

Lee, J. H. and Nan, A. (2012). Combination drug delivery approaches in metastatic breast cancer. *Journal of Drug Delivery*, 2012, 915375.

Li, Z., Su, K., Cheng, B., and Deng, Y. (2010). Organically modified MCM-type material preparation and its usage in controlled amoxicillin delivery. *Journal of Colloid and Interface Science*, 342(2), 607–613.

Lim, T. Y., Poh, C. K., and Wang, W. (2009). Poly(lactic-co- glycolic acid) as a controlled release delivery device. *Journal of Materials Science: Materials in Medicine*, 20(8), 1669–1675.

Lim, Y. B., Kim, T., Lee, J. W., Kim, S. M., and Kim, H. J. (2002). Self-assembled ternary complex of cationic Den-drimer, Cucurbituril, and DNA: noncovalent strategy in developing a gene delivery carrier. *Bioconjugate Chemistry*, 13(6), 1181–1185.

Lin, V. (2008). *Mesopourus Silica Closeup*. [Online Image]. Retrieved from https://en.wikipedia.org/wiki/File:Mesopourus_silica_closeup.jpg

Lin, W., Huang, Y. W., Zhou, X. D., and Ma, Y. (2006). In vitro toxicity of silica nanoparticles in human lung cancer cells. *Toxicology and Applied Pharmacology*, 217(3), 252–259.

Liu, X., and Sun, J. (2010). Endothelial cells dysfunction induced by silica nanoparticles through oxidative stress via JNK/P53 and NF-kB pathways. *Biomaterials*, 31(32), 8198–8209.

Liu, D., Yang, F., Xiong, F., and Gu, N. (2016). The smart drug delivery system and its clinical potential. *Theranostics*, 6(9) 1306–1323.

Lo, C. L, Lin, K. M., and Hsiue, G. H. (2005). Preparation and characterization of intelligent core-shell nanoparticles based on poly(D,L-lactide)-g-poly(N-isopropyl acrylamide-co-methacrylic acid). *Journal of Controlled Release*, 104(3), 477–488.

Luo, D. and Saltzman, W. M. (2000). Synthetic DNA delivery systems. *Nature Biotechnology*, 18(1), 33–37.

Luo, G., Yu, X., Jin, C., Yang, F., Fu, D., Long, J., and Lu, W. (2010). LyP-1-conjugated nanoparticles for targeting drug delivery to lymphatic metastatic tumors. *International Journal of Pharmaceutics*, 385(1–2), 150–156.

Luo, X., Matranga, C., Tan, S., Alba, N., and Cui, X. T. (2011). Carbon nanotube nanoreservior for controlled release of anti-inflammatory dexamethasone. *Biomaterials*, 32(26), 6316–6323.

Mao, H. Q., Roy, K., Troung-Le, V. L., Janes, K. A., Lin, K. Y., Wang, J. and Leong, K. W. (2001). Chitosan-DNA nanoparticles as gene carriers: synthesis, characterization and transfection efficiency. *Journal of Controlled Release*, 70(3), 399–421.

Martínez, A., Iglesias, I., Lozano, R., Teijon, J. M., and Blanco, M. D. (2011). Synthesis and characterization of thiolated alginate-albumin nanoparticles stabilized by disulfide bonds. Evaluation as drug delivery systems. *Carbohydrate Polymers*, 83(3), 1311–1321.

Masserini, M. (2013). Nanoparticles for brain drug delivery. *ISRN Biochemistry*

Medina, S. H., Tekumalla, V., Chevliakov, M. V., Shewach, D. S., and Ensminger, W. D. (2011). N-Acetylgalactosamine- functionalized dendrimers as hepatic cancer cell-targeted Carriers. *Biomaterials*, 32(17), 4118–4129.

Menjoge, A. R., Kannan, R. M., and Tomalia, D. A. (2010). Dendrimer-based drug and imaging conjugates: Design considerations for nanomedical applications. *Drug Discovery Today*, 15(5–6), 171–185.

Meyer, R. A., and Green, J. J. (2016). Shaping the future of nanomedicine: Anisotropy in polymeric nanoparticle design. *WIREs Nanomed Nanobiotechnol*, 8(2), 191–207. doi:10.1002/wnan.1348

Michalet, X., Pinaud, F. F., Bentolila, L. A., Tsay, J. M., Doose, S., Li, J. J., and Weiss, S. (2005). Quantum dots for live cells, in vivo imaging, and diagnostics. *Science*, 307(5709), 538–544.

Missirlis, D., Tirelli, N., and Hubbell, J. A. (2005). Amphiphilic hydrogel nanoparticles. Preparation, characterization, and preliminary assessment as new colloidal drug carriers. *Langmuir*, 21(6), 2605–2613.

Mody, V. V., Siwale, R., Singh, A., and Mody, H. R. (2010). Introduction to metallic nanoparticles. *Journal of Pharmacy and Bioallied Sciences*, 2(4), 2010, 282–289.

Monteiro-Riviere, N. A., Nemanich, R. J., Inman, A. O., Wang, Y. Y., and Riviere, J. E. (2005). Multi-walled carbon nanotube interactions with human epidermal keratinocytes. *Toxicology Letters*, 155(3), 377–384.

Mora-Huertas, C. E., Fessi, H., and Elaissari, A. (2010). Polymer-based nanocapsules for drug delivery. *International Journal of Pharmaceutics*, 385(1–2), 113–142.

Morimoto, N., Endo, T., Ohtomi, M., Iwasaki, Y., and Akiyoshi, K. (2005). *Macromolecular Bioscience*, 5(8), 710–716.

Müller, R. H., and Keck, C. M. (2004). Challenges and solutions for the delivery of biotech drugs a review of drug nanocrystal technology and lipid nanoparticles. *Journal of Biotechnology*, 113(1–3), 151–170.

Murakami, T., Sawada, H., Tamura, G., Yudasaka, M., Iijima, S., and Tsuchida, K. (2008). Water-dispersed single-wall carbon nanohorns as drug carriers for local cancer chemotherapy. *Nanomedicine*, 3(4), 453–463.

Nasongkla, N., Shuai, X., Ai, H., Weinberg, B. D., Pink, J., Boothman, D. A., and Gao, J. M. (2004).cRGD-functionalized polymer micelles for targeted doxorubicin delivery. *Angewandte Chemie International Edition*, 43(46), 6323–6327.

Nelson, A. L. (2010). Antibody fragments: hope and hype. *MAbs*, 2(1), 77–83.

Neuberger, T., Schopf, B., Hofmann, H., Hofmann, M., and von Rechenberg, B. (2005). Superpara-magnetic nanoparticles for biomedical applications: possibilities and limitations of a new drug delivery system. *The Journal of Magnetism and Magnetic Materials*, 293(1), 483–496.

Niemeyer, C. M. (2003). Functional hybrid devices of proteins and inorganic nanoparticles. *Angewandte Chemie*, 42(47), 5796–5800.

Olivier, J. C. (2005). Drug transport to brain with targeted nanoparticles. *NeuroRx*, 2(1), 108–119.

Olivier, J. C., Huertas, R., Lee, H. J., Calon, F., and Pardridge, W. M. (2002). Synthesis of pegylated immunonanoparticles. *Pharmaceutical Research*, 19(8), 1137–1143.

Pan, B. F., Gao, F., and Gu, H. C. (2005). Dendrimer modified magnetite nanoparticles for protein immobilization. *Journal of Colloid and Interface Science*, 284(1), 1–6.

Park, E. J., and Park, K. (2009). Oxidative stress and pro-inflammatory responses induced by silica nanoparticles in vivo and in vitro. *Toxicology Letters*, 184(1), 18–25.

Peracchia M. T., Fattal, E., Desmaële, D., Besnard, M., Noël, J. P., Gomis, J. M., and Couvreur, P. (1999). Stealth® PEGylated polycyanoacrylate nanoparticles for intravenous administration and splenic targeting. *Journal of Controlled Release*, 60(1), 121–128.

Perry, J. L., Martin, C. R., and Stewart, J. D. (2011). Drug-delivery strategies by using template-synthesized nanotubes. *Chemistry*, 17(23), 6296–6302.

Popovici, R. F., Seftel, E. M., Mihai, G. D., Popovici, E., and Voicu, V. A. (2011). Controlled drug delivery system based on ordered mesoporous silica matrices of captopril as angiotensinconverting enzyme inhibitor drug. *Journal of Pharmaceutical Sciences*, 100(2), 704–714.

Putnam, D., Gentry, C. A., Pack, D. W., and Langer, R. (2001). Polymer-based gene delivery with low cytotoxicity by a unique balance of side-chain termini. *Proceedings of the National Academy of Sciences of the United States of America*, 98(3), 1200–1205.

Rao, J. P., and Geckeler, K. E. (2011). Polymer nanoparticles: preparation techniques and size-control parameters. *Progress in Polymer Science*, 36(7), 887–913.

Ravi Kumar, M., Hellermann, G., Lockey, R. F., and Mohapatra, S. S. (2004). Nanoparticle-mediated gene delivery: state of the art. *Expert Opinion on Biological Therapy*, 4(8), 1213–1224.

Reichert, J. M. (2011). Antibody-based therapeutics to watch in 2011. *MAbs*, 3(1), 76–99.

Rejinold, N. S., Chennazhi, K. P., Nair, S. V., Tamura, H., and Jayakumar, R. (2011). Biodegradable and thermo-sensitive chitosan-g-poly(N-vinylcaprolactam) nanoparticles as a 5-fluorouracil carrier. *Carbohydrate Polymers*, 83(2), 776–786.

Remy, J. S., Abdallah, B., Zanta, M. A., Boussif, O., Behr, J. -P., and Demeneix, B. (1998). Gene transfer with lipospermines and polyethylenimines. *Advanced Drug Delivery Reviews*, 30(1–3), 85–95.

Rojahn, S. Y. (2012). Fine-tuning Nanotech to Target Cancer. Retrieved from MIT Technology Review website: https://www.technologyreview.com/s/427821/fine-tuning-nanotech-to-target-cancer/.

Ruponen, M., Honkakoski, P., Ronkko, S., Pelkonen, J., Tammi, M., and Urtti, A. (2003). Extracellular and intracellular barriers in non-viral gene delivery. *Journal of Controlled Release*, 93(2), 213–217.

Sanna, V. (2014). Targeted therapy using nanotechnology: Focus on cancer. *International Journal of Nanomedicine*. 9, 467–483.

Saraog, G. K., Gupta, P., Gupta, U. D., Jain, N. K., and Agrawal, G. P. (2010). Gelatin nanocarriers as potential vectors for effective management of tuberculosis. *International Journal of Pharmaceutics*, 385(1–2), 143–149.

Selvakannan, P. R., Mandal, S., Phadtare, S., Gole, A., Pasricha, R., Adyanthaya, S. D., and Murali, S. (2004). Water-dispersible tryptophanprotected gold nanoparticles prepared by the spontaneous reduction of aqueous chloroaurate ions by the amino acid. *Journal of Colloid and Interface Science*, 269(1), 97–102.

Sesana, A. M., Monti-Rocha, R., Vinhas, S. A., Morais, C. G., and Dietze, R. (2011). In vitro activity of Amphotericin B Cochleates against Leishmania Chagasi. *Memórias do Instituto Oswaldo Cruz*, 106(2), 251–253.

Shenoy, D., Little, S., Langer, R., and Amiji, M. (2005). Poly(ethylene oxide)-modified poly(beta-amino ester) nanoparticles as a pH-sensitive system for tumor-targeted delivery of hydrophobic drugs: Part 2. In vivo distribution and tumor localization studies. *Pharmaceutical Research*, 22(12), 2107–2114.

Shiba, K., Yudasaka, M., and Iijima, S. (2006). Carbon nanohorns as a novel drug carrier. *Nihon Rinsho*, 64(2), 239–246.

Shin, U. S., Yoon, I. K., Lee, G. S., Jang, W. C., Knowles, J. C., and Kim, H. W. (2011). Carbon nanotubes in nanocomposites and hybrids with hydroxyapatite for bone replacements. *Journal of Tissue Science and Engineering*, 2011, 674287.

Shin, Y., Chang, J. H., Liu, J., Williford, R., Shin, Y.-K., and Exarhos, G. J. (2001). Hybrid nanogels for sustainable positive thermosensitive drug release. *The Journal of Controlled Release*, 73(1), 1–6.

Skowron, S. (2006). *Cyclodextrin*. [Online Image]. Retrieved from https://commons.wikimedia.org/wiki/File:Cyclodextrin.svg

Song, X., Zhao, Y., Wu, W., Bi, Y., and Cai, Z. (2008). PLGA nanoparticles simultaneously loaded with vincristine sulfate and verapamil hydrochloride: systematic study of particle size and drug entrapment efficiency. *International Journal of Pharmaceutics*, 350(1-2), 320–329.

Stayton, P. S., Hoffman, A. S., Murthy, N., Lackey, C., Cheung, C., Tan, P., and Press, O. W. (2000). Molecular engineering of proteins and polymers for targeting and intracellular delivery of therapeutics. *Journal of Controlled Release*, 65(1–2), 203–220.

Sugisaki, K., Usui, T., Nishiyama, N., Jang, W. D., and Yanagi, Y. (2008). Photodynamic therapy for corneal neovascularization using polymeric micelles encapsulating Den-drimer porphyrins. *Investigative Ophthalmology & Visual Science*, 49(3), 894–899.

Tamer, U., Gundogdu, Y., Boyaci, I. H., and Pekmez, K. (2010). Synthesis of magnetic core-shell Fe_3O_4-Au, nanoparticles for biomolecule immobilization and detection. *Journal of Nanoparticle Research*, 12(4), 1187–1196.

Taratula, O., Garbuzenko, O. B., Kirkpatrick, P., Pandya, I., and Savla, R. (2009). Surface-engineered targeted PPI Dendrimer for efficient intracellular and intratumoral siRNA delivery. *Journal of Controlled Release*, 140(3), 284–293.

Thaci, B., Ulasov, I. V., Wainwright, D. A., and Lesniak, M. S. (2011). The challenge for gene therapy: Innate immune response to adenoviruses. *Oncotarget*, 2(3), 113–121.

Tomalia, D. A. (2005). Birth of a new macromolecular architecture: dendrimers as quantized building blocks for nanoscale synthetic organic chemistry. *Progress in Polymer Science*, 30(3–4), 294–324.

Torchilin, V. (2011). Tumor delivery of macromolecular drugs based on the EPR effect. *Advanced Drug Delivery Reviews*, 63(3), 131–135.

Torchilin, V. (2008). *Multifunctional Pharmaceutical Nanocarriers*. Springer Science.

Trementozzi, A. (2013). *SolidLipidNanoparticle*. [Online Image]. Retrieved from https://commons.wikimedia.org/wiki/File:SolidLipidNanoparticle.jpg

Velazquez, A. J., Carnahan, M. A., Kristinsson, J., Stinnett, S. and Grinstaff, S.W. (2004). New dendritic adhesives for sutureless ophthalmic surgical procedures: In vitro studies of corneal laceration repair. *Archives of Ophthalmology*, 122(6), 867–870.

Valencia, P. M., Pridgen, E. M., Rhee, M., Langer, R., Farokhzad, O. C., and Karnik, R. (2013). Microfluidic platform for combinatorial synthesis and optimization of targeted nanoparticles for cancer therapy. *ACS Nano*, 7(12), 10671–10680.

van Vlerken, L. E., Duan, Z., Little, S. R., Seiden, M. V., and Amiji, M. M. (2008). Biodistribution and pharmacokinetic analysis of paclitaxel and ceramide administered in multifunctional polymer-blend nanoparticles in drug resistant breast cancer model. *Molecular Pharmaceutics*, 5(4), 516–526.

Vij, N., Min, T., Marasigan, R., Belcher, C. N., and Mazur, S. (2010). Development of PEGylated PLGA nanoparticle for controlled and sustained drug Delivery in cystic fibrosis. *Journal of Nanobiotechnology*, 8, 22.

Vinogradov, S. V., Bronich, T. K., and Kabanov, A. V. (2002). Nanosized cationic hydrogels for drug delivery: preparation, properties and interactions with cells. *Advanced Drug Delivery*, 54(1), 135–147.

Vinogradov, S. V., Zeman, A. D., Batrakova, E. V., and Kabanov, A. V. (2005). Polyplex Nanogel formulations for drug delivery of cytotoxic nucleoside analogs. *Journal of Controlled Release*, 107(1), 143–157.

Wei, L., Hu, N., and Zhang, Y. (2010). Synthesis of polymer-mesoporous silica nanocomposites. *Materials*, 3(7), 4066–4079.

Whitehead, K. A., Langer, R., and Anderson, D. G. (2009). Knocking down barriers: advances in siRNA delivery. *Nature Reviews Drug Discovery*, 8(6), 129–138.

Wightman, L., Kircheis, R., Rössler, V., Carotta, S., Ruzicka, R., Kursa, M., and Wagner, E. (2001). Different behavior of branched and linear polyethylenimine for gene delivery in vitro and in vivo. *Journal of Gene Medicine*, 3(4), 362–372.

Wilczewska, A. Z., Niemirowicz, K., Markiewicz, K. H., and Car, H. (2012). Nanoparticles as drug delivery systems. *Pharmacological Reports,* 64(5), 1020–1037.

Wu, W., Chen, B., Cheng, J., Wang, J., Xu, W., Liu, L., and Li, J. Biocompatibility of Fe_3O_4/DNR magnetic nanoparticles in the treatment of hematologic malignancies. *International Journal of Nanomedicine*, 5, 1079–1084.

Wu, A. M., and Senter, P. D. (2005). Arming antibodies: Prospects and challenges for immunoconjugates. *Nature Biotechnology*, 23(9), 1137–1146.

Wu, C., Yang, Z., Peng, X., Tan, Y., Chen, M., Zhu, X, and Xu, Y. (2011). Optimization of the preparation process for an oral phytantriol-based amphotericin B cubosomes. *Journal of Nanomaterials*, 2011, 1–10. doi:10.1155/2011/308016

Xu, X., Ho, W., Zhang, X. Bertrand, N., and Farokhzad, O. (2015). Cancer nanomedicine: From targeted delivery to combination therapy. *Trends in Molecular Medicine* 21(4), 223–232.

Yallapu, M. M., Othman, S. F., Curtis, E. T., Gupta, B. K., Jaggi, M., and Chauhan, S. C. (2011). Multi-functional magnetic nanoparticles for magnetic resonance imaging and cancer therapy. *Biomaterials*, 32(7), 1890–1905.

Yang, J., Xu, C., Wang, C., and Kopeček, J. (2006). Refolding hydrogels self-assembled from N-(2- hydroxypropyl) methacrylamide graft copolymers by antiparallel coiled-coil formation, *Biomacromolecules*, 7(4), 1187–1195.

Yang, Z., Peng, X., Tan, Y., Chen, M., and Zhu, X. (2011). Optimization of the preparation process for an oral phytantriol-based amphotericin B cubosomes. *Journal of Nanomaterials*, 10.

Yu, K. O., Grabinski, C. M., Schrand, A. M., Murdock, R. C., Wang, W., Gu, B., and Hussain, S. M. (2009). Toxicity of amorphous silica nanoparticles in mouse keratinocytes. *Journal of Nanoparticle Research*, 11(1), 15–24.

Zhang, B., Chen, Q., Tang, H., Xie, Q., Ma, M., Tan, L., and Yao, S. (2010). Characterization of and biomolecule immobilization on the biocompatible multi-walled carbon nanotubes generated by functionalization with polyamidoamine dendrimers. *Colloids and Surfaces B: Biointerfaces*, 80(1), 18–25.

Zhang C. Y. Yang, Y. Q., Huang, T. X., Zhao, B., Guo, X. D., Wang, J. F., and Zhang, L. J. (2012). Self-assembled pH-responsive MPEG-b-(PLA-co-PAE) block copolymer micelles for anticancer drug delivery. *Biomaterials*, 33(26), 6273–6283.

Zhang, J.-S., Li, S., and Huang, L. (2003). Cationic liposome-protamine-DNA complexes for gene delivery. *Methods in Enzymology*, 373, 332–342.

Zhu, Y. (2014). A robust graft-to strategy to form multifunctional and stealth zwitterionic polymer-coated mesoporous silica nanoparticles. *Biomacromolecules*, 15(5), 1845–1851.

Zhuang, J., Jiwpanich, S., Deepak, V. D., and Thayumanavan, S. (2012). Facile preparation of nanogels using activated ester containing polymers. *ACS Macro Letters*, 1(1), 175–179.

10 Tissue Engineering and Regenerative Medicine

10.1 INTRODUCTION

Life science over the past few decades has moved forward at a rapid pace, and many do not fully realize the recentness of this acceleration. Tissue engineering and regenerative medicine are at the forefront of this advancement, representing a change in the technology of healthcare. While discussing the phenomena associated, it is important to remember that it has not been long since the biology of a living cell was observed, much less understood. It is important, then, to note that the technologies discussed in this paper represent a new subject in the field of science, developing over the past 20 or so years. It is also important to observe the multidisciplinary nature of the science (Van Blitterswijk, 2008), and the requirement of input from chemical, biological, mechanical, and engineering perspectives.

As the reader will be frequently reminded, tissue engineering and regenerative medicine are new disciplines. It is widely agreed that the origin of these disciplines lies somewhere in the collaboration between biomaterials, cell biologists, surgeons, and engineers (Brown, 2013). This coming together was for the purpose of generating therapeutic technologies, moving towards a distant dream of tissue regeneration; although in the sense provided may be synonymous with repair or replacement.

To describe the goals of regenerative medicine is to explain the limitations of existing prosthetic implants and living tissue grafts/cadaveric transplants. Implantable prosthetic devices have had an immensely successful history in many clinical and reconstructive surgical disciplines (Walker et al., 2007). Despite their many advantages, they still suffer the key limitation of never being more than a temporary substitute; they never work better than the day they are implanted. They are always foreign, artificial devices, which the body tolerates, until they wear out or clog up (Brown, 2013).

Overcoming the need to constantly treat and repair damaged human parts has for some time now been at the forefront of medicinal studies. The conventional practices are no doubt important and treating patients with various diseases in the best manner available is of course a societal benefit. However, advancements are becoming increasingly necessary, and progress in the medical field is coming.

10.2 TRADITIONAL MEDICINAL PRACTICES

The introductions of most tissue engineering research articles begin with a description of the growing need for new methods in medicine. This is by no means a criticism of the medicinal field, its practitioners, or those involved with it. To not marvel

by Yaser Dahman and Filip Jelic

at the miracles possible through modern medicine is to not understand how far the human species has come. There is no place that technological breakthroughs can be traced easier than with respect to medicine and doctoral practices. One need only point to the increasing life expectancy to describe the success of the medicinal field.

That success is, in fact, perhaps one of the reasons why a need for novel medicinal practices has emerged. To explain this reasoning, one needs to examine the current medicinal practice of living tissue grafts and transplants, from heart to liver, to skin, cornea, or tendon. They are fabulous methods of restoring certain functionalities to individuals who otherwise might suffer from a serious decrease in life quality, or even death. It would be difficult to label the practice anything other than a success as far as medicine is concerned. The responsibility of that success is being felt, however, in what some researchers call a "worst case scenario": the more successful the procedure is, the worse their problems become – for example, as kidney and heart transplants became successful and immune-suppression was better managed, the waiting list for donors became inexorably longer (Vacanti and Vacanti, 2007). As humans live longer and age better, suitable donors become more and more scarce and this only gets worse.

Another example of this ticking time bomb is the story of the prosthetic hip replacement. As a successful operation that has been around for a long time, there are more and more patients across an increasing age range that are demanding it. This has, in turn, resulted in the number of individuals living longer active lives spiraling up for many years. This would be fine, except for the problem that no matter how well prosthetics are made, they will always eventually wear out. As a consequence, the number of patients needing much more complex but less successful revision surgery to remove or replace worn implants has spiraled up (Brown, 2013). What this represents, in numbers, is a major healthcare cost, a problem that governments would prefer not to have.

The advent of superior medicine has brought with it a demand for more of it. As Figure 10.1 illustrates, the number of transplants has stayed fairly steady since the early 1990s. However, there is an ever-increasing disparity between the number of donors, which has stayed steady, and the number of patients waiting for one.

Looking beyond the scope of bridging the patient-donor gap, there are specific practices or areas of medicine that would seem to require upgrades. Bone ailments, for example, are an interesting case. Bone has a high regenerative capacity, especially in younger individuals, and a majority of fractures will heal without major intervention. However, larger bone defects can supersede that ability to regenerate and would require surgical intervention. Here, the current gold standard treatment is the use of a procedure called autografting, involving the harvest of "donor" bone from a non-load-bearing site in the patient and transplanting it into the defect site (Stevens, 2008). This brings with it a plethora of complications, most notably the loss of some function as the transplanted bone may not be capable of load bearing to the extent required.

Perhaps the most glaring of examples where current treatments might be falling short concerns the human spine. Most notably, issues in this area can be contributed to degenerative disc disease (DDD), which affect the intervertebral disc (IVD) between adjacent vertebral bodies (Yeganegi, 2009). DDD contributes to the pathogenesis of lower back pain, which has become an endemic problem placing a big

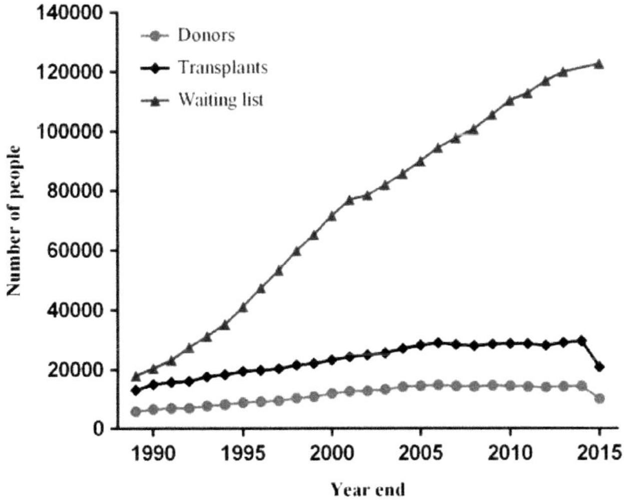

FIGURE 10.1 Increasing number of patients on the donor waiting list, compared to the number of transplants performed and the number of live donors (Almela et al., 2016).

economic burden on the health care system (Hudson et al., 2013). Current treatments range from conservative management to invasive procedures. They seek to eliminate pain generated by the ruptured or herniated discs but do not attempt to restore function or structure (Xu et al., 2014). The gold standard in DDD treatment is intervertebral fusion. Despite its frequency, the procedure often fails to completely alleviate patients of their back pain (Yeganegi, 2009).

As is often the case in tissue engineering *needs* scenarios, the restoration of function and integrity of a particular tissue is often lacking in current medicinal treatment methods. Modern medicine has come far enough to be able to alleviate patients of symptoms from what might be ailing them, but even that selected route is often temporary. What medicinal practices are lacking is the ability to fully restore or regenerate a tissue back to its functioning best, and truly heal a patient suffering from any type of degeneration.

10.3 TISSUE ENGINEERING

The term "tissue engineering" was officially coined at a National Science Foundation workshop in 1988. It was defined as "the application of principles and methods of engineering and life sciences towards the fundamental understanding of structure–function relationships in normal and pathological mammalian tissues and the development of biological substitutes to restore, maintain, or improve tissue function" (O'Brien, 2011). The field of tissue engineering (Figure 10.2) is highly multidisciplinary, drawing on experts from medicine, materials science, genetics, and all types of engineering. This paper will attempt to describe the field of tissue engineering and its advancements by breaking it down into its major components: cells, scaffolds, and signals (Figure 10.3).

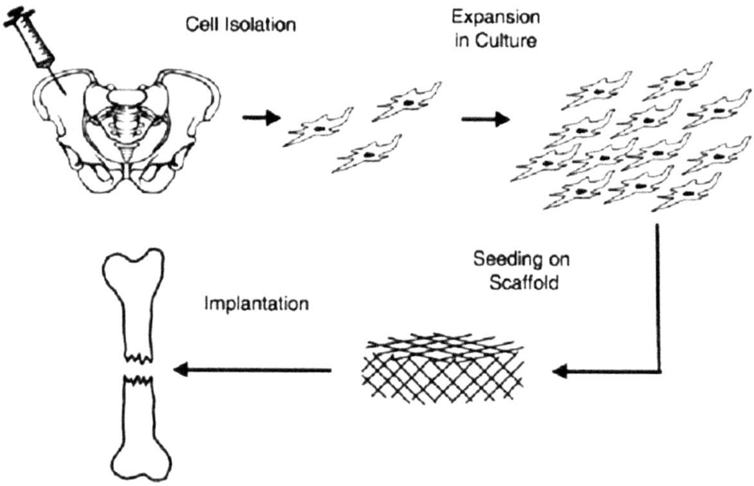

FIGURE 10.2 The complete, "ideal", tissue-engineering paradigm (Porter, Ruckh, and Popat, 2009).

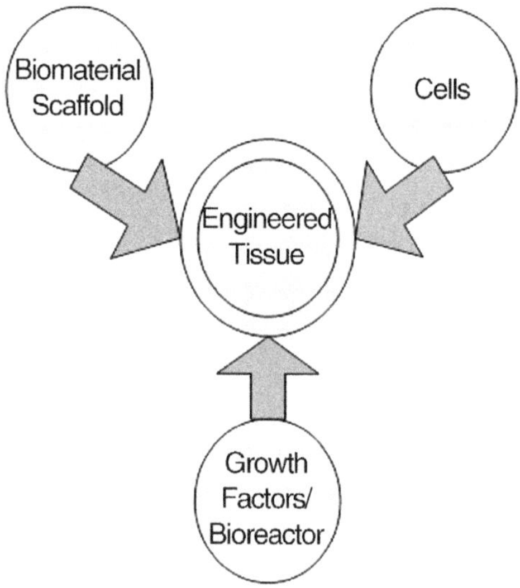

FIGURE 10.3 The tissue engineering triad: cells, signals, and the scaffold template for tissue formation (O'Brien, 2011).

10.3.1 Cells

Whatever the tissue engineering approach, whether it is replacement, repair, or regeneration, a critical issue in the design process is that of cell source. Where the cells come from and how they are employed are the first steps in the design process for treatment or therapy (Nerem, 2011). In addressing the issue, several considerations need to be evaluated:

- Will the source of the cells be endogenous (from the host) or exogenous (outside the host)?
- Whether to employ an autologous or allogeneic cell strategy or even a xenogeneic strategy.
- Whether to use undifferentiated stem cells, progenitor cells, or fully differentiated somatic cells.

- Differences with respect to the age of the donor and receptor, or with the state of disease of the host organ.
- Differences in sex of donor/receptor that must be taken into account.

It is difficult to pinpoint any one source of cells as a viable option for design. There has been research performed all over the map, using different types of stem cells (Nerem, 2011) and cells extracted from the native tissue (Salih, 2013).

10.3.1.1 Cell Sourcing

Since the advent of tissue engineering, there has been remarkable progress in the field. However, there remain many challenges that need to be addressed before the field can be exploited for widespread clinical applications. At the forefront of all the challenges are those related to cells, as they constitute the main pillar of tissue engineering strategies, from cell injection, cell induction, or cell-seeded scaffolds (Neel et al., 2014). Effective clinical application will require cells to be easily procurable, scalable in-vitro, and robust in culture and implantation (Almela et al., 2016). It is necessary that they integrate functionally with the recipient tissue, being non-immunogenic and safe, neither contaminated by pathogens nor tumorigenic (Vacanti, 2006).

Acquiring cells represents a major challenge linked with tissue engineering of every part of the body, regardless of tissue type or strategy applied. For clinical applications, it is possible to have cell origins that are either autogenic (from the same patient) or allogeneic (of a different, same-species donor) (Almela et al., 2016). Further, the type of cell extracted can be described in terms of maturity: differentiated or stem cells (Birla, 2014). Each type of cell source presents its own challenges and potential compromises to its reliability.

Allogeneic cells are donor-dependent and are associated with genetic mismatches and the high likeliness of immune rejection (Birla, 2014). The latter issue would require a further step in any type of implantation process to suppress any immune response. Autogenic cells eliminate the need for potential donor matching and remove immunological risk. However, they present their own set of issues. Firstly, the harvesting biopsy may yield limited quantities, if any, of viable cells (Birla, 2014). This is often the case when the source is a diseased organ. Secondly, the target tissue may be inaccessible for direct biopsy (as is the case for sourcing from the heart valve) or it may not be practical to perform a biopsy at all (spinal cord) (Palakkan et al., 2013).

When it comes to cell maturity, it is important to remember that all mature differentiated cells have a finite life span. Cell viability gradually declines until the growth is irreversibly halted as cells begin the process of senescence. The result is that, past a certain point in differentiation, the limited number of population doublings may be insufficient to provide a clinically relevant tissue (Trainor et al., 2014). The use of somatic cells is then limited in patients past a certain age, and the requirement of a donor becomes necessary. Another option in tissue engineering strategy is the use of stem cells. In tissue engineering applications, this has extended to adult stem cells (ASCs), embryonic stem cells (ESCs), and induced pluripotent stem cells (iPSCs) (Bhagavati, 2015). Stem cells have a unique capacity to self-renew and the ability to differentiate into various specialized cell types. They provide tissue engineering applications a certain degree of flexibility not available through the use of somatic cells. However, this flexibility does not come without its own complications. A summary of allogeneic and autogenic cells is provided in Table 10.1 above.

Adult stem cells (ASCs), while multipotent, have a finite potential for expansion. They are capable of generating cell types of the pertinent tissue, but not all cell lineages (Bhagavati, 2015). In addition, the availability of adult stem cells is extremely limited, e.g. the proportion of mesenchymal stem cells (MSCs) in bone marrow is 1: 100 000 nucleated cells (Choudhery et al., 2014), making them logistically difficult to obtain. Furthermore, an increase in donor age has a negative impact on cell viability, proliferation, and differentiation. MSCs from an older donor display more senescent features when compared to MSCs isolated from a younger one (Li et al., 2015).

Embryonic stem cells, on the other hand, are indefinitely proliferative and pluripotent. They are able to differentiate to all tissue types (Almela et al., 2016). ESCs

TABLE 10.1
Challenges Associated with Cultivation of Autologous/Allogenic Cells, Either Differentiated or Stem

	Autogenic	Allogenic
Differentiated	Associated with donor morbidity Requires extra surgical step Possible inaccessible areas for harvest Culture can be time consuming Very Expensive Finite life span	Dependent on finding a donor Immune system rejection if incompatible
Stem	*ASCs* Limited differentiation lineages Function is donor age-dependent Limited availability *iPSCs* Tumorigenicity Immunogenicity	*ASCs* Limited differentiation lineages Function is donor age-dependent Immune system rejection if incompatible *ESCs* Ethical issues Immunogenicity Tumorigenicity Difficulties in optimizing differentiation and culture purification

Source: Almela et al. (2016).

offer a solution to the cell scarcity that impedes tissue fabrication. There are several complications associated with the use of ESCs however. Firstly, the ethical and moral controversies surrounding the sacrifice of viable embryos to isolate ESCs have sparked a long ongoing debate among scientists and others with interest in the matter (Doerflinger, 2010). While this paper will not delve deeper into the issue, there are numerous resources available for arguments both for and against the use of ESCs in general. Secondly, a complication of ESC treatment is the potential rejection and immunogenic response of hosts into which such therapies are implanted. Accelerated in-vitro differentiation mechanisms of ESCs, with respect to the long period of normal development, will cause a lack of immune inhibition ligands and retention of residual antigens within the differentiated cells that are eventually recognized as foreign antigens (Martello and Smith, 2014). The final issue with ESCs stems from the difficulty in identifying the correct mixture of factors which drive the differentiation pathway to the desired lineage. This often leads to yields that are not 100% pure population (Almela et al., 2016). This incomplete differentiation poses a problem with respect to culture heterogeneity and requires an additional step to purify the differentiated cells and remove the undifferentiated cells, which are considered contaminants. Failure in this regard gives rise to another serious issue: the potential risk of tumorigenicity (Martello and Smith, 2014). The reason for this concern is that stem and cancer cells share many similar cellular and signalling pathways (Almela et al., 2016).

Attempting to overcome the issues associated with ESCs, induced pluripotent stem cells have been derived through cell reprogramming. By delivering several transcription factors, somatic cells are *dedifferentiated* and turned back to an ESC-like state (Takahashi and Yamanaka, 2006). The resulting iPSCs have been assumed to be as pluripotent as ESCs while being readily accessible. Without the ethical concerns governing the use of ESCs, the major issue associated with using iPSCs has centred on their proneness to tumorigenicity (Takahashi et al., 2007). Compared to ESCs, iPSCs are more prone to tumorigenicity. This is a consequence of multiple mutations resulting in genetic and epigenetic instability, the imperfect induction process, and the reprogramming genes themselves (Yu and Thomson, 2014). Also, while iPSCs used in treatment are generally autogenic, their use has shown to provoke an immune response similar to that of ESCs (Yu and Thomson, 2014). This is due to the epigenetic modifications, which can cause inappropriate gene and antigen expression. While the immune response is not as severe as that of ESCs, it is possible that engrafted aberrant cells might exploit this immune evasion and eventually result in tumour development (Almela et al., 2016).

Sources of cells can range even more than the type of cell extracted for tissue engineering applications. A trend in tissue engineering regarding cell usage is to derive multipotent cells and culture them in a manner that allows them to differentiate along the lineage desired for a particular application. While there remain application-specific limitations, the general modus operandi follows this path. Beyond the usage of embryonic stem cells, extracting from adult tissue leads to one popular area of multipotent cells used in numerous tissue engineering experiments: skeletal stem cells.

The ability of bone to remodel, along with the capacity of bone tissue to heal and regenerate following damage, supports the concept of a stem cell population within post-natal bone marrow. Bone marrow consists of hematopoietic and stromal compartments, and it has long been acknowledged as a source of hematopoietic stem cells (HSCs). Bone marrow cell suspensions include both HSCs and stromal cells. Stromal cells form stromal tissue which tends to function as a scaffold, composed of a cell network that provides physical and functional support to HSCs. Stromal cells are able to adhere to tissue culture plastic, while extraction of HSCs is made easier by the fact that they are non-adherent and can be readily removed from stromal cell cultures by simple washes (Black et al., 2015).

Another example of multipotent cell extraction exists in dental pulp stem cells (DPSCs). This line of cells is capable of differentiating into multiple cell lines providing an alternative source for tissue engineering applications (Song et al., 2016). DPSCs have the potential to differentiate into a smooth muscle phenotype in-vitro through treatment with various growth agents (Liu et al., 2013). They are able to maintain high proliferation abilities over a long period of time making them useful and beneficial for tissue engineering applications (Yoshiba et al., 2012).

A major issue in tissue engineering is providing a sufficient blood supply in the initial phase post-implantation. Vascularization must be established; otherwise, the implant must rely on diffusion for the supply of nutrients and the removal of waste. This can lead to nutrient limitations, resulting in improper integration or even death of the implant. Early on in the technology, most techniques to promote vascularization targeted delivery of angiogenic growth factors, relying on the ingrowth of host endothelial cells, without much success (Brown, 2013).

Establishing a blood vessel network by either vasculogenesis or angiogenesis is essential. It allows for cellular gaseous exchange, a supply of nutrients and removal of waste during new tissue formation (Black et al., 2015). The success of any ex-vivo graft is dependent on vascular invasion within the graft to ensure the cell's environmental requirements are met; new tissue formation within the implanted construct will only occur if there is rapid integration with the host vasculature (Buranawat, 2013). Without a functional vascular supply, a tissue engineered construct will ultimately be prone to failure.

Individual cells in-situ require a supply of oxygen and nutrients, along with waste removal processes. The delivery of growth factors and cytokines to the surrounding cells occurs by diffusion to and from a network of capillaries, which run throughout the tissue. What is also of concern is the time required for a tissue engineering construct to vascularize once implanted. For larger implants such as bone, the slow spontaneous vascular ingrowth may not be effective enough in covering the construct in time for proper cell function (Buranawat, 2013).

One strategy for getting over this issue is to *prevascularize* the tissue engineered construct ex-vivo prior to implantation (Buranawat, 2013). This type of approach drastically decreases the time that is needed to vascularize, as host vessels only have to grow into the outer regions of an implant. The goal of prevascularization is to create new vasculature via vasculogenesis inside an engineered scaffold prior to implantation. It can be achieved by culturing specific cell types in the presence of factors that ensure phenotypic stability and expression while encouraging interaction with the supporting scaffold (Cenni et al., 2009).

This strategy to enhance early vascularization has been studied. One method was to add endothelial cells to engineered muscle tissue, leading to the formation of a vascular network inside the muscle tissue in-vitro (Levenberg et al., 2005). The study, novel at the time, reported muscle tissue integrating properly with the surrounding tissue, while also indicating that the added vascular structure did not have a negative effect on the differentiation of the muscle tissue. The prevascular network fused with the blood vessels of the host and became functional vessels transporting blood. This study, performed back in 2005, showed that in-vitro prevascularization could be a successful strategy to improve vascularization after implantation.

10.3.2 Scaffolds

While cells are the important producers and directors of tissues, responsible for the synthesis and maintenance of tissue, it is the extracellular matrix (ECM) that makes up a substantial part of the tissue volume. It comprises a complex network of macromolecules, a variety of proteins (including several collagen types) secreted locally and organized in a fashion allowing cells to link to one another while also receiving the nutrients they require to do their job (Salih, 2013). The sourcing of cells, complicated in its own nature, requires little to no design on the part of tissue engineering scientists. Crafting a platform that mimics the complex nature of the ECM is much more of a thoughtful activity.

Perhaps the most important, and most challenging, aspect of tissue engineering is the design and implementation of a proper scaffold for any particular application. Considering that the goal of tissue engineering is regeneration or replacement of damaged tissue, the ideal process is to combine cells from the body with highly porous scaffold biomaterials, acting as templates for tissue regeneration in order to guide the growth of new tissue (O'Brien, 2011).

Tissue engineering has seen numerous scaffolds produced from a variety of biomaterials through a plethora of fabrication techniques, all in an attempt to regenerate different tissues or even organs of the body (Agarwal et al., 2009). Regardless of the type of tissue, there are a number of key considerations when designing a scaffold for a particular application:

- Biocompatibility. The very first criterion for scaffolds. When designing a scaffold for tissue engineering, it must be biocompatible; cells must adhere, function normally, and migrate to the surface and begin to proliferate and create ECM. Once that is accomplished and implantation has taken place, the scaffold or construct must not elicit an immune reaction and must prevent rejection by the body (O'Brien, 2011).
- Biodegradability. Scaffold constructs should be designed with biodegradability in mind, as the end goal of the entire procedure is to allow the body's own cells to replace the implant over time and continue functioning normally (Okamoto and Baiju, 2013). The by-products of this degradation should be non-toxic and allowed to exit the body without interfering with routine function (O'Brien, 2011).
- Mechanical Properties. In an ideal scenario, the scaffold used for a tissue engineering construct will have the same mechanical properties as

the anatomical site into which it is to be implanted (Walser and Ferguson, 2016). From a practical perspective, the scaffold structure must also be strong enough to allow for surgical handling during implantation (O'Brien, 2011).

- Architecture. Scaffolds should be designed such that there is an interconnected pore structure and high level of porosity, as this will allow for cellular penetration and easy diffusion of nutrients to cells and the ECM (O'Brien, 2011). It will also allow for the removal of waste products out of the scaffold.

Tissue engineering uses scaffolds to provide temporal support for cells to regenerate new extracellular matrices to replace tissue that has been destroyed by injury, disease, or congenital defects. A natural ECM separates different tissues, forms a supportive meshwork around cells, and provides anchorage to the cells (Agarwal et al., 2009). It is important for manufactured scaffolds to mimic these properties, temporarily, while degrading at a sufficient rate to allow for replacement by newly fabricated ECM. The many considerations in tissue engineering scaffolds make it a complicated science, with much research being devoted specifically to their design. It is important then to consider the biomaterial to be used in constructing the right scaffold and then to find the ideal method of fabrication and set up.

In tissue engineering applications, there are three main groups of biomaterials used in the fabrication of scaffolds: ceramics, synthetic polymers, and natural polymers (O'Brien, 2011). Each has their own specific advantages/disadvantages, and it is becoming increasingly common to see scaffolds comprised of two or more different phases.

Ceramics, while not generally used in soft tissue regeneration, have been applied extensively in bone tissue engineering. Examples include hydroxyapatite (HAp) (Bose et al., 2012) and tri-calcium phosphate (TCP) (O'Brien, 2011), used in numerous applications. They are typically characterized by high mechanical stiffness (Young's modulus), very low elasticity, and a hard-brittle surface (O'Brien, 2011). From a bone perspective, they exhibit excellent biocompatibility due to their chemical and structural similarity to the mineral phase of native bone. The interactions of osteogenic cells with ceramics are important for bone regeneration, and ceramic structures have shown an ability to enhance osteoblast differentiation and proliferation (Hench, 1998). Various ceramics have been used already in dental and orthopedic surgery to fill bone defects; however, their clinical application for tissue engineering has been limited thus far due to their brittleness, difficulty of shaping, and the fact that tissue engineered bone constructs have yet to show the ability to bear the mechanical loading needed for remodelling (Van Landuyt et al., 1995).

There have been numerous synthetic polymers used in attempting to produce functional scaffolds, including polystyrene, poly-l-lactic acid (PLLA), polyglycolic acid (PGA), and poly(D,L-lactic-co-glycolic acid) (PLGA) (O'Brien, 2011). These materials have generally shown a desirable ability to be fabricated with a specifically tailored architecture, as their flexibility allows for easier shaping than ceramics (Okamoto and Baiju, 2013). Synthetic polymers also show favourable degradation characteristics as the property can be adjusted by varying the polymer itself or its composition (Lu et al., 2000). There are drawbacks to using synthetic polymers in tissue engineering scaffolds, mainly stemming from the risk of rejection due to reduced bioactivity (O'Brien, 2011). There are also concerns over the degradation processes of PLLA and PGA as they degrade via hydrolysis, producing carbon dioxide and therefore lowering the local pH, a phenomenon which can result in tissue necrosis (Liu et al., 2006).

The use of biological materials as scaffold biomaterials has been common as well. Collagen, various proteoglycans, alginate-based substrates, and chitosan have all been used in the production of scaffolds for tissue engineering (O'Brien, 2011). Unlike synthetic polymers, these natural polymer structures are biologically active and tend to promote excellent cell adhesion and growth (Kim et al., 2006). They are also biodegradable and allow host cells over time to produce their own extracellular matrix to replace the degraded scaffold (Damadzeh et al., 2010). However, fabricating scaffolds from natural polymers presents its own set of challenges to overcome. For one, it is difficult to fabricate a scaffold with a homogenous and reproducible structure (Freiss and Schlapp, 2006). Secondly, the scaffolds tend to have poor mechanical properties, limiting their use in any type of load-bearing application (O'Brien, 2011).

With each of the described scaffold materials having specific disadvantages, there has been much research devoted to the development of composite scaffolds comprising more than one phase (O'Brien, 2011). Numerous research papers have been published attempting to introduce ceramics into polymer-based scaffolds, while others have combined synthetic polymers with natural ones (Wu et al., 2007). This new trend in scaffolding has opened the science to new possibilities and points of research.

An example of this composite technology can be found in bone tissue engineering. A number of different scaffolding materials have been used in the field ranging from calcium phosphates (Bose et al., 2012) to natural and synthetic hydroxyapatite (Knowles, 2003). Successful research combining collagen (Type I) and glycosaminoglycan (GAG) to produce a highly porous collagen-GAG scaffold (O'Brien, 2011) has also been done. While O'Brien et al. have demonstrated the success of these CG scaffolds for bone repair in minimally weight-bearing regions of the body (Lyons et al., 2010), in order to facilitate repair of regions where scaffolds are subjected to higher levels of loading, they have introduced a ceramic phase to the scaffold for strengthening. The experiments have led to the development of a series of highly porous biomimetic collagen-hydroxyapatite (CHAp) scaffolds, based on the two primary constituents of bone (O'Brien, 2011).

There are other examples of hydroxyapatite being used in bone tissue engineering scaffolds as composites based on HAp particles and biopolymers show good osteoconductivity, osteoinductivity, biodegradability, and high mechanical strength (Okamoto and Baiju, 2013). One example is the composite scaffold including poly(l-lactic acid) and HAp (Nejati et al., 2008). Other successful researchers have combined HAp with poly(-caprolactone) (PCL) (Rezwan et al., 2006). The reasoning behind the composition is to take advantage of the HAp's and polymers' mechanical and physiological properties in order to meet the requirements of host tissue (Bose et al., 2012). These new advanced techniques have led to vast improvements in the field of bone tissue engineering, reaching ever closer to clinical trials.

There are other interesting developments in the use of materials in tissue engineering scaffolds. For more complex tissue, traditional scaffolding methods seem to have limitations in their usability as viable constructs for tissue replacement or repair. One interesting example of the complicated nature of the body is the IVD (Figure 10.4). IVDs, making up one fourth of the spinal column's length, are fibrocartilagenous cushions serving as the spine's shock absorbing system, protecting the vertebrae, brain, and other structures that make up a human spine. The discs are responsible for the spine's motion, mainly extension and flexion. It is composed of two distinct tissues, with the annulus fibrosus enclosing the nucleus pulposus. The annulus fibrosus resembles a strong radial tire, with its structure made up of lamellae; concentric sheets of collagen fibers connected to vertebral end plates and oriented at various angles. The nucleus pulposus contains a hydrogel-like matter that resists compression and allows it to absorb water at generous amounts – similar in make up to the annulus fibrosus (water, collagen, and proteoglycans), varying only in the composition of each element, meant to differentiate in the water retention capabilities of each separate layer. (Spine Doctor, 2017).

Tissue engineering has provided the treatment of IVD degeneration new hopes in treatment. However, the complicated nature of the tissue has made it difficult to fabricate suitable scaffolds for much more than in-vitro studies of cell culture. An ideal scaffold in this scenario is particularly complicated. Efforts in producing annulus fibrosus tissue in-vitro have involved various polymeric scaffolds (Yeganegi, 2009) organized in a variety of manners. Different types of scaffolds have been used as suitable implants to achieve mechanical stability, all showing limited success.

An interesting development, then, in the field of annulus fibrosus and IVD tissue engineering came about when researchers began to decellularize native tissue and use it as a scaffold. This has been researched by Xu et al. (2014), who proposed that decellularized tissue may be, perhaps, the ideal scaffold for annulus fibrosus tissue engineering. The reasoning behind the method is that annulus fibrosus, with its complicated structure and complex components distributed unevenly, would be unfeasibly difficult to fabricate artificially. In their attempt to develop an ideal technology,

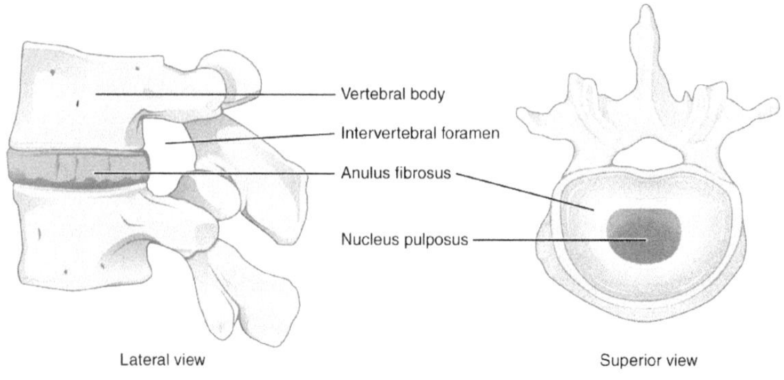

FIGURE 10.4 Diagram of the IVD in-between vertebrae in the spinal column (Betts et al., n.d.).

they were able to remove the native cells and preserve tissue-specific ECM components to closely match the natural tissue.

This decellularization technology has been studied before for other, perhaps *less* complicated, tissue. Decellularized matrices have been used in the tissue engineering of blood vessels (Boer et al., 2011), bladder (Cheng et al., 2010), bone (Dong et al., 2009), and cartilage (Kheir et al., 2011). With varying levels of success, the new technology requires more research. The development of decellularization techniques has led to more research being devoted to those types of scaffolds.

When it comes to scaffolding in tissue engineering, what often gets overlooked are the design methods for scaffolds. While the choice in biomaterials is well studied and constantly the focal point of many research articles, there is little more than a mention of the method of fabrication for the particular scaffold in question. It is a fair point, as most fabrication techniques were readily developed prior to the field of tissue engineering becoming a real focus of medicinal research. It is worth noting, however, that the field of tissue engineering has taken some of these pre-existing techniques and really found a use for them in scaffold design.

When discussing scaffold fabrication methods, the focus rests primarily on designing polymeric scaffolds. Numerous techniques have been used to mould and shape polymers in the past to varying degrees of success. For example, selective laser sintering has been used in fabricating a hybrid poly(-caprolactone)/hydroxyapatite polymer scaffold (Wiria et al., 2007) and in the formation of a poly(vinyl alcohol) and hydroxyapatite composite scaffold (Wiria et al., 2008). There have also been extrusion-based scaffolds developed using rapid prototyping techniques that have been well studied (Hogue et al., 2011), and stereolithography techniques used for developing bone tissue engineering scaffolds (Lee et al., 2007).

There are other standard methods of producing scaffolds suitable for tissue engineering applications. These conventional methodologies known for producing porous tissue engineered scaffolds (molecular self-assembly, phase separation, etc.) have their share of drawbacks. The major problem with these methods is the difficulty of generating highly porous scaffolds, and their incapability in mimicking the native ECM structure (Agarwal et al., 2009). A more modern technique, appearing in numerous research papers over the past decade or so, promises better results when dealing with both issues: electrospinning (Figure 10.5).

Electrospinning is a versatile technique for fabricating micro- or nanofibers of a variety of materials with polymer or polymer composite bases (Wang and Wang, 2014). It produces non-woven meshes containing fibers ranging in diameter from tens of nanometers up to a micrometric scale (Lanutti et al., 2007). While traditional methods can produce fibers as fine as 10–100 μm, for the generation of fibers down to 15 nm, electrospinning is an extremely useful technique (Agarwal et al., 2009).

The technique is gaining in popularity with researchers using the fabrication method in the design of chitosan-poly(vinyl alcohol) nanofibers (Agrawal and Pramanik, 2016) and numerous other oriented nanofibrous membranes for tissue engineering applications (Walser and Ferguson, 2016). The advances of the technique have led to the development of scaffolds that possess adequate biodegradability characteristics and mechanical properties that are desirable for tissue engineering applications (Agrawal and Pramanik, 2016).

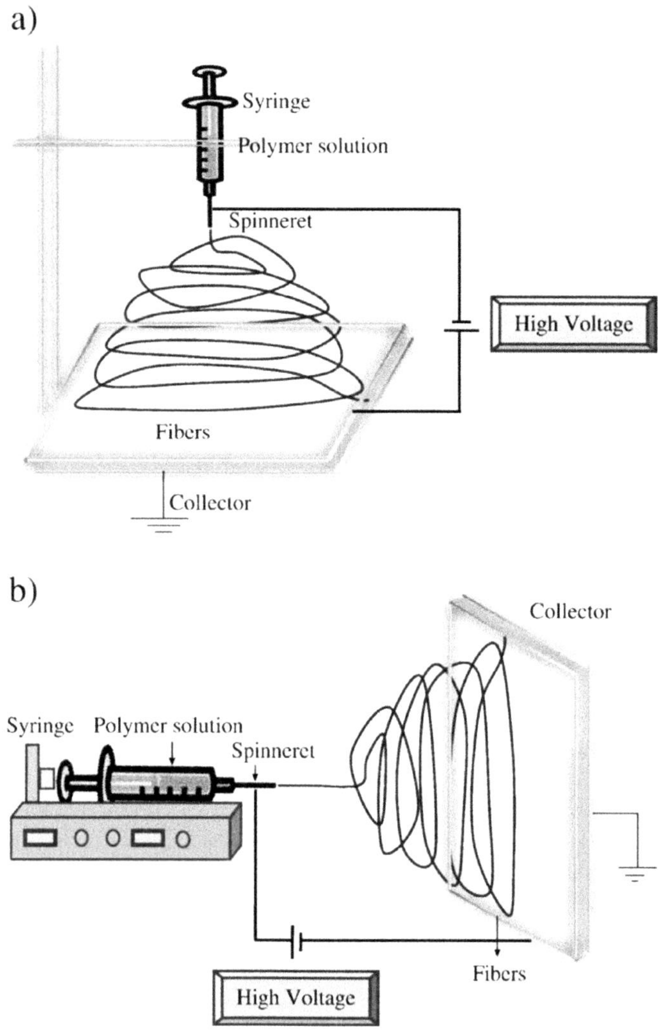

FIGURE 10.5 Electrospinning technique. An electrical charge draws a fine line (typically micro- or nanoscale) of fibers from a liquid. When a sufficiently high voltage is applied to the liquid droplet, the body of liquid becomes charged, and electrostatic repulsion counteracts the surface tension of the liquid – a stream of liquid erupts from the surface, until it is finally deposited on the grounded collector (Bhardwaj and Kundu, 2010).

10.3.3 Signals

Much cell behaviour is regulated by adhesion to an ECM. This includes cell proliferation, differentiation, apoptosis, and, most importantly, gene expression (Salih, 2013). It is this gene expression that leads to protein assembly, or disassembly, and function. Cell adhesion firstly involves the binding and clustering of integrin receptors to an immobilized matrix, and active spreading of cells against the substrate (Chen et al., 1997). Integrin signalling, cell shape, and cytoskeletal structure coordinate to control endothelial cell proliferation, differentiation, and apoptosis. It is well understood that adhesion is not a simple signalling event determined by binding of integrin to its ligand, but a rather complex interplay between biochemical signals of integrins and structural changes associated with cell spreading (Roberts et al., 1998). What is important to note here is that cells respond to the mechanical and biochemical changes in the ECM through the crosstalk between integrins and the actin cytoskeleton (Dike et al., 1999). They have a unique ability to respond to the molecular composition and physical properties of the ECM and integrate both mechanical and chemical signals through direct association with the cytoskeleton.

In-vivo, the ECM serves as a local storage reservoir for growth factors. There are many proteins present within the ECM that have dual binding sites: one for cell adhesion, and the other for growth factors (Salih, 2013). Once the cell has adhered, it allows for a local concentration of the growth factors near to their cell-surface receptors and cell adhesion sites, forming a higher concentration of their signal. Examples of such are vascular endothelial growth factor (VEGF) and fibroblast growth factor (FGF) (Patel et al., 2007). For tissue engineering applications, scaffolds must be able to mimic this behaviour once implanted until the newly produced ECM is in place and is able to take over.

Many external forces and prompts control cell differentiation. There are a number of environmental factors that play a role in the production of an extracellular matrix. While the type of cell used and the scaffold it is based in are imperative in tissue engineering design, it would be unwise to not consider the signalling for cell development equally. These signals can come in a variety of different techniques, whether biochemical, mechanical, growth factors, etc.

While chemical gradients and genetic regulation play vital roles in morphogenesis, it has become clear that the expression of genetic markers, while necessary, is not sufficient in the explanation of differentiation (Bernstein, 2011). There has been a rise in evidence which suggests that epigenetic factors include mechanical and structural cues that play essential roles in both embryogenesis and organogenesis. A better understanding of the mechanisms of mechanical interaction between the cells and their immediate microenvironment is necessary for appropriately directing the development of cells and the desired ECM products.

Mechanical forces have long been implemented in the regulation of many physiologic and pathologic processes. Cells respond to mechanical signals depending upon their substrate/environmental material properties and the surroundings it interacts with. These mechanical cues are converted into biochemical signals at the molecular level to provide and ensure the correct cellular response (Clause et al., 2011). With the rise in awareness of cell response to mechanical stimulation, it became an essential consideration in designing biomaterials and scaffolds with appropriate mechanical properties.

Stem cell differentiation, in particular, is regulated by a diverse array of extracellular cues. There has been much evidence recently suggesting that mechanical interactions between the ECM and cell-surface receptors as well as physical interactions between neighbouring cells play important roles in stem cell self-renewal and differentiation (Sheehy and Parker, 2011). Researchers are also becoming aware that the ECM affects cellular behaviour through many physical mechanisms, such as ECM geometry, elasticity, and propagation of mechanical signals to intracellular compartments. There has been much effort in developing biomaterials that exploit cellular microenvironments in order to guide cells to desired phenotypes and organization into functional tissue. Figure 10.6 compares the influencing factors on cells and their impact on cell behaviour.

Older approaches to the design of scaffolds included mainly the specification of material properties which were appropriate for mature tissues (Singh et al., 2008). Modern approaches used now aim to exploit endogenous tissue engineering strategies, like those occurring during development and recapitulated during post-natal wound healing, while providing mechanical stability via exogenous means (Knothe Tate et al., 2007). Thus, the modern approach of delivering tissue engineering signals is to incorporate them into scaffold design. With templates specifically designed to mimic the environment of condensed mesenchyme during development, tissue scaffolds now are serving as delivery devices for extrinsic cues; this means incorporating biochemical and mechanical signals (Singh et al., 2008).

A novel approach has recently emerged as perhaps the optimal method of drug delivery in tissue engineering applications. It deals with the formation of gradients in controlling signal delivery and input systems. Gradients in cellular and extracellular architecture, as well as in mechanical properties are apparent in native tissues. What has been proposed then is for tissue engineers to take a cue from nature and incorporate this gradient structure into their design strategies. Indeed, this has been a growing trend, as Singh et al. (2008) reviewed when discussing potential approaches to gradient methodology, offering a stark contrast to traditional approaches of homogenous signal delivery of cells and growth factors using isotropic scaffolds.

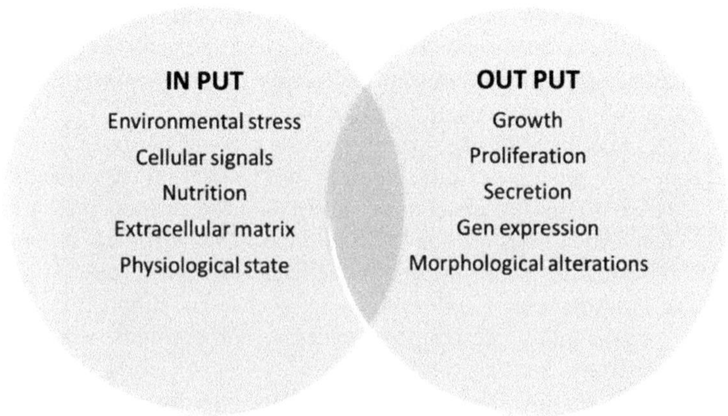

FIGURE 10.6 The influencing factors on cells and their impact in determining cell behaviour (Almela et al., 2016).

10.4 REGENERATIVE MEDICINE

Compared to tissue engineering, regenerative medicine seems to be more difficult to define; it was used earlier but was always more broadly described. Most scientists now view it as the field of stem cell research, where stem cells drive embryonic formation or are inductively organized to induce a response to regenerate tissue (Meyer et al., 2009). It seems that a rigid definition of regenerative medicine is not yet available, while the principle approaches defining the science are still being outlined. Formalities aside, stem cells, the center of regenerative medicine, hold great promise for the future of the field.

The field of stem cells has recently become one of the most intensely studied areas in medicine. This is mainly due to the fact that there are no suitable medical devices or therapies to cure diseases associated with cell or tissue loss (Somasundaram, 2015). It is in stem cells, which have the innate ability to replicate themselves and differentiate into other cell types, where there lies potential for the clinical application of tissue regeneration, repair, or transplant in humans. Ever since the discovery of stem cells, scientists have dreamed of using them to repair damaged tissue or create new organs, leading to technologies that will revolutionize the field of medicine.

Stem cell research is currently being fuelled by the expectation that it will lead to therapeutic strategies that have the potential to become a viable alternative to short-term medication and surgical procedures for the treatment of degenerative diseases (Chacko and Kuppusamy, 2015). In this paper, the focus of stem cells thus far has been their role and use in tissue engineering principles. There are other applications of stem cells not necessarily defined under the umbrella of tissue engineering pertaining to stem cell treatment and usage. The main one of these applications is indeed stem cell therapy.

10.4.1 Stem Cell Transplantation

Medicinal scientists have come to realize the therapeutic potency of stem cells – it has come far enough to completely change the landscape of medical research and become the backbone of the regenerative medicine field. The Canadian Stem Cell Foundation has taken a lead in defining four ways that stem cells are used for clinical benefit (Canadian Stem Cell Foundation, 2014):

- Stem Cell Transplantation
- Cellular Therapy
- Tissue Regeneration
- Drug Delivery

Cellular therapy and tissue regeneration are fields of research, which rely heavily on the ability of stem cells to multiply indefinitely and become cells required by specific parts of the body. Most research here is focused on ex-vivo studies and largely falls under tissue engineering regimes in attempting to replace tissues or organs that are damaged or degenerated.

Stem cell transplantation, however, is an interesting new field of medicinal research that differs slightly from the tissue engineering aspects of other regenerative medicine approaches. Stem cell therapy is the use of stem cells to prevent or treat a disease or degenerative condition (National Cancer Institute, 2013). Stem cell transplantation, in basic terms, is the infusion of healthy stem cells to replace diseased or damaged ones (Murnaghan, 2016). If successful, the healthy stem cells will integrate into the body and give rise to more cells that are able to perform the necessary functions of a specific tissue.

Biomedical science has entered a new era using cellular therapy to treat a wider and wider spectrum of human diseases, ranging from nerve injury to myocardial infarction and more (Leavitt, 2011). Hematopoietic stem cell transplantation has, over the past 50 years, been used to advance treatment technologies of all types. An example of this is the early successful transplant of patients that suffered from immunodeficiency syndromes (Linker, 2003). While the field remains relatively new, with clinical successes occurring sparsely over the course of the past half-century, there is promise. Time is still required to fully understand the mechanisms and tendencies of stem cells. While it is envisioned that many cellular therapies will be developed through some type of modification of autologous cells, it is also well understood that it will not happen tomorrow (Leavitt, 2011).

Where these treatments are used most are in cancer patients. The primary method of treating most types of cancer, particularly leukemia, is chemotherapy. A patient receives this therapy or some type of radiation treatment to destroy cancer cells, but unfortunately, healthy cells are also destroyed in the process (Murnaghan, 2016). A stem cell transplant is then a viable option to replace the lost or damaged cells with new, functional ones – it can provide red and white blood cells, as well as platelets that are important for metabolism, clotting, and immunity. Therefore, bone marrow transplants and blood stem cell transplants are most commonly used in the treatment of leukemia and lymphoma (National Cancer Institute, 2013).

10.5 FUTURE OF THE TECHNOLOGY

There can be no questioning that major breakthroughs have taken place, with economic activity within the tissue engineering sector growing exponentially each year (Lysaght et al., 2008) and an increasing number of products entering the marketplace and clinical trials (O'Brien, 2011). While progress is evident, significant research is required in many areas of the field.

There are many clinicians that will question the general procedural approach to tissue engineering in particular. The fact that most applications currently being researched or already developed require an in-vitro testing phase brings about questions of the efficiency of the process (O'Brien, 2011). It is an approach that requires two separate procedures, meaning delays in treatment during culturing. From a commercial perspective, this approach also poses problems due to the prolonged regulatory process required before such a construct can be approved for clinical use. It is, unfortunately, an inherent flaw for some tissues; cartilage, for example, which doesn't have the ability to regenerate itself when damaged, requires long-term in-vitro engineering as a viable solution to damaged or degenerated areas (Bernstein, 2011).

There are some tissues, such as bone, that have an intrinsic ability to repair, remodel, and regenerate (O'Brien, 2011). The task of tissue engineering and regenerative medicine, then, is to harness this innate regenerative capacity to really push the field forward and bring about a new age in the field of medicine.

10.6 CONCLUSION

The future of tissue engineering and regenerative medicine holds the promise of custom-made medical solutions for injured or diseased patients. As the science of tissue engineering and regenerative medicine becomes established as a central discipline in biomedicine, it is interesting to speculate where the field will head. Both areas have matured to the point that their research and clinical aspects can be conceptually categorized. It is a growing hope, then, that regenerative medicine can be introduced to the world as the new age of medicine. Work is required, but the hope only grows.

REFERENCES

Agarwal, S., Wendorff, J. H., and Greiner, A. (2009). Progress in the field of electrospinning for tissue engineering applications. *Advanced Materials*, 21(32–33), 3343–3351.

Agrawal, P., and Pramanik, K. (2016). Chitosan-poly(vinyl alcohol) nanofibers by free surface electrospinning for tissue engineering applications. *Tissue Engineering and Regenerative Medicine*, 13(5), 485–497.

Almela, T., Brook, I. M., and Moharamzadeh, K. (2016). The significance of cell-related challenges in the clinical applciation of tissue engineering. *Journal of Biomedical Materials Research Part A*, 104(A), 3157–3163.

Bernstein, H. S. (Ed.) (2011). *Tissue Engineering in Regenerative Medicine*. Springer Science.

Betts, J. G., Desaix, P., Johnson, E., Johnson, J. E., Korol, O., Kruse, D. and Young, K. A. (n.d.). Axial Skeleton. In *Anatomy and Physiology*. Retrieved from https://cnx.org/contents/FPtK1zmh@6.27:fEI3C8Ot@6/Preface.

Bhagavati, S. (2015). Stem cell therapy: Challenges ahead. *Indian Journal of Pediatrics*, 82(3), 286–291.

Bhardwaj, N., and Kundu, S. C. (2010). Electrospinning: A fascinating fiber fabrication technique. *Biotechnology Advances*, 28(3), 325–347. doi:10.1016/j.biotechadv.2010.01.004

Birla, R. (2014). *Introduction to Tissue Engineering: Applications and Challenges*. IEEE Press.

Black, C. R. M., Goriainov, V., Gibbs, D., Kanczler, J., Tare, R. S., and Oreffo, R. O. C. (2015). Bone tissue engineering. *Current Molecular Biology Reports*, 1(3), 132–140.

Boer, U., Lohrenz, A., Klingenberg, M., Pich, A., Haverich, A., and Wilhelmi M. (2011). The effect of detergent based decellularization procedures on cellular proteins and immunogenicity in equine carotid artery grafts. *Biomaterials*, 32(36), 9730–9737.

Bose, S., Roy, M. , and Bandyopadhyay, A. (2012). Recent advances in bone tissue engineering scaffolds. *Trends in Biotechnology*, 30(10), 546–554.

Brown, R. A. (2013). *Extreme Tissue Engineering: Concepts and Strategies for Tissue Fabrication*. Wiley-Blackwell.

Buranawat, B. K. (2013). Vascularization of tissue-engineered constructs. In V. Salih (Ed.), *Standardisation in Cell and Tissue Engineering: Methods and Protocols*. Woodhead Publishing.

Canadian Stem Cell Foundation (2014). *Clinical Trials*. Retrieved March 26, 2017, from Stem Cell Foundation: http://stemcellfoundation.ca/en/toward-treatments/clinical-trials/.

Cenni, E., Ciapetti, G., Granchi, D., Fotia, C., Perut, F., Giunti, A., and Baldini, N. (2009). Endothelial cells incubated with platelet-rich plasma express PDGF-B and ICAM-1 and induce bone marrow stromal cell migration. *Journal of Orthopaedic Research*, 27(11), 1492–1498.

Chacko, S. M., and Kuppusamy, P. (2015). Considerations for culturing and preserving adult stem cells for therapeutics. In I. Somasundaram (Ed.), *Stem Cell Therapy for Organ Failure*. Springer India.

Chen, C. S., Mrksich, M., Huang, S., Whitesides, G. M., and Ingber, D. E. (1997). Geometric control of cell life and death. *Science*, 276(5317), 1425–1428.

Cheng, H. L., Loai, Y., Beaumont, M., and Farhat, W. A. (2010). The acellular matrix (ACM) for bladder tissue engineering: A quantitative magnetic resonance imaging study. *Magnetic Resonance in Medicine*, 64(2), 341–348.

Choudhery, M. B., Badowski, M., Muise, A., Pierce, J., and Harris, D. T. (2014). Donor age negatively impacts adipose tissue-derived mesenchymal stem cell expansion and differentiation. *Journal of Translational Medicine*, 12(8).

Clause, K., Lam, J., Segura, T., and Barker, T. H. (2011). Directing cell fate through biomaterial microenvironments. In H. Bernstein (Ed.), *Tissue Engineering in Regenerative Medicine*. Springer Science.

Damadzeh, B., Jabari, H., Skrifvars, M., Airola, K., Moritz, N., Vallittu, P. K. (2010). Effect of ceramic filter content on the mechanical and thermal behaviour of poly-L-lactic acid and poly-L-lactic-co-glycolic acid composites for medical applications. *Journal of Materials Science: Materials in Medicine*, 21(9), 2523–2531.

Dike, L. E., Chen, C. S., Mrkisch, M., Tien, J., Whitesides, G. M., and Ingber, D. E. (1999). Geometric control of switching between growth, apoptosis, and differentiation during angiogenesis using micropatterned substrates. *In Vitro Cellular & Developmental in Biological Animals*, 35(8), 441–448.

Doerflinger, R. M. (2010). Old and new ethics in the stem cell debate. *The Journal of Law, Medicine & Ethics*, 38(2), 212–219.

Dong, S. W., Ying, D. J., Duan, X. J., Xie, Z., Yu, Z. J., Zhu, C. H., and Sun, J. S. (2009). Bone regeneration using an acellular extracellular matrix and bone marrow mesenchymal stem cells expressing Cbfa1. *Bioscience, Biotechnology, and Biochemistry*, 73(10), 2226–2233.

Freiss, W., and Schlapp, M. (2006). Sterilization of gentamicin containing collagen/PLGA microparticle composites. *European Journal of Pharmaceutics and Biopharmaceutics*, 63(2), 176–187.

Hench, L. (1998). Bioceramics. *Journal of the American Ceramic Society*, 81(7), 1705–1727.

Hogue, M. E., Chuan, Y. L., and Pashby, I. (2011). Extrusion based rapid prototyping technique: An advanced platform for tissue engineering scaffold fabrication. *Biopolymers*, 97(2), 83–93.

Hudson, K., Alimi, M., Grunert, P., Härtl, R., and Bonassar, L. J. (2013). Recent advances in biological therapies for disc degeneration: tissue engineering of the annulus fibrosus, nucleus pulposus, and whole intervertebral discs. *Current Opinion in Biotechnology*, 24(5), 872–879.

Kheir, E., Stapleton, T., Shaw, D., Jin, Z., Ingham, E., and Fisher, J. (2011). Development and characterization of an acellular porcine cartilage bone matrix for use in tissue engineering. *Journal of Biomedical Materials Research A*, 99(2), 283–294.

Kim, S., Park, M. S., Jeon, O., Choi, C. Y., and Kim, B. (2006). Poly(lactide-co-glycolide)/hydroxyapatite composite scaffolds for bone tissue engineering. *Biomaterials*, 27(8), 1399–1409.

Knothe Tate, M. L., Ritzman, T. F., Schneider, E., and Knothe, U. R. (2007). Testing of a new one-stage bone-transport surgical procedure exploiting the periosteum for the repair of long-bone defects. *Journal of Bone and Joint Surgery American*, 89(2), (307).

Knowles, J. C. (2003). Phosphate based glasses for biomedical application. *Journal of Materials Chemistry*, 13(10), 2395–2401.

Lanutti, J., D. Renekerb, T. Ma, D. Tomaskod, and D. Farsone (2007). Electrospinning for tissue engineering scaffolds. *Material Science and Engineering: C*, 27(3), 504–509.

Leavitt, A. D. (2011). Hematopoietic stem cell Transplantation: Reflections on yesterday and thoughts for tomorrow. In H. S. Bernstein, *Tissue Engineering in Regenerative Medicine*. Humana Press.

Lee, K. W., Wang, S., Fox, B. C., Ritman, E. L., Yaszemski, M. J., and Lu, L. (2007). Poly(propylene fumarate) bone tissue engineering scaffold fabrication using stereolithography: effects of resin formulations and laser parameters. *Biomacromolecules*, 8(4), 1077-1084.

Levenberg, S., Rouwkema, J., Macdonald, M., Garfein, E. S., Kohane, D., Darland, D. C., and Langer, R. (2005). Engineering vascularized skeletal muscle tissue. *Nature Biotechnology*, 23, 879–884.

Li, C., Wei, G., Wen, G., Qi, B., Xu, L., and Tao, S. (2015). Donor age and cell passage affect osteogenic ability of rat bone marrow mesenchymal stem cells. *Cell Biochemistry and Biophysics*, 72(2), 543–549.

Linker, C. A. (2003). Autologous stem cell transplantation for acute myeloid leukemia. *Bone Marrow Transplantation*, 31(9), 731–738.

Liu, H.-L., Liu, S.-J., Xiao, Z.-L., Chen, Q.-Y., and Yang, D.-W. (2006). Excess molar enthalpies of binary mixtures for (tributyl phosphate + methanol/ethanol) at 298.15K. *Journal of Thermal Analysis and Calorimetry*, 85(3), 541-544.

Liu, Y., Deng, B., Zhao, Y., Xie, S., and Nie, R. (2013). Differentiated markers in undifferentiated cells: expression of smooth muscle contractile proteins in multipotent bone marrow mesenchymal stem cells. *Development Growth and Differentiation*, 55(5), 591–605.

Lu, L., Peter, S. J., Lyman, M. D., Lai, H. L., Leite, S. M., Tamada, J. A., and Mikos, A. G. (2000). In vitro and in vivo degradation of porous poly(DL-lactic-co-glycolic acid) foams. *Biomaterials*, 21(18), 1837–1845.

Lyons, F. G., Al-Munajjed, A. A., Kieran, S. M., Toner, M. E., Murphy, C. M., Duffy, G. P., and O'Brien, F. J. (2010). The healing of bony defects by cell-free collagen-based scaffolds compared to stem cell-seeded tissue engineered constructs. *Biomaterials*, 31(35), 9232–9249.

Lysaght, M. J., Jaklenec, A., and Deweerd, E. (2008). Great expectations: Private sector activity in tissue engineering, regenerative medicine, and stem cell therapeutics. *Tissue Engineering - Part A*, 14(2), 305–315.

Martello, G., and Smith, A. (2014). The nature of embryonic stem cells. *Annual Review of Cell and Developmental Biology*, 30, 647–675.

Meyer, U., Meyer, T., Handschel, J., and Wiesmann, H. P. (2009). *Fundamentals of Tissue Engineering and Regenerative Medicine*. Springer.

Murnaghan, I. (2016). Why Perform a Stem Cell Transplant. Retrieved March 23, 2017, from Explore Stem Cells: http://www.explorestemcells.co.uk/WhyPerformStemCellTransplant.html.

National Cancer Institute. (2013, August 12). *Blood-Forming Stem Cell Transplants*. Retrieved March 26, 2017, from National Cancer Institute: https://www.cancer.gov/about-cancer/treatment/types/stem-cell-transplant/stem-cell-fact-sheet#q3.

Neel, E. A. A., Chrzanowski, W., Salih, V. M., Kim, H.-W., and Knowles, J. C. (2014). Tissue engineering in dentistry. *Journal of Dentistry*, 42(8), 915–928.

Nejati, E., Mirzadeh, H., and Zandi, M. (2008). Synthesis and characterization of nano-hydroxyapatite rods/poly(l-lactic acid) composite scaffolds for bone tissue engineering. *Composites Part A: Applied Science and Manufacturing*, 39(10), 1589–1596.

Nerem, R. M. (2011). Tissue engineering: From basic biology to cell-based applications. In L. L. Song (Ed.), *Stem Cell and Tissue Engineering*. World Scientific.
O'Brien, F. J. (2011). Biomaterials & scaffolds for tissue engineering. *Materials Today*, 14(3), 88–95.
Okamoto, M., and Baiju, J. (2013). Synthetic biopolymer nanocomposites for tissue engineering scaffolds. *Progress in Polymer Science*, 38(10–11), 1487–1503.
Palakkan, A. A., Hay, D. C., Anil Kumar, P. R., Kumary, T. V., and Ross, J. A. (2013). Liver tissue engineering and cell sources: Issues and challenges. *Liver International*, 33(5), 666–676.
Patel, V. N., Knox, S. M., Likar, K. M., Lathrop, C. A., Hossain, R., Eftekhari, S., and Hoffman, M. P. (2007). Heparanase cleavage of perlecan heparan sulfate modulates FGF10 activity during ex vivo submandibular gland branching morphogenesis. *Developmental Cell*, 134(23), 4177–4186.
Porter, J. R., Ruckh, T. T., and Popat, K. C. (2009). Bone tissue engineering: A review in bone biomimetics and drug delivery strategies. *Biotechnology Progress*, 25(6), 1539–1560. doi:10.1002/btpr.246
Rezwan, K., Chen, Q. Z., Blaker, J. J., and Boccaccini, A. R. (2006). Biodegradable and bioactive porous polymer/inorganic composite scaffolds for bone tissue engineering. *Biomaterials*, 27(18), 3413–3431.
Roberts, C., Chen, C. S., Mrksich, M., Martichonok, V., Ingber, D. E., and Whitesides, G. M. (1998). Using self-assembled monolayers presenting GRGD and EG30H groups to characterize long-term attachment of bovine capillary endothelial cells to surfaces. *Journal of the American Chemistry Society*, 120, 6548–6555.
Salih, V. (2013). Fundamentals of cell and matrix biology for tissue engineering. In V. Salih (Ed.), *Standardisation in Cell and Tissue Engineering: Methods and Protocols*. Woodhead Publishing.
Sheehy, S. P., and Parker, K. K. (2011). The role of mechanical forces in guiding tissue differentiation. In H. Bernstein (Ed.), *Tissue Engineering in Regenerative Medicine*. Springer Science.
Singh, M., Berkland, C., and Detamore, M. S. (2008). Strategies and applications for incorporating physical and chemical gradients in tissue engineering. *Tissue Engineering: Part B*, 14(4), 341–366.
Somasundaram, I. (2015). *Stem Cell Therapy for Organ Failure*. Springer India.
Song, B., Jiang, W., Alraies, A., Liu, Q., Gudla, V., Oni, J., and Agarwal, M. (2016). Bladder smooth muscle cells differentiation from dental pulp stem cells: Future potential for bladder tissue engineering. *Stem Cells International*.
Spine Doctor. (2017). Intervertebral Discs. Retrieved March 25, 2017, from Spine Universe: https://www.spineuniverse.com/anatomy/intervertebral-discs.
Stevens, M. M. (2008). Biomaterials for bone tissue engineering. *Materials Today*, 11(5), 18–25.
Takahashi, K., Tanabe, K., Ohnuki, M., Narita, M., Ichisaka, T., Tomoda, K., and Yamanaka, S. (2007). Induction of pluripotent stem cells from adult human fibroblasts by defined factors. *Cell*, 131(5), 861–872.
Takahashi, K., and Yamanaka, S. (2006). Induction of pluripotent stem cells from mouse embryonic and adult fibroblast cultures by defined factors. *Cell*, 126(4), 663–676.
Trainor, N., Pietak, A., and Smith, T. (2014). Rethinking clinical delivery of adult stem cell therapies. *Nature Biotechnology*, 32(8), 729–735.
Vacanti, C. (2006). The history of tissue engineering. *Journal of Cellular and Molecular Medicine*, 569–576.
Vacanti, J., and Vacanti, C. A. (2007). The history and scope of tissue engineering. In R. Lanza, J. Vacanti, and R. Langer (Eds.), *Principles of Tissue Engineering* (3rd ed.). Academic Press.
Van Blitterswijk, C. (Ed.) (2008). *Tissue Engineering*. Elsevier.

Van Landuyt, P., Li, F. , Keustermans, J. P. Streydio, J. M., Delannay F., and Munting, E. (1995). The influence of high sintering temperatures on the mechanical properties of hydroxylapatite. *Journal of Materials Science: Materials in Medicine*, 6(1), 8–13.

Walker, J. M., Hauser, H., and Fussenegger, M. (Eds.) (2007). *Methods in Molecular Medicine: Tissue Engineering*. Humana Press.

Walser, J., and Ferguson, S. J. (2016). Oriented nanofibrous membranes for tissue engineering applications: Electrospinning with secondary field control. *Journal of the Mechanical Behavior of Biomedical Materials*, 58, 188–198.

Wang, C., and Wang, M. (2014). Electrospun multifunctional tissue engineering scaffolds. *Frontiers of Material Science*, 8(1), 3–19.

Wiria, F. E., Chua, C. K., Leong, K. F., Quah, Z. Y., Chandrasekaran, M., and Lee, M. W. (2008). Improved biocomposite development of poly(vinyl alcohol) and hydroxyapatite for tissue engineering scaffold fabrication using laser sintering. *Journal of Materials Science: Materials in Medicine*, 19(3), 989–986.

Wiria, F. E., Leong, K. F., Chua, C. K., and Liu, Y. (2007). Poly-ε-caprolactone/hydroxy-apatite for tissue engineering scaffold fabrication via selective laser sintering. *Acta Biomaterialia*, 3(1), 1–12.

Wu, W., Feng, X., Mao, T., Feng X., Ouyang, M. H., Zhao, G., and Chen, F. (2007). Engineering of human tracheal tissue with collagen-enforced poly-lactic-glycolic acid non-woven mesh: A preliminary study in nude mice. *British Journal of Oral and Maxillofacial Surgery*, 45(4), 272–278.

Xu, H., Xu, B., Yang, Q., Li, X., Ma, X., Xia, Q., and Zhang, Y. (2014). Comparison of decellularization protocols for preparing a decellularized porcine annulus fibrosus scaffold. *PLoS One*, 9(1).

Yeganegi, M. (2009). *Characterization of a Biodegradable Electrospun Polyurethane Nanofiber Scaffold Suitable for Annulus Fibrosus Tissue Engineering*. ProQuest Dissertations Publishing.

Yoshiba, N., Yoshiba, K., Ohkura, N., Shigetani, Y., Takei, E., Hosoya, A., and Nakamura, H. (2012). Immunohistochemical analysis of two stem cell markers of α-smooth muscle actin and STRO-1 during wound healing of human dental pulp. *Histochemistry and Cell Biology*, 138(4), 583–592.

Yu, J., and Thomson, J. A. (2014). Induced pluripotent stem cells. In R. Lanza, R. Langer, and J. Vacanti (Eds.), *Principles of Tissue Engineering* (4th ed.). Elsevier/Academic Press.

11 Applications of Biomaterials in Hard Tissue Replacement

11.1 BACKGROUND

A variety of filling and repairing materials have been utilized to restore the function and configuration of a deficient part of a living body. Typical filling and repairing materials for living bodies include artificial bones and analogues such as artificial dental roots and crowns as well as artificial joints. They are usually recognized as living hard tissue replacements. These living hard tissue replacements are required to be mechanically strong, tough, and stable in living bodies and should have a high affinity for living bodies. An additional important factor is the comfort of a replacement's shape, because a living hard tissue replacement has to be a custom finished part conforming to an individual patient's deficient body part. The biological affinity used in this context means how a living hard tissue replacement adapts itself to and merges or assimilates with the surrounding living tissue where the replacement is embedded or implanted. Thus, a material having a high biological affinity is scarcely recognized as xenobiotic by the surrounding tissue. Particularly when such material is used as an artificial bone, it can promote osteogenesis from the surrounding bone to eventually form a firm bond between itself and the bone tissue. Among the currently available artificial bone materials, those featuring high mechanical strength and in-vivo stability are metals such as titanium and zirconium, alloys containing such metals, and ceramics such as alumina, silicon nitride, and zirconia. Materials with high mechanical strength and in-vivo stability, however, have low biological affinity, that is, they are unlikely to assimilate with living tissue, resulting in an extended cure time and poor adherence to the living tissue. In addition, they must be extracted and removed by surgical operations after they have performed their duties. Typical of known materials having high biological affinity are calcium phosphate-based ceramic materials including apatite, especially hydroxyapatite and tricalcium phosphate. Apatite has the best biological affinity as understood from the fact that bone is essentially composed of apatite if organic components are excluded. As previously described, most living hard tissue replacements have a particular shape and a complex profile depending on individual patients. In particular, oral surgical implants such as crowns and dental roots widely differ in the shape depending on individual patients and sites. In the prior art, an implant of desired shape is generally prepared from calcium phosphate-based ceramic material, typically apatite by moulding the

---

by Yaser Dahman and Ragvendra Pratap Singh

material by an injection moulding or casting technique, followed by sintering and reshaping. These manufacturing methods, however, have many drawbacks including lack of dimensional precision, difficulty in changing the shape, surface defects induced by reshaping, and losses of strength due to stresses.

Further, it is critical for implants to adhere to living bones. For promoting adherence, it is very important to control the surface nature of implants. However, ordinary processing techniques have limited freedom to control the surface nature, for example, to achieve a mirror finish or a rough surface.

An overview in 1988, the National Science Foundation Workshop first invented the term "Tissue Engineering" to refer to "the application of principles and methods of engineering and life sciences to understanding basic structure–function relationships in normal and pathological mammalian tissues and the development of biological substitutes to restore, maintain, or improve tissue function" (Skalak and Fox, 1988). Since then, as a new discipline and potential medical field, tissue engineering has made rapid progress that promises to eliminate re-operations by using biodegradable biological substitutes, solve problems of the immune rejection of implants, infections, or transmission of diseases associated with allografts and xenografts, and the shortage of organ donation, initiate the process of natural regeneration with biological substitutes to repair (Zippel, 2010; Wang, 2006), and to replace the lost or damaged tissues, i.e. to provide long-term solutions, offering potential treatments for the currently untreatable medical conditions. Tissue engineering strategies involve the combination of living cells with a natural/synthetic support or scaffold to develop a biological substitute or a three-dimensional living construct that is structurally, mechanically, and functionally equal to or superior to the tissue to be replaced (Zippel, 2010; Kim et al., 2005). Developing such a tissue engineering construction requires careful selection of three major components: scaffolding, signalling factors, and cells, generally called the triad of tissue engineering (Chan and Leong, 2008). Porous three-dimensional scaffolds are usually seeded with cells and occasionally with signalling molecules or biophysical stimuli in the form of a bioreactor (Martin, 2004). These cellular scaffolds remain subjected to a pre-implantation differentiation culture in-vitro to synthesize tissues and, before transplantation or direct implantion on the injured site, using body systems where tissue regeneration is induced in-vivo.

11.2 TISSUE OF THE BODY

Hard tissue is tissue which is mineralized and has a firm intercellular matrix. The hard tissues of humans are bone, tooth enamel, dentin, and cementum.

11.2.1 Enamel

Enamel is the hardest substance in the human body and contains the highest percentage of minerals, 96%, with water and organic matter composing the rest. The primary mineral is hydroxyapatite, which is crystalline calcium phosphate (Vugman et al., 1995). The enamel is formed on the tooth as the tooth develops into the gum, before it enters the mouth. Once fully formed, it does not contain any blood vessels or nerves (Staines et al., 1981). The enamel does not contain collagen, as it is found in other hard tissues such as dentin and bone, but it contains two classes of unique proteins: amelogenins and enamelin. Although the role of these proteins is not fully understood, it is believed that they assist in the development of enamel by providing a framework for minerals to be formed among other functions. Once mature, enamel is almost totally devoid of softer organic matter. Enamel is avascular, has no nervous feed, and is not renewed, but it is not a static tissue as it may undergo changes in mineralization. Dentin by weight consists of 70% mineral hydroxyapatite, 20% organic matter and 10% water. It is yellow in appearance which greatly affects the colour of a tooth due to the translucency of the enamel. Dentine, less mineralized and less fragile than enamel, is necessary for enamel support. Dentine rates about 3 on the Mohs scale of mineral hardness.

11.2.2 Cementum

Cementum is slightly softer than dentin and includes about 45–50wt.% inorganic matter, e.g. hydroxyapatite, and 50–55wt.% organic matter and water (Fractures (Broken Bones), n.d.). The organic part is mainly composed of collagen and proteoglycans. The membrane is avascular, receiving its nutrition through its own imbricated cells of the surrounding vascular periodontal ligament.

11.2.3 Bone

Bone is a rigid organ that forms part of the vertebral skeleton. The bones support and protect the various organs of the body, produce red and white blood cells, store minerals, and allow mobility. The bone tissue is a kind of dense connective tissue. Bones have a variety of shapes and sizes and have a complex internal and external structure. They are lightweight but strong and hard and serve multiple functions. Mineralized bone tissue or bone tissue is of two types: cortical and spongy, giving it rigidity and a three-dimensional internal coral structure. Additional kinds of tissue present in bones includes marrow, endosteum, periosteum, nerves, blood vessels, and cartilage.

11.3 HUMAN BONE SYSTEM

The human skeleton contains 206 bones. Each of them has a different size and length. Figure 11.1 shows the human skeleton. The human skeleton has six main functions:

- The skeleton provides the framework to support our body allow us to maintain the body shape.
- The bones of the skeleton provide the attachment surface for muscle, tendons, and ligaments.
- Our movement is dependent on the skeletal muscle. Without the skeleton to give leverage, movement would be greatly restricted.
- The skeleton protects many vital organs.
- The skeleton is the site of hematopoiesis – the generation of blood cell.

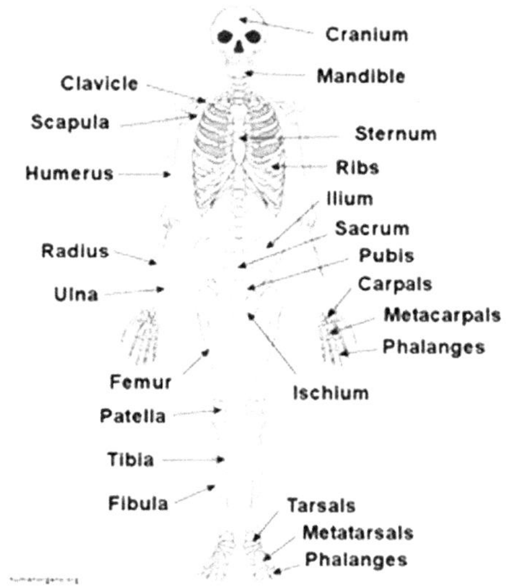

FIGURE 11.1 Human skeleton (Villarreal, 2007).

Bone also serves as a mineral storage deposit in which nutrients can be stored and retrieved. Calcium, especially, can be released by dissolution of bone tissue under the control of 1,25-dihydroxyvitamin D3 during periods of low calcium intake.

11.3.1 Bone Characteristic

Bone is one of the hardest structures of the human body. Bones come in a variety of shapes and have a complex internal and external structure, allowing them to be lightweight yet strong and hard while fulfilling their many other functions. Bone has special characteristics. Bone is anisotropic. Its modulus is dependent upon the direction of loading, and the mechanical properties dependent upon the direction of loading. Bone is weaker in shear than in tension and compression. Bone is viscoelastic. Its force deformation characteristics are dependent upon the rate of loading. Stress–strain character is dependent upon the rate of applied strain (time dependent) Figure 11.2 shows bending, compression, and torsion occurring on the bone.

11.3.2 Mechanical Properties of Bone

Compact bone has a porosity of 5–30% and a cancellous bone of about 30–90% of the proportion of volume occupied by any mineralized tissue. A key requirement in bone is the resistance to compression, and the most important factor in compressive strength is the degree of mineralization. Loss of mineralization results in an increased risk of fracture. The compressive strength of cortical bone varies. In humans, the compressive strength is about 200 MPa (megaPascal) for the femur; the

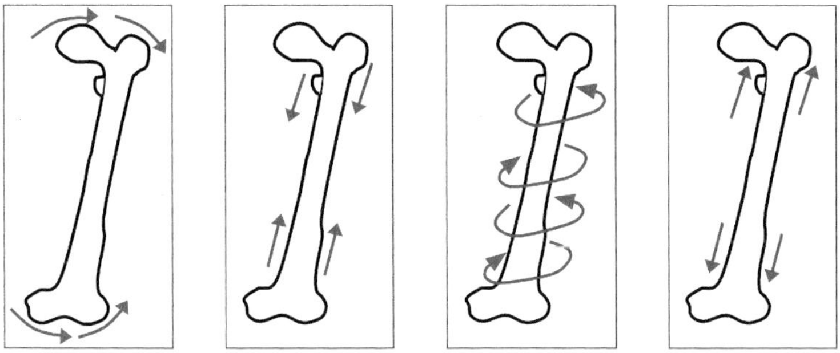

FIGURE 11.2 Bending, compression, and torsion occur on the bone (Bin Pubin, 2007).

elastic modulus of compression is about 17 GPa (gigaPascal). Compressive strength is much lower and the results obtained have varied depending on the location of the bone. Compression resistances of 0.15–27 MPa and an elastic modulus of 50 350 MPa have been reported for cancellous bone. Table 11.1 and Table 11.2 show the mechanical properties of human bone.

TABLE 11.1
Young's Modulus, Compressive Strength, Tensile Strength, Density and Fracture Toughness of Human Bone

	Young's Modulus (GPa)	Compressive Strength (MPa)	Tensile Strength (MPa)	Density (g/cm^3)	Fracture Toughness ($MPam^{1/2}$)
Cortical Bone	3.8–11.7	88–164	82–114	1.7–2.0	2–12
Cancellous	0.2–0.5	23	10–20	–	–

Source: Bin Pubin (2007).

TABLE 11.2
Mechanical Properties of Human Bone

Cortical Bone	Compressive Strength, MPa	131–224 longitudinal
	Tensile strength, MPa	106–133 transverse
	Shear strength, MPa	80–172 longitudinal
	Elastic Modulus, GPa	51–56 transverse
		53–70
		11–20 longitudinal
Cancellous bone	Tissue compressive strength, MPa	0.5–50
	Tissue elastic modulus, MPa	5–150
	Material elastic modulus, GPa	1–11

Source: Bin Pubin (2007).

11.3.3 Bone Fracture

A bone fracture is a medical condition in which a bone becomes cracked, burst or bisected as a result of physical trauma. Bone fracture can also occur due to certain medical conditions that weaken bones, such as osteoporosis or certain types of cancer. A broken bone is not always defined as a fracture, especially since a fracture is not always defined as a broken bone. A broken bone is defined as a complete rupture of the bone, as opposed to a fracture covering any type of crack or rupture in the bone. Fatigue and impact loads are the most common reasons for bone fracture. Figure 11.3 shows the types of bone fracture.

11.4 FRACTURE FIXATION

Fractured bones need to be fixed surgically for their healing and proper functioning as early as possible.

- Temporary internal fixation devices
- Wire (twisted and knotted)/pins
- Screws
- Nails or Rods
- Plates
- Permanent joint replacement
- Hip
- Ankle
- Shoulder
- Elbow

Wire/pins: Wires are often used to join bones together. They are often used to hold pieces of bone too small to be attached with screws. In many cases, they are used together with other forms of internal fixation, but they can be used alone to treat small bone fractures, such as those found in the hand or foot. Wires are usually removed after a certain time but can be left permanently for some fractures.

Screws: Screws are used for internal fixation more often than any other type of implant. Although the screw is a simple device, there are different models based on the type of fracture and how the screw will be used. The screws are of different sizes for use with bones of different sizes. The screws can be used alone to maintain a fracture, as well as plates, stems, or nails. After bone care, the screws can be left in place or removed.

Nails or rods: In some long bone fractures, the best way to keep the bony pieces is to insert a stem or nail through the hollow center of the bone that normally contains the marrow. The screws at respective ends of the stem are used to prevent fracture after shortening or rotating, and to hold the stem in place until the fracture has healed. The stems and screws can be left in the bone after the healing is complete. This is the method used to treat the majority of fractures of the femur and the tibia.

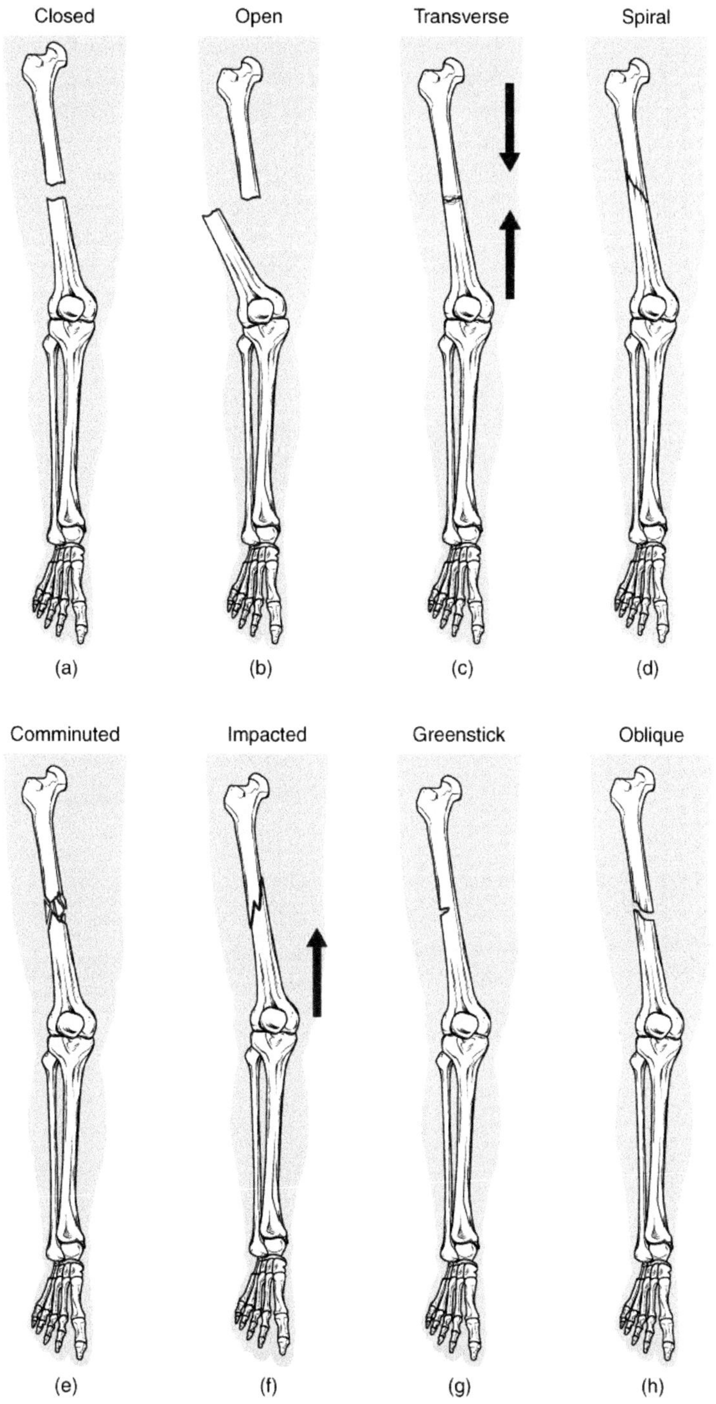

FIGURE 11.3 Types of bone fracture (Betts et al., n.d.).

Plates: Fracture fixation by bone plate ensures rigid immobilization by the fracture site and decreases the fracture gap, allowing primary bone healing by a new formation. The role of the bone plate and screws is to keep the fragments of the bone in position until the bone heals. The wide bone dynamic compression plate (DCP) is used when the actual bone is fractured or ruptured. The fixing of the plate conforms to the place of injury. The wide bone DCP is submerged in large bones like the humerus, tibia, and femur. The function of the plate is like an internal splint and compressor of the bone. The bone protects the plaque. Figure 11.4 through Figure 11.7 show the internal fixation of different fractures using bone plates.

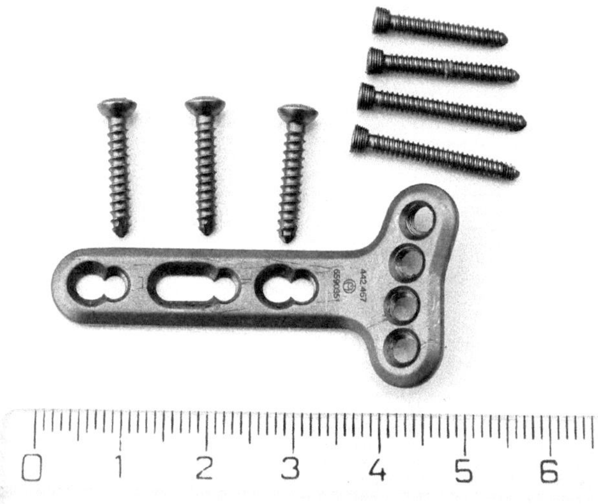

FIGURE 11.4 Simulation of internal fixation using bone plates (Hagens, 2011).

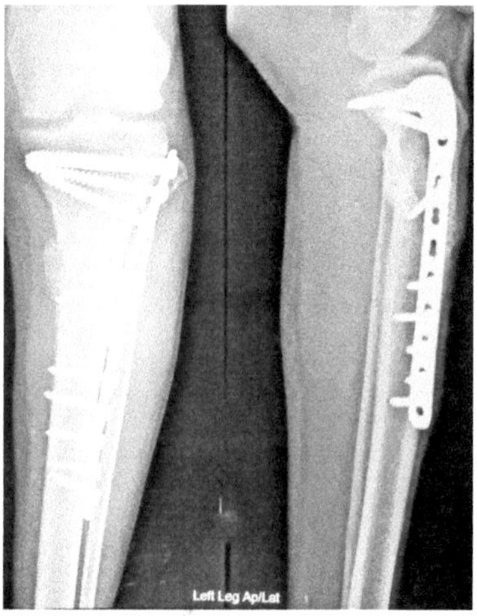

FIGURE 11.5 Internal fixation at humerus bone (Shah, 2014).

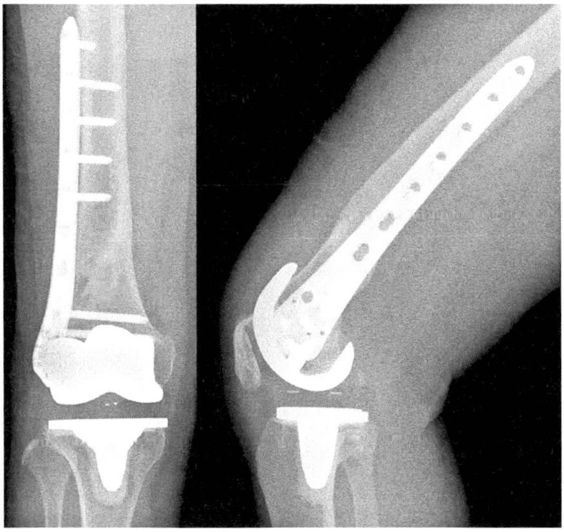

FIGURE 11.6 Internal fixation at human femur bone (Kim et al., 2015).

11.5 BONE TISSUE AND ANATOMY

Tissue Components:

Bone tissue, or osseous tissue, is a type of connective tissue that contains many calcium and phosphorous salts. Approximately 25% of the bone tissue is water, and another 25% is made up of protein fibers like collagen. The remaining 50% of the bone tissue is a mixture of mineral salts, primarily calcium and phosphorous.

Kinds of Bone Tissue:

There are two different types of bone tissue: compact and spongy bone.

Compact Bone:

- Is made up of concentric rings of matrices that surround central canals which contain blood vessels.
- Embedded in this bone tissue are small cave-like spaces called lacunae, which are connected to each other through small tunnels called canaliculi.
- The lacunae contain osteocyte cells. Osteocytes help maintain healthy bone tissue and are involved in the bone remodelling process.
- Compact bone has a Young's modulus of elasticity ranging from 17–20 GPa and compressive strength in the range of 131–224 MPa

Spongy Bone:

- Looks like an irregular latticework (or sponge) with lots of space.
- These spaces are filled with red bone marrow, which is the site of the formation of blood cells.
- Young's modulus and compressive strength for trabecular bones are 50–100 MPa and 5–10 MPa respectively.

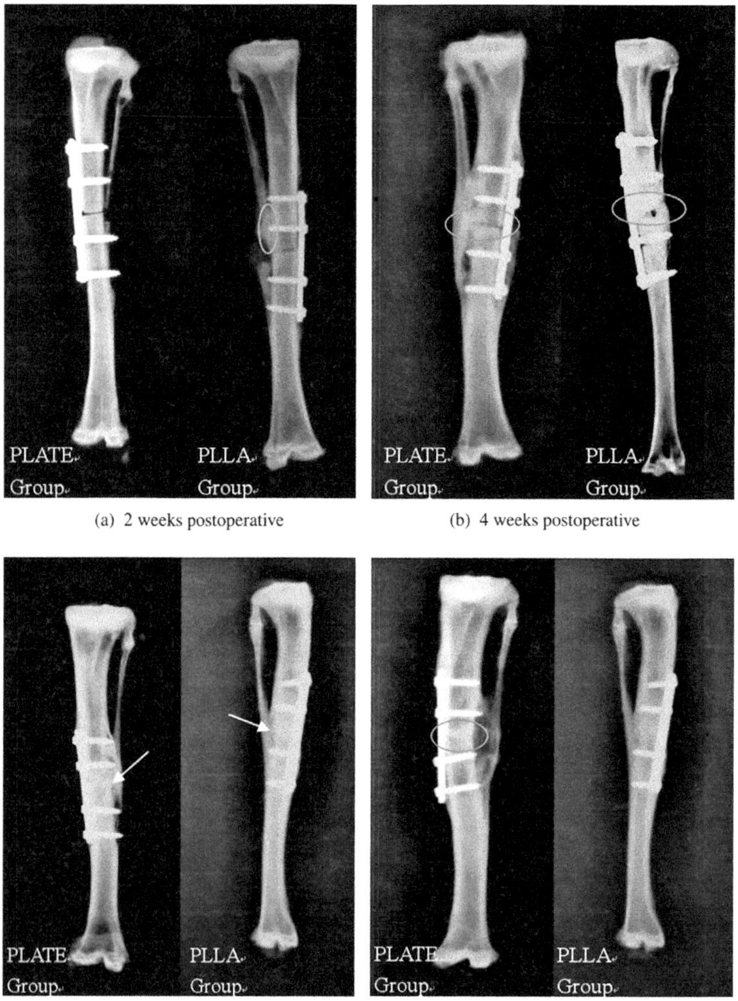

(a) 2 weeks postoperative

(b) 4 weeks postoperative

(c) 8 weeks postoperative

(d) 12 weeks postoperative

FIGURE 11.7 Internal fixation using bone plate (Fan et al., 2008).

Bone Anatomy:

Bones in the body are found in two basic forms, long bones and plates. Long bones are hollow inside. The inner space is called the medullary cavity; two membranes surround the outside of the bone, which is the periosteum. The inside of the bone lining the medullary cavity is the endosteum.

Bone Parts:

The epiphysis refers to the end of the bone, and the diaphysis refers to the main shaft of the bone. Each epiphysis is connected to an adjacent bone using a structure called a joint. The ends of each bone are covered with articular cartilage, which keeps the ends of the bones from grinding together when moving.

TABLE 11.3
Cells Involved in Bone Development and Maintenance

Cell Type	Function
Chondroblasts and chondrocytes	Cells that build and maintain cartilage
Osteoblasts	Young bone-forming cells that cause the development of the hard extracellular of the bone
Osteoclasts	Cells that dissolve bones

TABLE 11.4
Mechanical Properties of Bone

Property	Cortical Bone	Cancellous Bone
Tensile Strength (MPa)	50–100	10–100
Compressive Strength (MPa)	130–230	2–12
Young's Modulus (GPa)	7–30	0.02–0.5
Strain to failure (%)	1–3	5–7
Shear Strength (MPa)	53–70	
Shear Modulus (GPa)	3	

Source: Velasco, Narváez-Tovar, and Garzón-Alvarado (2015).

Table 11.3 summarizes the cells that are involved in the development and maintenance of bones. Table 11.4 lists the mechanical properties of bone.

11.6 DEVELOPING BIOACTIVE COMPOSITE MATERIALS

11.6.1 Bone and the Composite Strategy

In the development of new engineering materials, apart from the other properties required for specific applications, solid and rigid materials coupled with reasonable ductility are always targeted. In the development of new biomaterials for tissue replacement, the structure and properties of the tissue to be replaced, i.e. the biological model, must be considered, because if the properties of the new material are significantly different from those of the host tissue, the developing material will cause dynamic changes in the host tissue after implantation, as discussed by Wolff's law, and will therefore not achieve the objectives integrated into the original conceptual design. It is therefore essential to have a good understanding of biological models before developing new biomaterials.

Bone serves as a model for making new materials for the replacement of hard tissue. Bone is a typical composite material with a complex structure in which some levels of organization, from macro- to microscale, can be identified, as seen

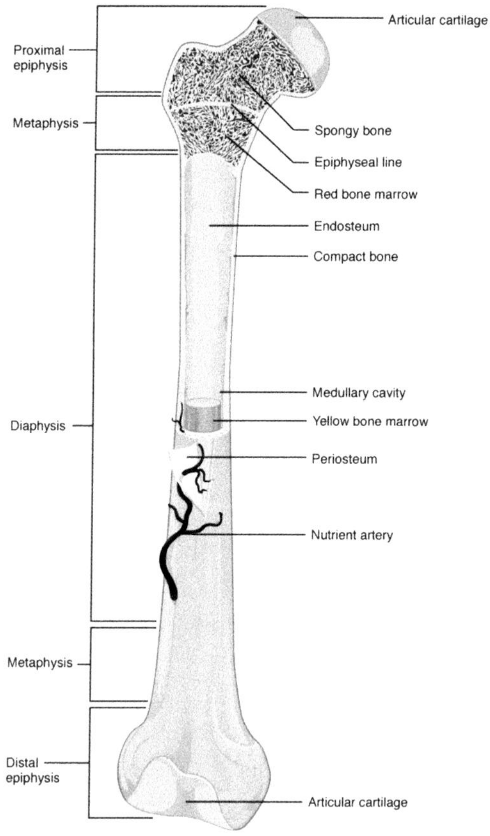

FIGURE 11.8 Structural organization in a human long bone (Betts et al., n.d.).

in Figure 11.8 (Betts et al., n.d.). Two levels of composite structure are considered when developing bone substitutes: firstly, bone-apatite-reinforced collagen forming individual lamellae at nm–mm scale, and secondly, interstitial bone reinforced by osteon on scale mm to mm. It is the microscopic apatite–collagen composite that provides the basis for the production of bioactive biopolymer composites as analogue biomaterials for bone replacement (Bonfield, 1987). The mechanical properties of bone have been well documented, which serve as a benchmark against which the mechanical performance of bone-analogous materials is evaluated. As an anisotropic material, cortical bone has a range of associated properties rather than a set of unique values (Bonfield, 1987): 7–30 GPa for Young's modulus, 50–150 MPa for tensile strength and 1–3% for elongation at fracture.

Since bone is an apatite–collagen composite material at the ultrastructural level, a polymer matrix composite containing a particulate bioactive component appears to be a natural choice for substitution of cortical bone. The bioactivity of the composite, which is reduced by the bioactive component in the composite, resolves to promote tissue growth adjacent to the implant and the formation of a strong bond between the tissue and the implant after implantation. The matrix polymer will provide ductility and other associated properties which are required of hard tissue replacement materials.

11.6.2 Biomaterials for Hard Tissue Repair

In a few years, the market for biomaterials in orthopedics is increasing at a high rate. While in the past, the materials needed for implantation had to be "bioinert", scientists have moved to the "bioactive", which interact with biological molecules or cells and regenerate damaged tissues. In the case of bone tissue engineering, the material should preferably be inductive, osteoinductive, osteoconductive, and capable of osseointegration, i.e. be able to integrate into the bone–bone surrounding.

Over the past two decades, many bone replacement materials have replaced the need to transplant autologous and allogeneic bones. In bioactive ceramics of the present day, biological or synthetic polymers, bioactive glasses, and composites of these materials are used. The most recent advances are in the development of bioactive materials for bone regeneration. Particular attention is paid to recent developments in bioactive glasses containing sintered Na, bioactive glasses containing borate, and elastomeric composites. Bioactive glasses are not new to bone tissue engineering, as their highly adjustable mechanical properties, their degradation rate, their ability to withstand bone, and the differentiation of osteoblasts from stem and progenitor cells are superior to other biomaterials than bioceramics. Although boride and other trace elements have a beneficial effect on bone remodelling, there is a high risk associated with the design of new composite materials. Elastomeric composites are superior to thermoplastic matrix composites and have excellent elastic properties, ideal for the replacement of a key elastic protein (collagen) in bone tissue. According to the paradigm of tissue engineering, these materials should be able to be resorbed and replaced by newly formed biological tissue of the body with the passage of time, and offer both mechanical integrity and flexibility in the environment of injured or damaged bones.

11.6.3 Bioactive Bioceramics

Bone apatite is one of the biological apatites that make up the mineral phase of calcified tissues in the body. The use of a synthetic compound similar to bone apatite is perceived as advantageous to the replacement of hard tissue compared to other synthetic materials. Consequently, there has been sustained interest over the past 20 years in hydroxyapatite (HAp, Ca_{10} $(PO_4)_6(OH)_2$), which resembles bone apatite and is a member of the calcium phosphate family that is part of the bioceramic bio cyclic group. HAp has excellent biocompatibility and is an osteon conductor (Ladizesky et al., 1998). It has been clinically used alone as a bioactive material in the form of a powder, a porous structure, or a dense body (Hench, 1993). However, the most widespread success of HAp is its use as a bioactive coating on total hip prostheses. Another attractive member of the family of calcium phosphates for medical applications is tricalcium phosphate (TCP, Ca_3 $(PO^4)_2$), which plays an important role as resorbable bioceramic. TCP was used for bone repair in the form of ceramic blocks, granules, or calcium phosphate cements (Bonner et al., 2002). HAp and TCP are weak biological ceramics and therefore cannot be used alone as major carrier implants in the human body.

Bioglass and A-W glass ceramics are also bioactive bioceramics that have been used successfully for tissue replacement. Bioglass is a family of bioactive glasses containing SiO_2, Na_2O, CaO, and P_2O_5 in specific proportions. A particular advantage of bioglass (45S5 bioglass) is its ability to bind to hard and soft tissue (Hench, 1991). The shortcoming of bioglass is a mechanical weakness and a low breaking strength due to an array of two-dimensional amorphous glass. The flexural strength of most bioglass compositions is in the range of 40–60 MPa, which is not suitable for major load applications. By heat treatment, suitable glass can be transformed into crystalline–crystalline composites containing crystalline phases of controlled sizes and contents. The resulting glass ceramic may have superior mechanical properties to the parental glass as well as to the sintered crystalline ceramic. The bioactive A-W glass ceramic is therefore made from the parent glass in the pseudo-ternary $3CaO{\cdot}P_2O_5$-$CaO{\cdot}SiO_2$-$MgO{\cdot}CaO{\cdot}2SiO_2$ systems, which is produced by the conventional quenching process. The bioactivity of the A-W glass ceramic is much higher than that of sintered HAp. The A-W glass ceramic has excellent mechanical properties and has therefore been used clinically for iliac prostheses and vertebrae and as intervertebral spacers (Kokubo, 1992). Bioceramics such as HAp, TCP, bioglass, and A-W glass ceramics can be used in the form of particles as bioactive reinforcement phases in bioactive tissue substitutes.

11.6.4 Synthetic Biodegradable Polymers

The most thoroughly researched synthetic biodegradable polymers are the poly (also known as polyesters). These synthetic polymers can be synthesized from a wide range of monomer units via aperture and condensation polymerization methods. The poly(hydroxyl acid) has an ester linkage, which is cleaved by hydrolysis, which results in a reduction in the molecular weight (Mw) of the polymer. However, this reduction in Mw does not decrease the mass of the implant materials. The degree of degradation of the polyesters depends on the exposed surface, the crystallinity, the initial Mw, and the ratio between the hydroxyl ions and the monomers (in the copolymers).

The most studied and most widely used polymers of the poly(hydroxy acid) class are poly(glycolic acid) (PGA), poly(lactic acid) (PLA) and their poly(lactic-co-glycolide) (PLGA) copolymer. In addition to PGA, these polymers are soluble in a variety of organic solvents and can therefore be treated by many solvent-based and heat-based methods. These polymers are considered suitable candidates for bone repair and regeneration applications because they are biocompatible and biodegradable in the human body. Biodegradation is mediated by hydrolytic degradation through the de-esterification process and the removal of the monomer by-products takes place through natural excretion pathways. All the polyesters, theoretically, can be rendered degradable by the esterification method. However, this is a chemically reversible process and only the aliphatic chains between the ester linkages can degrade within the time required to be useful for biomedical applications.

Implantable devices for internal fixation for fracture repair have been made using these polymers. They first generated interest three decades ago when polyesters were used for suture materials and remain one of the most widely used synthetic biodegradable polymers. When polyesters are used alone for the manufacture of devices, the mechanical properties of the highly porous scaffolds are relatively lower than those required for bone tissues engineering applications. They also reduce the local pH in-vivo due to degradation products, which, in turn, accelerates the rate of implant degradation to a degree that limits their clinical utility. Another disadvantage of this rapid disintegration is that acid degradation by-products (monomeric hydroxyl-carboxylic acids or oligomers) induce an inflammatory reaction.

11.6.4.1 Poly(Glycolic Acid)

Poly(glycolic acid) (PGA) (Figure 11.9) is a highly crystalline synthetic polymer (45% to 50% crystallinity) of glycolic acid (Sheikh et al., 2015). Due to high crystallinity, melting point (>200°C), tensile modulus, and controlled solubility, PGA was first employed for clinical use as sutures and as biomedical implants. PGA possesses a high degradation rate due to its hydrophilic nature and the mechanical strength of PGA after implantation for 14 days usually decreases by 50% and by ~90% after 28 days. The degradation product of PGA is hydroxyacetic acid and is either metabolized by the liver (with CO_2 and H_2O as final products) or discharged through the kidneys via the urine. Biodegradation, no aggregation, and a lack of cytotoxic response are the main advantages of using PGA as a degradable material.

PGA has been used as self-reinforced foam and is stiffer (Young's modulus of 12.5 GPa) than other degradable polymers for clinical use. Also, PGA loses its mass in 6–12 months due to in-vivo degradation. PGA has been evaluated as a biomaterial for the fabrication of devices used for the internal fixation of bone. Since PGA loses its strength after implantation with time, this limits its usefulness for load-bearing fractured segments. PGA has similarly been reinforced by amorphous carbonated-apatite and used as a bone replacement graft material, but it was observed that this material was only useful in small defects or non-loading-bearing situations.

11.6.4.2 Poly(Lactic Acid)

Poly(lactic acid) (PLA) (Figure 11.10) is an aliphatic thermoplastic polyester with linear polymeric chains and undergoes in-vivo biodegradability via enzymatic and hydrolytic pathways (Sheikh et al., 2015). PLA has excellent mechanical and thermal properties, is biocompatible and biodegradable, and has a renewable source, which makes it affordable and available for biomedical applications. Lactic acid is a chiral molecule and exists as two stereoisomeric forms, which result in distinct polymers

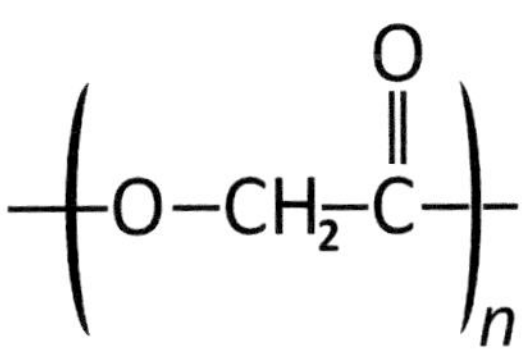

FIGURE 11.9 Structural formula of poly(glycolic acid) (Sheikh et al., 2015).

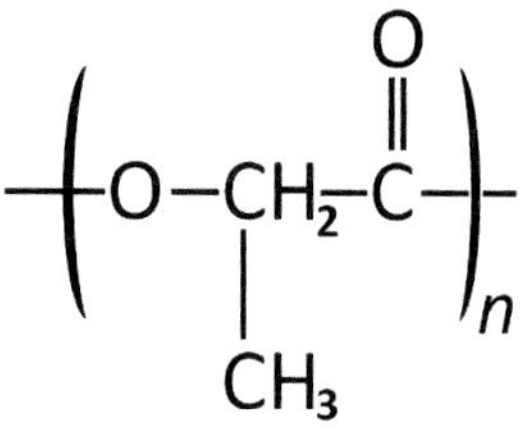

FIGURE 11.10 Structural formula of poly(lactic acid) (Sheikh et al., 2015).

based on morphology such as L-PLA, D-PLA, D, L-PLA, and meso-PLA. L-PLA and D-PLA are stereoregular, D, L-PLA is a racemic polymer (a mixture of L- and D-lactic acid), and meso-PLA is obtained from D, L-lactide. Crystalline L-PLA, which is resistant to hydrolysis, and amorphous D, L-PLA, which is more sensitive to hydrolysis, are mostly used for clinical applications. In-vivo, the lactic acid that is released by poly(L-lactic acid) (PLLA) degradation is converted into glycogen in the liver or incorporated into the tricarboxylic acid cycle and excreted from the lungs as water and carbon dioxide.

Scaffolds fabricated for bone tissue engineering applications need specific material properties (porous architecture, adequate porosity levels, and mechanical strength) and therefore L-PLA is preferred in orthopedic applications because it satisfies most of these requirements. PLLA has been investigated as a biomaterial and fabricated into scaffolds by utilizing salt leaching, phase separation, and gas induced foaming methods. These technologies and methods can be used to fabricate porous polymers having a porosity below 200 μm. However, they do not allow control over porosity in the 200–500 μm size range, which is imperative for new bone formation and vascular ingrowth. Precise extrusion manufacturing (PEM) is another method that has been used to produce PLLA scaffolds. The scaffold porosity was ~60%, but the effectiveness of adequate porosity distribution resulted in improved mechanical properties (~8 MPa compressive strength). Thermally induced phase separation (TIPS) is another technique that can be used successfully to fabricate highly porous scaffolds for bone tissue engineering. This technique utilizes dioxane as a solvent and can be used to create a composite structure having interconnected pores of PLLA with hydroxyapatite. The porosity obtained by this technique can be as high as 95%, with pore sizes ranging from a few microns to several hundred microns, along with an improvement in the mechanical properties (from ~6 MPa for PLLA alone to 11 MPa for the composite). The composite skeleton of the PLLA/HAp implant shows good bonding to bone structure. The compressive strength, porosity levels, and distribution and interfacial properties have been improved upon further by using micro- and nanosized HAp, which encourages molecular interactions and the formation of chemical linkages between the PLLA matrix and the inorganic fillers. PLA synthetic polymers have also been utilized to create a partially degradable bone graft for supporting weak bone in the proximal femur. This biomaterial comprises an outer elastic layer of D, L-PLA, HAp and calcium carbonate with an inner layer of titanium dip-coated into solutions of PLLA with suspended calcium salts. D, L-PLA, owing to its fast degradation rate, is strategically placed on the outside to promote biodegradation and replacement with new bone tissue. The PLLA degrades gradually and provides the biocompatible interface among the biological tissues and the inert metallic core of the implant, which provides the required mechanical stability. Table 11.5 lists a summary of PLA composites as well as their enhancements.

11.6.4.3 Poly(Lactide-co-glycolide)

PLGA is designed by the combination of lactic and glycolic acid. L- and D, L- have both been utilized for copolymerization and when used in the compositional range of 25%–75%, form an amorphous PLGA polymer. PLGA used in clinical applications has been shown to be biocompatible, non-cytotoxic, and non-inflammatory. Although PLGA has been extensively used in a variety of clinical applications, its use is limited in the field of orthopedics. The reason for this is probably the hydrophobic nature of PLGA, which does not support cell adhesion for promoting bone ingrowth. By altering the unit ratio of lactide to glycolide and the molecular weight (Mw), its biodegradation and mechanical properties can somewhat be controlled. Even with optimization, PLGA is not an ideal candidate to be used for load-bearing applications due to the low mechanical strength.

11.6.4.4 Poly(ε-Caprolactone)

Poly(ε-caprolactone) (PCL) (Figure 11.11) is an aliphatic polyester that is semi-crystalline and can be processed in various forms due to its being highly soluble in a variety of organic solvents (Sheikh et al., 2015). PCL is a polymer that has a very high thermal stability when compared with other aliphatic polymers. The decomposition

TABLE 11.5
Composites of PLA and Their Enhanced Properties

Composite of PLA	Enhancement	Reference
n-HAp/PLA	Porosity, protein adhesion, bioactivity	Kothapalli et al. (2005)
Sol-gel bioactive glass/PLA	Hydrophobicity, retention of mechanical properties for longer duration	Sepulveda et al. (2001)
PLA/CDHAp	Biodegradability, cell adhesion, and growth capability	Zhou et al. (2011)
PLLA/PLA	Lower crystallinity, faster hydrolysis, and degradation	La Carrubba et al. (2008)
SR-PDLLA/PLLA	Mechanical strength, retention of strength for longer duration	Majola et al. (1992)
Collagen/PLA	Tensile strength, fibrous tissue ingrowths	Liao et al. (2004)
PLA/Starch	Young's modulus, water absorption	Avérous (2004)
PLA/Modified TiO_2	Improved crystallinity, improved mechanical properties, reduced complex viscosity during moulding	Mahshid et al. (2011)
PLA/Organo-montmorillonite	Thermal properties were greatly enhanced	Depan, Kumar, and Singh, (2009)

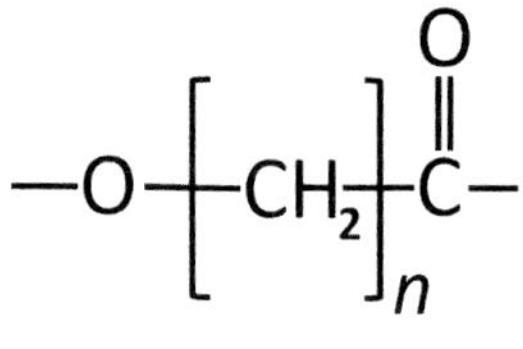

FIGURE 11.11 Structure formula of poly(ε-Caprolactone) (Sheikh et al., 2015).

temperature (Td) of PCL is 350°C, while the Td of aliphatic polyesters is usually between 235°C and 255°C.

PCL is a biodegradable and biocompatible polymer that used for bone repair and treatment of bone defects. However, PCL is not an ideal biomaterial for these purposes due to its low degradation rate and lower mechanical properties. The fusion and diffusion technique was used to enhance PCL with HAp. Using this method, the polymer is completely melted and the HAp particles used as reinforcing fillers are dispersed in the polymer matrix. The particle size used to make the composites with this technique is very important. It was observed that HAp particles with a size range of 3 to 8 μm imparted higher compression resistances to the composite materials. Although the addition of fillers improves the compressive strength, increasing the content of the filler more than a certain level makes these PCL/HAp composites too fragile for clinical use.

11.6.4.5 Benzyl Ester of Hyaluronic Acid

The benzyl esters of hyaluronic acid are also known as HYAFF-11; they demonstrate good degradation and their degradation products are non-toxic. The degradation time varies from 1–2 weeks to 2–3 months and occurs by hydrolysis via ester linkages. Degradation depends on the degree of esterification with de-esterified HYAFF-11 being more soluble and resembling the precursor hyaluronic acid. HYAFF-11 has been studied for use in bone tissue engineering and vascular graft preparation applications. HYAFF-11 was enhanced by α-tricalcium phosphate (α-TCP) to form a hydrogel. It has been seen that the compressive strength is improved from ~3 MPa for pure HYAFF-11 to ~17 MPa for the hydrogel. This increase in the value of the compressive strength, being closer to the supposed bone strength, suggests that these HYAFF-11-based hydrogels can be used as bio-resorbable bone fillers for orthopedic and oral maxillofacial applications.

11.6.4.6 Poly-Para-Dioxanone

Poly-para-dioxanone (PDS) (Figure 11.12) is a polymer consisting of repeated ether–ester units (Sheikh et al., 2015). PDS is obtained by ring-opening polymerization of

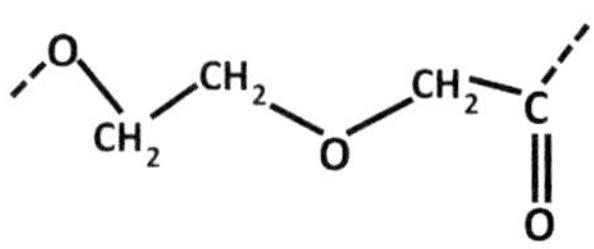

FIGURE 11.12 Structural formula of poly-para-dioxanone (Sheikh et al., 2015).

the para-dioxanone monomer. PDS is a polyester used in the field of medicine in the form of films, laminates, moulded products, foams, adhesives, and surface coatings. Due to its excellent biocompatibility, biodegradation, and flexibility, PDS has been studied for use in tissue regeneration and fracture repair applications. The PDS used for internal fixation of fractures has been shown to be completely biodegradable in bone tissue. The PDS can be resorbed completely in-vivo within 5–7 months by altering its crystallinity, molecular weight, and melting temperature.

11.7 FACTORS INFLUENCING BIOACTIVE COMPOSITES

A composite material consists of two chemically distinct phases (metallic, ceramic, or polymeric), which are separated by interfaces. A composite is designed to have a combination of the best characteristics of each of the component materials. The classification of engineering composite materials is based on matrix materials (metals, ceramics, and polymers) or reinforcement sizes/shapes (particles, mustache/short fibers, and continuous fibers) (Hull and Clyne, 1996; Schwartz, 1992). Most engineering composite materials are developed to provide unique mechanical properties such as strength, stiffness, toughness, and fatigue resistance. For biomedical composites, although excellent mechanical performance is desirable and often targeted for improvement, the biocompatibility of the material is paramount. Biological compatibility is more important than mechanical compatibility. Being composed of two or more types of materials, composites have an increased likelihood of causing unfavourable tissue reactions. However, bioactive and resistant composites have the advantage of overcoming the fragility problems of bulk bioceramics while maintaining a bioactive response in-vivo. The classification of biomedical composites can be based on matrix materials or on the bioactivity of composites (at least one of the constituent materials of a composite must be bioactive, which may render the composite bioactive; in some cases, two or all of the components are bioactive). Using the matrix material as the basis of classification, there are three types of biomedical composites:

Polymer matrix composites, e.g. carbon/PEEK (carbon reinforced polyether ether ketone), HAp/HDPE (Hydroxyapatite-reinforced High-density polyethylene).
Metal matrix composites, e.g. HAp/Ti (Hydroxyapatite-reinforced Titanium), HAp/Ti-6Al-4V (Hydroxyapatite-reinforced Titanium alloy).
Ceramic matrix composites, e.g. stainless steel/HAp, glass/HAp.

Using the bioactivity of composites as the basis for classification, there are also three types of biomedical composites:

Bioinert composites, e.g. carbon/carbon, carbon/PEEK.
Bioactive composites, e.g. stainless steel/bioglass, HAp/HDPE, HAp/Ti-6Al-4V.
Bioresorbable composites, e.g. TCP/PLA (Tricalcium phosphate-reinforced Polylactic acid), TCP/PHB (Tricalcium phosphate-reinforced Polyhydroxybutyrate).

Two types of reinforcements are normally used for biomedical composites: fibers and particles. With few exceptions, fibers and particles in biomedical composites are harder and stronger than the matrix and thus reinforce composites. Because the reinforcement and the matrix interact with each other in different ways in different composite systems, composites must be treated individually. The properties of biomedical composites are strongly affected by a number of factors, some of which are listed below:

- Reinforcement shape, size, and size distribution
- Reinforcement properties and volume percentage
- Bioactivity of the reinforcement (or the matrix)
- Matrix properties (molecular weight, grain size, etc.)
- Distribution of the reinforcement in the matrix
- Reinforcement–matrix interfacial state

Among these factors, the properties of the constituent materials are major influencing factors. However, factors such as the composite architecture (percentage of reinforcement, distribution, orientation, etc.) and the strength–matrix bond condition also play an important role. By carefully supervising these factors, the mechanical and biological performance of bioactive composites can be designed to meet various clinical requirements. Brief discussions on the major factors of polymers filled with bioactive particles are presented in this section. The physical characteristics (shape, size, etc.) of the reinforcement are very important in determining the mechanical properties of a composite. In the idealized situation for the mathematical modelling of the mechanical behaviour of a particulate composite, it is normally assumed that the reinforcement has a spherical shape (Figure 11.13a). In fact, the bioactive reinforcing particles may have an irregular, flat, or acicular shape. The HAp particles in the commercially available spray-dried powders may have an irregular shape, shown in (Figure 11.13d), which are composed of highly bonded HAp crystallites. This type of irregular shape is preferred over the spherical shape since the molten polymer can penetrate the hollows on the surface of the particle during the high temperature composite treatment and thus form a mechanical locking with the particle at ambient or body temperature. With the smooth surface of the sphere, the particles do not provide such a locking mechanism and, in the absence of chemical bonding between the polymer and the particle, will be dissociated from the polymer when a tensile stress is applied. Commercially

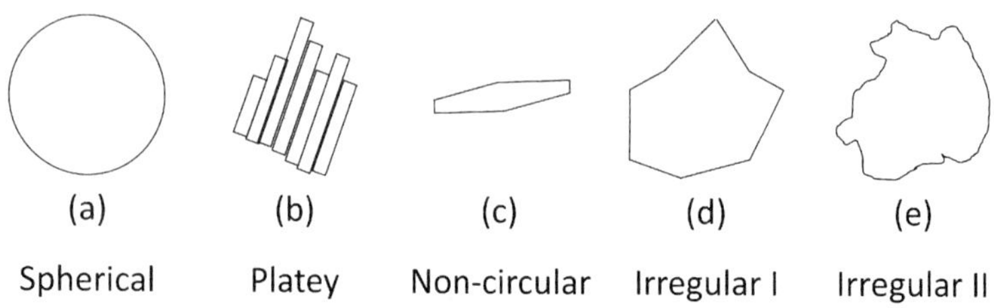

FIGURE 11.13 Shape of bioceramics for biomedical composites.

produced TCP particles can also take this irregular shape but have micro/nanopores in the particles. Therefore, under sufficient shear stress during the composite heat treatment, the porous TCP particles can fragment into smaller particle fragments. If the bioactive (or vitroceramic reinforcing) glass particles are made by the conventional glass making process (i.e. melting and tempering), the glass particles assume the shape shown (Figure 11.13c), which has sharp angles. The particles of this form cause stress concentrations in the composites around the cutting edges and are therefore not preferred. An additional milling process may be required to remove (or at least reduce) the pointed ends of the glass particles prior to composite processing. The plate shape (Figure 11.13b) is not usually encountered for particles in bioactive composites. When the calcium phosphate particles produced by the precipitation method are directly used for the composites, the particles of nanometric size generally have an acicular shape (Figure 11.13c). In such a situation, the aspect ratio (i.e. width/length ratio) of the particles is an important parameter and the orientation of the acicular particles should be considered. By using conventional plastics processing technology to produce bioactive composites, the average size of the bioactive particles (primary particles) normally varies from several micrometers to tens of micrometers. The fine ceramic particles tend to combine between strongly bonded aggregates, which can still unite to produce even larger structures, commonly called "agglomerates". To form high-quality, high-performance ceramic polymer composites, agglomerates or aggregates of particles must be decomposed during the composite treatment into primary particles (i.e. the smallest particulate parts of the minor component present in the powder of ceramic manufactured or as-received) are sufficiently dispersed in the polymer matrix (Figure 11.14). The dispersion of the condensed state particles (Figure 11.14a) in the intermediate state (Figure 11.14b) may not be sufficient because the contact points of the particles provide crack initiation sites or act to improve the propagation of the cracks, resulting in premature failure of the composite when the composite is subjected to mechanical stresses. Ideally, the particles present in the composite should be in a dispersed state, as indicated in (Figure 11.14c). Therefore, specially designed processing equipment is often required which produces sufficiently large shear forces to overcome various particle adhesion forces during the composite melt processing

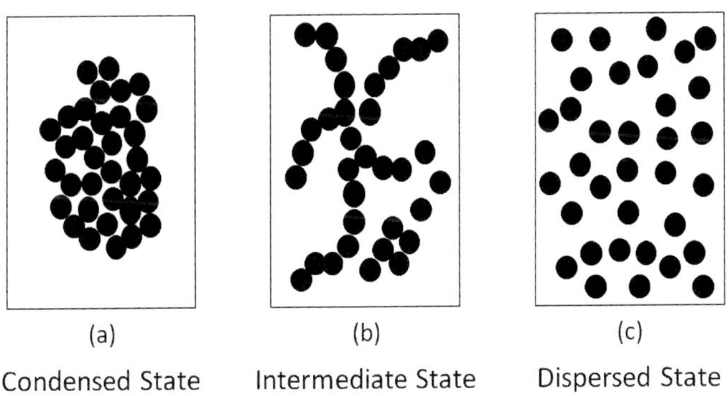

FIGURE 11.14 Possible distributions of bioceramic particles in biomedical composites.

so that agglomerates or aggregates of particles can be reduced to primary particles and the primary particles can be uniformly distributed in the composite (Ess, 1987) and (Wang, 1994).

Even with machines well designed to disperse rigid particles in a soft matrix, the input energy has to be carefully controlled. Inadequate energy input does not result in breakage of agglomerates or aggregates of particles (Figure 11.15). On the other hand, excessive energy input can cause fragmentation of the primary particles, which may not induce beneficial effects and accompanied heat generation, resulting in thermal degradation of the polymer. In addition to the operating conditions of the machine, which include the design of the machine (mainly for the generation of high shear forces), energy input (rotor or screw speed), temperature, and pressure, other factors such as the characteristics of the ceramic particles (i.e. particle morphology, size, etc.), inter-particle attraction, surface treatment of particles, and the volume fraction of particles in the composite can significantly affect the dispersion and distribution of the particles in the composite. The surface treatment of bioceramic particles may not alleviate the processing difficulties and may not necessarily lead to an improved dispersion of the particles (Wang, 2001). Analysis of the efficiency of a manufacturing process with respect to the dispersion and distribution of particles in a composite can be assisted using scanning electron microscopy and image analysis techniques (Wang, 1996). At different stages of processing, the analysis of the polished samples is imaged and the images analysed. It should be borne in mind that polished samples provide only two-dimensional diameters (circular equivalent) of ceramic particles. To determine the three-dimensional average volume diameter of these particles (spherical equivalent) in the composite, stereology as well as image analysis should be used [85,86]. To produce bioactive bone substitute materials, the bioactive phase of a particular composite must exceed a certain volume fraction. Below this volume fraction, even if the bioactive phase is incorporated into

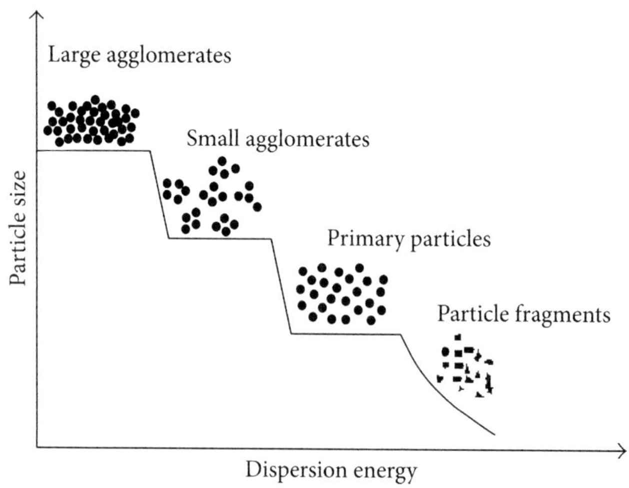

FIGURE 11.15 Schematic diagram showing the relationship between the average sizes of dispersed particles and the dispersion energy required. (Njuguna, Vanli, and Liang, 2015).

the matrix, the composite may not possess the desired bioactivity. It has been demonstrated that for the hydroxyapatite-reinforced high-density polyethylene composite, the critical HAp volume percentage is about 20%, above which bone apposition could occur on a composite implant. Similarly, for other composites to be useful bone replacement materials, the bioceramic content in these composites should be greater than the minimum amount(s) (probably about 20vol.%). In this regard, with respect to particulate-filled polymers, bioactive composites containing 20vol.% or more of bioceramic are highly filled polymer systems. In the plastics industry, it is generally recognized that highly charged high-quality polymers are very difficult to produce unless specially designed machines are used and considerable experience in the processing of plastics has been gained. Even in the face of problems such as highly filled polymers, which are highly viscous at their processing temperatures, a reasonably uniform distribution of the bioceramic particles in the composite must be ensured. Structural defects such as micro- and macropores and cracks can often be present in moulded parts because air is trapped in the mouldings and the differences in physical properties between the bioceramic and the polymer. However, such defects are obviously intolerable for medical devices that are intended to improve the quality of life of patients without their premature failure in service.

The packing behaviour of bioceramic particles in the polymer matrix is an important factor in the understanding and also the design of bioactive composites. For every filled polymer system, there is a maximum volume of particles that can be incorporated before a continuous network of the particles is formed and voids. The packing behaviour of particulate materials depends largely on particle size, shape, and surface characteristics. For a theoretical analysis, it is assumed that the particles have a mono-modal size distribution with a sharp peak (Curve (a) in Figure 11.16) or even just have one uniform size, which makes it difficult to achieve high packing density. In reality, most as-produced (or as-received) ceramic powders have large size distributions, sometimes with a small particle size range, as in Curve (b) in (Figure 11.16). The other ceramic powders may exhibit bimodal size distributions

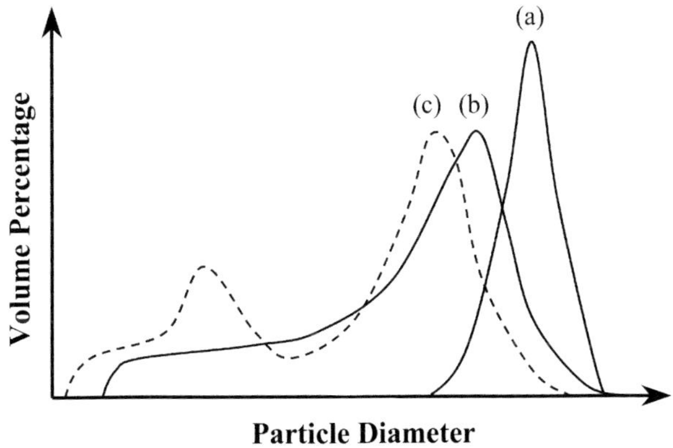

FIGURE 11.16 Schematic diagram showing the particle size distribution of particulate reinforcement (a) mono-modal size distribution (b) mono-modal size distribution with a long tail end (c) bimodal size distribution. (Wang, 2003).

shown in Curve (c) in (Figure 11.16). The two-dimensional distribution of the particles is more efficient (Figure 11.17), as the small particles can occupy the space between large particles, a high bioceramic content per unit volume in the composite. The present invention relates to a method for the production of ceramic particles in the form of a ceramic material. The interphase between the reinforcement and the matrix in composites determines the mechanical behaviour and properties of composite materials and their impact on the structure of the composite material and the interfacial bonding, and avoiding catastrophic failure. For engineering composite materials, the silane coupling agents are often used for glass fibers in fiber-reinforced plastics (FRPs) in order to provide a strong chemical link between the oxide groups and the polymer molecules of the resin. In most bioactive composites, chemical bonding does not exist and the interfacial bond strength totally depends on the mechanical interlock between the bioceramic particles and the polymer matrix. In a theoretical analysis, it is shown that if the composite is under tension, high stress concentrations develop at the poles of spherical particles (Figure 11.18). In the polar area, when the tensile stress exceeds the low interfacial strength provided by the locking mechanism, debonding of the bioceramic particle from the polymer matrix inevitably takes place. To prevent (or delay) the debonding process, it appears necessary to provide chemical links between bioactive particles and the matrix polymer for biomedical composites.

The hard bioceramic particles in composites not only provide the reinforcement but also render the composite bioactive when there is a sufficient amount of the particles in the composite. For achieving the reinforcing effect, factors such as the size, shape, and mechanical properties of the particles need to be considered. For example, the Young's modulus values of bulk HAp, bioglass and A-W glass-ceramic are 80–120 GPa, 30–35 GPa, and 118 GPa, respectively. If a high bioactivity level is desired for achieving a strong bond between the composite implant and the host tissue within a short period, bioceramics exhibit high degrees of bioactivity. Different bioceramics have their own characteristics, and the judicial selection of a particular bioceramic for the composite is based on the clinical requirement, the composite production route, and sometimes the cost involved. When the degradation behaviour of the polymer is biodegradable, the molecular weight of the polymer can be modified. It is necessary to ensure that the materials used in the composite materials are

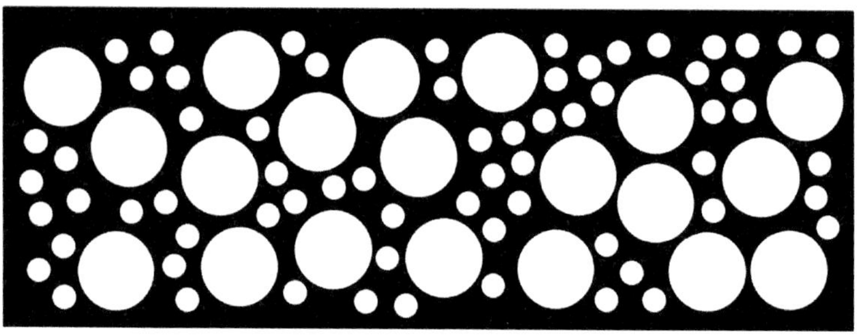

FIGURE 11.17 Schematic diagram showing the distribution in a polymer matrix of bio ceramic particles of a bimodal size distribution.

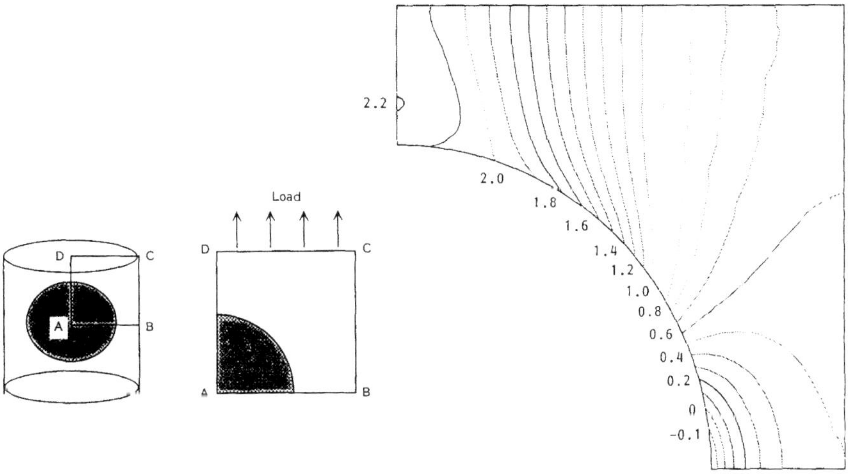

FIGURE 11.18 Stress concentration around spherical bio ceramic particles in bioactive composites (Guild and Bonfield, 1993).

of the highest quality. However, the thermally unstable, thermally non-defective, thermally non-decomposable polymerization of the thermoplastic polymer is not possible.

11.8 BIOACTIVE COMPOSITES FOR HARD TISSUE

Since the pioneering work in the bone field of using hydroxyapatite as a bioactive phase and a reinforcement of high-density polyethylene to produce a bone analogue, a number of bioactive composite systems consisting of bioceramics and biomedical polymers have been studied. In this section, only a few systems presented in (Figure 11.19) are briefly reviewed. There are other bioceramic polymer

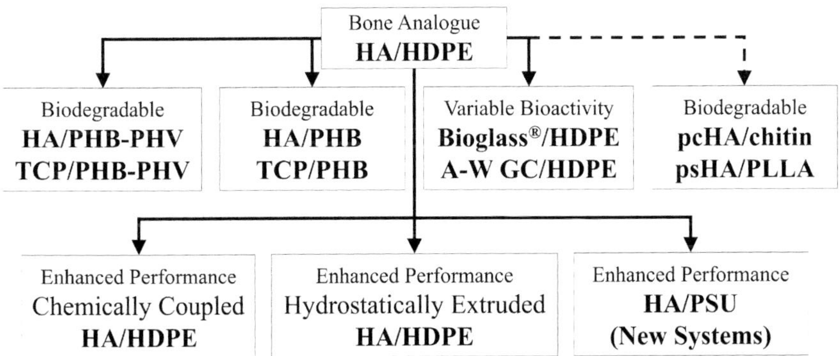

FIGURE 11.19 Development of bioactive composites for medical applications (Min Wang, 2003).

systems that have been or are being investigated for tissue replacement. The particular combination of a bioceramic with a polymer for a composite is based on important factors that control composite performance and composite production, most of which have been discussed in the previous sections. The production of bone analogues using polymers as templates has been lengthy compared to the production of bioactive composites for tissue replacement using metal matrices or ceramic matrices. In the case of bioactive metal matrix composites, metallic matrices provide the necessary strength and toughness. In the case of ceramic matrix composites, bioactive ceramics are most likely to be matrices and the incorporation of vitreous material or metal fibers results in the ceramic being cured. Bioactive metal matrix composites and bioactive ceramic matrix composites have their drawbacks and disadvantages as tissue replacement materials. It is nevertheless interesting to study the possibilities of developing these materials for their intended applications. The use of bioactivity of bioceramic particles in composites for tissue replacement led to research into the production of new materials such as bioactive bone cement and bioactive dental materials. These new materials, with the incorporation of bioceramic particles, could induce or improve adjacent tissue formation and finally establish a strong bond with the newly formed tissue. Tissue engineering has emerged in recent years as a promising and viable way to solve the problems of tissue loss and organ failure. One of the major problems in tissue engineering is the development of biodegradable scaffolds suitable for cell seeding and subsequent tissue growth. There are a number of candidate polymers for tissue scaffolds and various techniques have been used to produce polymer scaffolds. With appropriate modifications, some commonly used manufacturing techniques have been used to make bioactive scaffolds, which contain bioceramic particles, for tissue engineering applications. These bioactive scaffolds must improve cell adhesion and tissue formation while also possessing higher strength and stiffness than their polymer homologues in the initial stages of cell seeding and subsequent tissue growth.

11.8.1 Hydroxyapatite-Reinforced High-Density Polyethylene (HAp/HDPE)

The hydroxyapatite-reinforced high-density polyethylene (HAp/HDPE) composite (also known as HAPEXTM from 1995, when Smith & Nephew Richards Inc. introduced its mid-sized composite ear implant series) is the first composite bioceramic and biopolymer designed to mimic structure and associated bone properties, which resulted in the research and development of other bioactive composites (Figure 11.19) using the same rationale. The main advantages of using polyethylene as a matrix material are as follows: it is a proven biocompatible polymer widely used in orthopedics; it is a ductile polymer which allows the incorporation of a large number of bioceramic particles in the system; the polymer having a high bioceramic particle content can still be melt processed using current plastics technology; and HDPE is a linear polymer whose molecular chains can be aligned for the improvement of properties when advanced processing technology is used. Development of HAp/HDPE compositions involved the use of calcined bone ash (CBA), commercially available synthetic Hap and synthetic Hap produced in-house, and may involve the use of carbonaceous apatite or other substituted apatites that are more similar to bone apatite, and therefore more bioactive than HAp. Different polyethylene and different grades of HDPE have also been used for scientific research and product development. The HAp/HDPE system takes progressed according to the modern development track of a new biomaterial, conceptual design, production of materials, evaluation (physical, mechanical, and biological), protection of intellectual property (IP), clinical trial, industrial involvement, regulatory approval, and end-use in medical devices for patients. Some research activities on HAp/HDPE composites continue, focusing on a few issues that will help to better understand the system and optimize composites. HAp/HDPE composites containing up to 45vol.% (i.e. 73wt.%) of HAp can be routinely achieved by standard procedures. A twin-screw extruder or internal mixer can be used to efficiently compound the materials. The use of two roll mills appeared inadequate because of their inability to cope with HAp high-density fractions, and also due to degradation of the polymer. Composite plates as thick as 20 mm can be made by compression moulding using composite powders. These plates were free of voids, as revealed by X-ray radiographs. Rheological studies revealed that the combination of particulate HAp into HDPE caused an increase in the viscosity of the composites by their processing temperatures and that there were treatment windows for HAp/HDPE composites containing various amounts of HAp. The swelling rate of the pieces of the HAp/HDPE composite was reduced when the HAp content was increased.

It was demonstrated that following the standardized production procedure the difference between the actual HAp mass percentages in the produced composites and the desired amounts of HAp in the composites was negligible and therefore HAp/HDPE of good compositions had been made. Microscopic examination of the composites revealed that the HAp particles were well dispersed in the HDPE and that the composites had a uniform distribution of HAp particles (Figure 11.20a). Structural analysis using stereology indicated that the high shear forces generated during the composition process separated the agglomerates of HAp particles into unit particles in the polymer matrix. The average particle diameter of the HAp particles in the moulded HAp/HDPE was almost identical to the average particle size of the HAp powder used for composites production. It was found that the average molecular weight of the HDPE was slightly decreased during the composite production process, with the rate of decrease depending on the HAp volume fraction. The incorporation of HAp particles also resulted in a decrease in the degree of crystallinity of the HDPE, with composites of higher HAp content having lower crystallinity degrees for the polymer matrix. Various aspects of the mechanical performance of HAp/HDPE composites have been studied. By varying the amount of HAp in the composite, a range of mechanical properties can be obtained. An increase in the HAp volume percentage results in an increase in the Young's modulus, shear modulus, storage modulus (in dynamic mechanical analysis (DMA)), microhardness, and HAp/HDPE tensile strength, with corresponding decreases in deformation to fracture and attenuation of energy for fracture. The particle morphology and mean particle size of HAp had an impact on the mechanical properties of HAp/HDPE composites. HAp/HDPE with 45vol.% of HAp has a Young's modulus value of 5.54

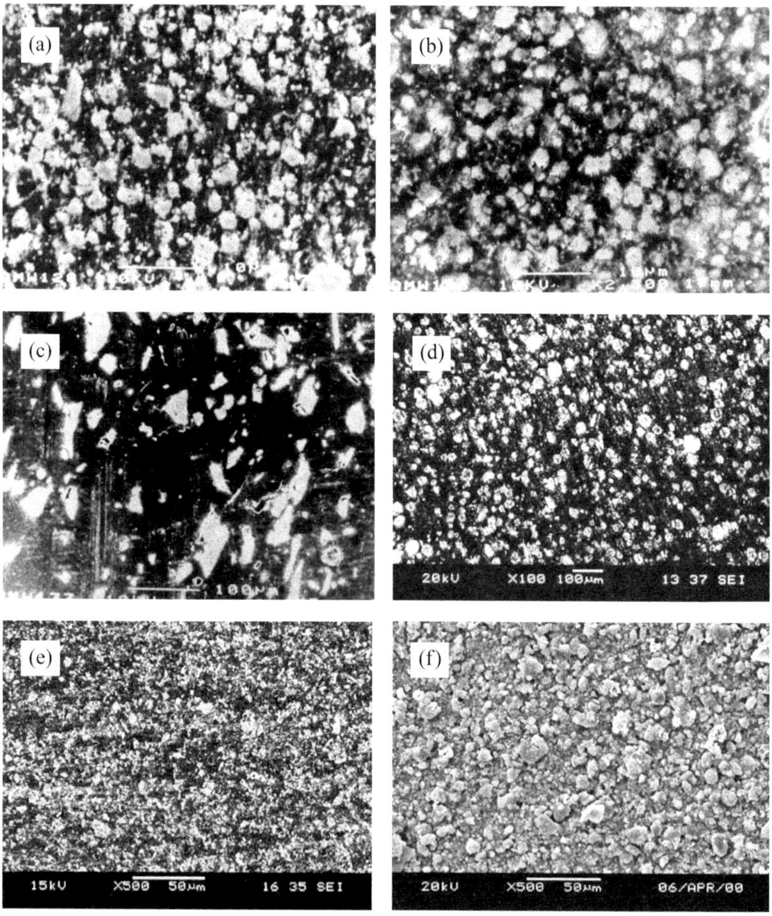

FIGURE 11.20 Microstructure of various bioactive composites (a) HAp/HDPE, (b) hydrostatically extruded HAp/HDPE, (c) Bioglass/HDPE, (d) HAp/PSU, (e) TCP/PHB (f) pcHA/chitin (Wang, 2003).

GPa, which approaches the lower limit for human cortical bone. Examination of the fracture surfaces of HAp/HDPE composites suggested that in the composites there was only a mechanical bond between the HAp particles and the HDPE matrix resulting from the HDPE removal from around the individual HAp particles during the heat treatment. In the aqueous environment, water absorption by HAp/HDPE composites was low (0.1% at 37°C) and the strength and modulus of the composites were not significantly affected after prolonged contact with physiological fluids. The incorporation of HAp particles into the HDPE improved short-term creep resistance when the samples were subjected to similar stresses and an increase in the volume fraction of HAp increased the creep resistance. However, failure of the composites could occur at long periods due to delamination at the HAp/HDPE interface. In general, the fatigue life of HDPE and the HAp/HDPE composite was reduced with increasing shear stress in biaxial stress. The HAp/HDPE composite is not suitable for articulated surface implants.

The in-vitro and in-vivo biological performance of HAp/HDPE composites has also been thoroughly evaluated. In in-vitro experiments using primary cultures of human osteoblastic cells, it was observed that the osteoblast cells attached to the HAp particles of the composite and then proliferated, which clearly showed the biocompatibility and bioactivity of the HAp/HDPE composites. In in-vivo experiments using adult white rabbits from New Zealand, it was demonstrated that after 6 months' implantation into the lateral femoral condyle 40% of the composite surface of the implant was covered with a newly formed bone, exhibiting a good osteotic conductivity of the composites. The biological performance (i.e. bioactivity) of the composites depended on the HAp volume percentage of the composites. HAp/HDPE composites were first used for implants of periosteal orbital soils in the correction of bulk sockets and in the reconstruction of the orbital floor after trauma. Postoperative clinical examinations reported good patient satisfaction and computerized tomograms of patients revealed the integration of implants at the orbital floor 6 months after implantation. Over the previous few years, HAp/HDPE composite ear implants have been developed, having the combined advantage of bioactivity, flexibility, and cutting ability of composites, and satisfactory clinical results have been achieved.

11.8.2 Chemically Coupled HAp/HDPEXTM

As there is only a mechanical bond between HAp particles and the HDPE matrix in HAp/HDPE composites and theoretical analysis has shown that debonding of HAp particles from the matrix can take place at polar areas, silane surface treatment of HAp particles and acrylic acid grafting of polyethylene were investigated for improving the reinforcement–matrix bonding of the composites. Only limited improvements in tensile strength and ductility were achieved while the Young's modulus was slightly decreased. It was observed that the chemical bond established between HAp and HDPE delayed the debonding process but could not prevent de-bonding which caused the eventual failure of the composites.

11.8.3 Hydrostatically Extruded HAp/HDPEXTM

Hydrostatic extrusion is one technology that can be used to align polymer chains, and thereby the mechanical properties of the polymer can be significantly improved. This technique was used to align the polyethylene chains in the HAp/HDPE composites, which is possible due to the linear molecular structure of the polymer. Creating that progressive extrusion ratios consume led to a higher modulus and strength of HAp/HDPE composites, which are within the limits for the mechanical properties of cortical bone (Table 11.6). The HAp/HDPE fracture strain was also substantially increased by hydrostatic extrusion. The extruded HAp/HDPE containing 40% by volume of HAp had a fracture strain that was much higher than that of human cortical bone (9.4% vs 1–3%). Hydrostatic extrusion did not modify the uniform distribution of the HAp particles in the composites (Figure 11.20b) and the bioactivity of

TABLE 11.6
Mechanical Properties of Hydrostatically Extruded HAp/HAp–PE Composites

Extrusion Ratio	HAp Volume (%)	Young's Modulus (GPa)	Tensile Strength (MPa)	Flexural Modulus (GPa)	Flexural Strength (MPa)
1:1	0	0.65	17.9	1.1	23
5:1	0	2.59	61.2	2.2	52
8:1	0	4.08	158.2	2.2	48
1:1	40	4.29	20.7	4.7	32
5:1	40	5.89	64.8	7.2	73
8:1	40	9.91	91.2	9.0	88

the composites was retained after extrusion. Therefore, HAp/HDPE subsequently processed by hydrostatic extrusion has great potential for major load applications.

11.8.4 Hydroxyapatite-Reinforced Polysulfone

In addition to polyethylene, there are a few other biomedical polymers that could be used for the production of bone-like materials. Polysulfone (PSU) is an amorphous polymer, which has a high specific strength and modulus. To grow bioactive composites for carrier prostheses, PSU might be a better choice for the matrix of a composite than HDPE because its strength and modulus are significantly higher, which may provide a higher level of mechanical properties for composites. Other favourable properties of PSU include low creep rate, oxidation resistance, excellent hydrolysis resistance or molecular weight reduction, and stability in aqueous inorganic acids, alkaline, and saline solutions, and biological inertia. In addition, PSU has a high radiation resistance to beta, gamma, and infrared radiation and X-rays, and can be sterilized by steam. Therefore, the HAp/PSU composite was developed as a new hard tissue replacement material. Production of the HAp/PSU composite follows the same procedure as for HAp/HDPE composites. The HAp/PSU composite containing up to 40vol.% of HAp was produced. The HAp particles were also well dispersed in the PSU matrix (Figure 11.20d) and the desired amount of HAp in the composite was confirmed. The density close to the theoretical value was obtained for the composite, which indicates a structure without vacuum. Rheological examination revealed that the HAp/PSU composite exhibited pseudoplastic flow behaviour at treating temperatures. With an increase in HAp content, the stiffness of the HAp/PSU composite also increased. The mechanical properties of the HAp/PSU composite are in the lower limit of the bone. As with HAp/HDPE composites, in the biaxial fatigue test, torsional stress considerably reduced the fatigue life of the HAp/PSU composite. It has been found that the HAp/PSU composite is not suitable for articulated surface implants.

11.8.5 Bioglass-Reinforced High-Density Polyethylene

In order to establish a stronger bond between the implant and the tissue in a shorter period of time, glass or ceramics, which are more bioactive than HAp such as organic glass or A-W glass ceramic could be used as bioactive, phase-in composites. Following implantation, bioglass implants can elicit specific physiological responses, including the supply of reactive silica to the surface, calcium and phosphate groups, and alkaline pH levels at the tissue interfaces, thus providing strong bioactivity and conditions to establish a strong tissue-implant bond. Using technology for HAp/HDPE composites, bioglass-reinforced polyethylene composites were produced. It was found that the bioglass particles were well dispersed and a reasonably homogeneous distribution of the particles in the polymer matrix was obtained (Figure 11.20c). The compound containing up to 30vol.% of bioglass exhibited elastic compliance levels, tensile strength, and breaking strain, comparable to those of soft connective tissue.

Compounds having bioglass volumes greater than 30vol.% possessed mechanical properties comparable to spongy bone. In in-vitro experiments using a simulated body fluid, it was found that it took longer for bone apatite to form on bioglass/HDPE composite surfaces than on HAp/HDPE composite surfaces, indicating higher bioactivity of the bioglass/HDPE composite. However, the mechanical properties of the bioglass/HDPE composite decreased over time in the aqueous environment. In in-vitro experiments using human osteoblast (HOB) cells, it was observed that the cells bind to the bioglass particles in the composite, indicating excellent biocompatibility and bioactivity of the composite. Recent transmission electron microscopy (TEM) examination of the interface between the HOB cells and the composite indicated a direct bond between the hydroxide carbonate hydroxide (HCH) layer, which formed on bioglass particles in-vitro and on the HOB cells.

11.8.6 A-W Glass Ceramic-Reinforced High-Density Polyethylene

Bioglass is highly bioactive, but its mechanical properties are low due to the two-dimensional amorphous glass array forming the glass. The A–W glass ceramic (AWGC) has excellent mechanical properties while possessing a high bioactivity. The particulate AWGC can be used as a stiffer reinforcement in the composite while providing the composite with much higher bioactivity than the HAp particles. Therefore, the established processing technology for HAp/HDPE composites was used for the production of AWGC/HDPE composite. As for HAp/HDPE and Bioglass/HDPE composites, a homogeneous distribution of the AWGC particles in the polyethylene matrix was achieved using the standard production procedure. The Young's modulus and the microhardness of the composite also increased with an increase in the volume fraction of AWGC while the tensile strength and the rupture stress decreased. Even with 40vol.% of AWGC particles, the composite still had considerable ductility. Current research on this composite system focuses on its mechanical properties and in-vitro bioactivity.

11.8.7 Calcium Phosphate-Reinforced Polyhydroxybutyrate

Biodegradable materials have attracted attention in the research and development of new biomaterials. These materials are designed to gradually degrade in the body and will eventually be replaced by newly formed material. They could provide time-varying mechanical properties and their use can ensure the complete dissolution of the implant, eliminate long-term biocompatibility problems, or avoid secondary surgical operations. The requirements for biodegradable materials are that they must degrade in the body at a rate that can be controlled, and degradation products should be non-toxic, biocompatible, and easily excreted. Following implantation in the body, biodegradable bone substitute material will have gradual decreases in strength and stiffness for a clinically determined optimal period. As the bone recovers itself, the external charge will be transferred from the biodegradation implant to the bone. This method provides the best biomaterial solution for temporary tissue replacement and eventual tissue regeneration if requirements for initial stiffness and strength and other short-term properties can be met. Composites made of bioactive (and bioresorbable) ceramics and biodegradable polymers have great promises for these purposes. Polyhydroxybutyrate (PHB) is a natural beta-hydroxy acid (linear polyester). Its ability to degrade and reabsorb in the human body environment makes it an appropriate candidate as a matrix for bioactive and biodegradable composite implants that will guide tissue growth and may eventually be replaced by newly formed tissue. Being thermoplastic, PHB can be processed using conventional manufacturing technologies such as extrusion, injection moulding or compression moulding. Therefore, using the technology for the HAp/PSU composite, the HAp and TCP particles were separately integrated with PHB to form composites for tissue replacement and regeneration applications. Particular bioceramics (HAp or TCP) could be distributed homogeneously in the PHB matrix for HAp/PHB and TCP/PHB composites (Figure 11.20e). The stiffness of the composites increased with an increase in the bioceramic content. In in-vitro experiments using simulated body fluid (SBF), bone apatite formed on HAp/PHB and TCP/PHB composites, which was indicative of the bioactivity of these materials in-vivo. With prolonged immersion in SBF (i.e. beyond 2 months), the HAp/PHB and TCP/PHB composites showed decreases in the storage modulus (from the dynamic mechanical analysis(DMA)), which indicates degradation of composites in the simulated body environment. The structure and mechanical properties of bone apatite formed in-vitro on HAp/PHB and TCP/PHB composites are similar to those formed on other bioactive materials.

11.8.8 Calcium Phosphate-Reinforced Chitin

Chitin is another natural polymer that can be used for biodegradable composites. It is an important constituent of the exoskeleton of crustaceans, molluscs, and insects. Chitin as a natural polymer is biodegradable because its b-1,4 glycosidic bonds are sensitive to lysozymes present in the human body. Poorly crystallized HAp (pcHAp), more bioactive and soluble than highly crystallized HA, was used as a bioactive and biodegradable phase for chitin. The pcHAp/chitin composite could be produced using the solution casting technique, with a homogeneous distribution of the pcHAp particles in the current composite (Figure 11.20f). The solution casting process did not change the crystal structure of the chitin. The results of the tensile tests revealed that the strength and modulus of pcHAp/chitin composite decreased with an increase in the amount of particulate pcHAp in the composite. In-vitro mineralization experiments have shown that the pcHAp particles render the bioactive composite and have considerably improved the composite's ability to induce the formation of bone apatite on its surface.

11.8.9 Bioactive and Biodegradable Scaffolds

The pore size of the scaffolds could be well controlled using porous particles of different sizes. The concentration of the polymer solution in scaffold production should be carefully selected as it has been found that with an increase in polymer concentration the pore interconnection has decreased with increasing wall thickness of the pores. Highly porous scaffolds containing 20wt.% of bioactive ceramic particles could be produced. The in-vitro experiments showed, of course, that the pcHAp particles increased the formation of bone apatite on the surface of PSHA/PLLA composite scaffolds when immersed in SBF. Scaffolding degradation in SBF was also observed. The introduction of bioactivity into biodegradable scaffolds by incorporating particulate biological ceramics can improve cell seeding and hence subsequent tissue growth. Scaffolding (production and selection) is only part of the "tissue engineering triad". The other two parts of the triad, namely cells and signalling (molecules), are equally important components that determine the ultimate success (or failure) of a tissue engineering strategy. The search for a suitable scaffold for a particular application is the construction technology that underpins the development of tissue engineering, and different application situations require scaffolding of different characteristics.

11.9 CHALLENGES AND FUTURE DIRECTIONS

The field of bone tissue engineering is at an exciting point, with enormous research activity focused on delivering new and improved biomimetic materials. The level of biological complexity that needs to be recapitulated within synthetic three-dimensional environments is still uncertain. Further elucidation of communication among the cells and of the complex interplay between cells and their matrix will support focus strategies to enable the presentation of biofactors in the correct context both chemically, temporally, and in terms of their distribution. Similarly, the clinical application of surface structuring approaches will require further understanding of the interactions occurring at the cell surface/substrate interface. Vascularization of large tissue constructs remains a significant challenge and some engineering-based approaches to try and overcome this have been discussed here. It is worth noting that advances in microsurgical techniques are also underway to allow reconstructive surgeons to generate so-called "axially vascularized" tissues that can overcome some of the existing problems in achieving rapid vascularization of implanted biomaterials. This highlights the importance of close interaction between the surgical and cell biology communities as we move from the bench closer to the bedside. The harvest of pluripotent mesenchymal cells from sources other than bone marrow, for example, from the periosteum or adipose tissue, also warrants consideration. Advances in materials processing are also having a positive impact on the field. In the body, bone often has a structurally important interface with other tissues such as cartilage and ligament/tendon, for which designed scaffolds can be used to create tissue interfaces. For example, computer-aided design and SFF polymer/ceramic composites have been used to create a construct for a bone-cartilage interface by seeding chondrocytes (cartilage cells) within the cartilage portion and BMP-7 transduced cells on the ceramic portion. The potential to combine three-dimensional printing of scaffolds with three-dimensional printing of cells and biologics, while currently challenging, will enable the development of new designer material/biofactor hybrids. Soft material routes similar sol-gel dispensation might also be a plan to incorporate biomolecules during scaffold fabrication, although this is still under development. It is likely that bio-functionalization strategies will continue to receive a well-deserved focus, as will approaches to better integrate micro- and nanoscale features into designed scaffolds. Developments in this field will find a wealth of applications in our ageing population.

11.10 CONCLUDING REMARKS

Hard tissues in the human body are natural composite materials and they serve as models in the development of tissue replacement materials. Over the last two decades, various bioactive composites have been investigated for tissue replacement and tissue regeneration purposes. Each of these composites has its distinctive characteristics and may be used in specific clinical situations. The successful clinical use of bioactive composites has paved the way for further developing this type of biomaterials for various applications. With new knowledge being gained of natural tissues and the human body and the advancement of composite science and technology, newer and better composite materials will become available for substituting diseased, damaged or worn out body parts. Natural tissues such as bone have the exceptional ability of self-repair. It remains a great challenge for man to produce what nature has made for us.

REFERENCES

Avérous, L. (2004). Biodegradable multiphase systems based on plasticized starch: A review. *Journal of Macromolecular Science, Part C*, 44(3), 231–274. doi:10.1081/MC-200029326

Betts, J. G., Desaix, P., Johnson, E., Johnson, J. E., Korol, O., Kruse, D., and Young, K. A. (n.d.). Bone tissue and the skeletal system. In *Anatomy and Physiology*. Retrieved from https://cnx.org/contents/FPtK1zmh@6.27:fEI3C8Ot@6/Preface.

Bin Pubin, M. K.A. (2007). *Designing a New Concept of Lc-DCP Broad Type Bone Plate* [Report]. Retrieved from Faculty of Mechanical Engineering Universiti Malaysia Pahang website: http://umpir.ump.edu.my/2325/1/MUHAMMAD_KAMAL_ASYRAF_BIN_PUBIN.PDF.

Bonfield, W. (1987). Materials for the replacement of osteoarthritic hip joints. *Metals and Materials*, 3, 712–716

Bonner, M., Saunders, L. S., Ward, I. M., Davies, G. W., Wang, M., Tanner, K. E., and Bonfield, W. (2002). Anisotropic mechanical properties of oriented HAPEXTM. *Journal of Materials Science*, 37(2), 325–334.

Chan, B. P., and Leong, K. W. (2008). Scaffolding in tissue engineering: General approaches and tissue-specific considerations. *European Spine Journal*, 17(S4), 467–479. doi:10.1007/s00586-008-0745-3

Depan, D., Kumar, A. P., and Singh, R. P. (2009). Cell proliferation and controlled drug release studies of nanohybrids based on chitosan-g-lactic acid and montmorillonite. *Acta Biomaterialia*, 5(1), 93–100. doi:10.1016/j.actbio.2008.08.007

Ess, J. W., and Hornsby, P. R. (1987). Twin-screw extrusion compounding of mineral filled thermoplastics: dispersive mixing effects. *Plastics and Rubber Processing Applications*, 8(3), 147–156.

Fan, Y., Xiu, K., Duan, H., and Zhang, M. (2008). Biomechanical and histological evaluation of the application of biodegradable poly- l -lactic cushion to the plate internal fixation for bone fracture healing. *Clinical Biomechanics*, 23, S7–S16. doi:10.1016/j.clinbiomech.2008.01.005

Fractures (Broken Bones). (n.d.). Retrieved from American Academy of Orthopaedic Surgeons website: https://orthoinfo.aaos.org/en/diseases--conditions/fractures-broken-bones

Guild, F. J., and Bonfield, W. (1993). Predictive modeling of hydroxyapatite-polyethylene composite. *Biomaterials*, 14(13), 985–993.

Hagens, W. (2011). *Titanium plaatje voor pols.* [Online Image]. Retrieved from https://commons.wikimedia.org/wiki/File:Titanium_plaatje_voor_pols.jpg

Hench, L. L., and Wilson, J. (Eds.) (1993). *An Introduction to Bioceramics*. World Scientific.

Hench, L. L. (1991). Bioceramics: From concept to clinic. *Journal of the American Ceramic Society*, 74(7),1487–1510.

Hull, D., and Clyne, T. W. (1996). *An Introduction to Composite Materials* (2nd ed.). Cambridge University Press.

Kim, W., Song, J. H., and Kim, J. (2015). Periprosthetic fractures of the distal femur following total knee arthroplasty: Even very distal fractures can be successfully treated using internal fixation. *International Orthopaedics*, 39(10), 1951–1957. doi:10.1007/s00264-015-2970-9

Kokubo, T. (1992). Bioactivity of glasses and glass–ceramics. In P. Ducheyne, T. Kokubo, and C.A. van Blitterswijk (Eds.), *Bone Bonding Biomaterials*. Reed Healthcare Communications.

Kothapalli, C. R., Shaw, M. T., and Wei, M. (2005). Biodegradable HA-PLA 3-D porous scaffolds: Effect of nano-sized filler content on scaffold properties. *Acta Biomaterialia*, 1(6), 653–662. doi:10.1016/j.actbio.2005.06.005

La Carrubba, V., Pavia, F. C., Brucato, V., and Piccarolo, S. (2008). PLLA/PLA scaffolds prepared via thermally induced phase separation (TIPS): Tuning of properties and biodegradability. *International Journal of Material Forming*, 1(S1), 619–622. doi:10.1007/s12289-008-0332-5

Ladizesky, N. H., Pirhonen, E. M., Appleyard, D. B., Ward, I. M., and Bonfield, W. (1998). Fibre reinforcement of ceramic/polymer composites for a major load-bearing bone substitute material. *Composites Science and Technology*, 58(3–4), 419–34.

Liao, S. S., Cui, F. Z., Zhang, W., and Feng, Q. L. (2004). Hierarchically biomimetic bone scaffold materials: Nano-HA/collagen/PLA composite. *Journal of Biomedical Materials Research*, 69B(2), 158–165. doi:10.1002/jbm.b.20035

Mahshid, S., Li, C., Mahshid, S. S., Askari, M., Dolati, A., Yang, L., and Cai, Q. (2011). Sensitive determination of dopamine in the presence of uric acid and ascorbic acid using TiO2 nanotubes modified with pd, pt and au nanoparticles. *The Analyst*, 136(11), 2322. doi:10.1039/c1an15021a

Majola, A., Vainionp, S., Rokkanen, P., Mikkola, H.-M., and Törmälä P. (1992). Absorbable self-reinforced polylactide (SR-PLA) composite rods for fracture fixation: Strength and strength retention in the bone and subcutaneous tissue of rabbits. *Journal of Materials Science: Materials in Medicine*, 3(1), 43–47. doi:10.1007/BF00702943

Njuguna, J., Vanli, O. A., and Liang, R. (2015). A review of spectral methods for dispersion characterization of carbon nanotubes in aqueous suspensions. *Journal of Spectroscopy*. 2015, 1–11.

Sepulveda, P., Jones, J. R., and Hench, L. L. (2001). Characterization of melt-derived 45S5 and sol-gel -derived 58S bioactive glasses. *Journal of Biomedical Materials Research*, 58(6), 734–740. doi:10.1002/jbm.10026

Shah, N. (2014). *X ray Internal Fixation Leg Fracture*. [Online Image]. Retrieved from https://commons.wikimedia.org/wiki/File:X_ray_internal_fixation_leg_fracture.jpg

Sheikh, Z., Najeeb, S., Khurshid, Z., Verma, V., Rashid, H., and Glogauer, M. (2015). Biodegradable materials for bone repair and tissue engineering applications. *Materials*, 8(9), 5744–5794. doi:10.3390/ma8095273

Skalak, R., and Fox, C. F. (1988). *Tissue Engineering: Proceedings of a Workshop, Held at Granlibakken, Lake Tahoe, California, February 26-29, 1988* (Vol. 107). Alan R. Liss: Chicago.

Staines, M., Robinson, W. H., and Hood, J. A. A. (1981). Spherical indentation of tooth enamel. *Journal of Materials Science*, 16(9), 2551–2556. doi:10.1007/BF01113595

Velasco, M. A., Narvaez-Tovar, C. A., and Garzón-Alvarado, D. A. (2015). Design, materials, and mechanobiology of biodegradable scaffolds for bone tissue engineering. *BioMed Research International*.

Villarreal, M.R. (2007). *Human Skeleton Front en*. [Online Image]. Retrieved from https://commons.wikimedia.org/wiki/File:Human_skeleton_front_en.svg

Vugman, N. V., Rossi, A. M., and Rigby, S. E. (1995). EPR dating CO2- sites in tooth enamel apatites by ENDOR and triple resonance. *Applied Radiation and Isotopes: Including Data, Instrumentation and Methods for Use in Agriculture, Industry and Medicine*, 46(5), 311–315. doi:10.1016/0969-8043(94)00154-R

Wang, M, Deb S, Tanner, K. E., and Bonfield, W. (1996). Hydroxyapatite–polyethylene composites for bone substitution: effects of silanation and polymer grafting. *Proceedings of the Seventh European Conference on Composite Materials*, Vol. 2, London.

Wang, M., Porter, D., and Bonfield, W. (1994). Processing, characterisation, and evaluation of hydroxyapatite reinforced polyethylene composites. *British Ceramic Transactions*, 93, 91–95.

Wang, M. (2001). Bioactive ceramic–polymer composites for bone replacement. *Proceedings of the 13th International Conference on Composite Materials*, Beijing.

Wang, M. (2003). Developing bioactive composite materials for tissue replacement. *Biomaterials*, 24(13), 2133–2151.

Zhou, H., Touny, A. H., and Bhaduri, S. B. (2011). Fabrication of novel PLA/CDHA bionanocomposite fibers for tissue engineering applications via electrospinning. *Journal of Materials Science: Materials in Medicine*, 22(5), 1183–1193. doi:10.1007/s10856-011-4295-6

Zippel, N., Schulze, M., and Tobiasch, E. (2010). Biomaterials and mesenchymal stem cells for regenerative medicine. *Recent Patents on Biotechnology*, 4(1), 1–22. doi:10.2174/187220810790069497

12 Applications of Biomaterials in Soft Tissue Replacement

12.1 INTRODUCTION

In the human body, soft tissues are found all over and they refer to tissues that connect, support, and surround organs and structures. The different types of soft tissue in the body include: fat, muscle, blood and lymph vessels, synovial fluid, fibrous tissue (ligaments and cartilages), and nervous tissue. These tissues have distinctive functions in the body that include structural support, protection, storage, transfer of nutrients and waste, and signal delivery. Soft tissue makes up many important structures in the body that serve important functions, and if they are damaged it can impair one's way of living. There are four types of tissue present in the body (Figure 12.1) and they are classified as connective, epithelial, muscular, and neural (Marieb, 2006).

Fat tissues, also known as adipose tissues, are composed of lipid filled cells referred to as adipocytes, along with fibroblasts and immune cells. They have an extensive network of collagenous extracellular matrices (ECM) that are infiltrated by blood vessels (Alkhouli, 2013). Adipose tissue can be found all over the body and are responsible for cushioning and protecting organs and structures, thermal insulation, as well as the storage of surplus triglycerides to be used as energy. Adipose tissue is a connective tissue that is characterized by high expandability, which is proportional to obesity. Triglycerides make up the majority of the volume of adipose tissue as well as approximately 85% of the tissue's weight (Alkhouli, 2013). Levels of collagen in adipose vary depending on the expandability of the tissues. Not many studies out there examine the mechanical properties of adipose tissue; most research seems to be focused on the behavioural side of adipose tissue when it undergoes expansion or compression (Alkhouli, 2013).

Muscle tissue functions by transforming chemical energy into mechanical energy. It can be differentiated into three types: skeletal, smooth, and cardiac (Marieb, 2006). Skeletal muscle tissues have long, striated, and multinucleate cells. They attach to bones and help in the movement and stabilization of the skeleton. Each muscle consists of many muscle fascicles (bundle of cells), and each fascicle consists of many muscle fibers (cell). Each muscle fiber consists of many myofibrils, which consist of the functional unit of the muscle (actin and myosin). Smooth muscle tissue consists of short, spindle-shaped, non-striated cells. Smooth muscle is involuntary, and can be found in the organs of the visceral region of the body such as stomach and intestines where they allow these organs to contract and expand. Cardiac muscle tissue

by Yaser Dahman and Zaid Dwaik

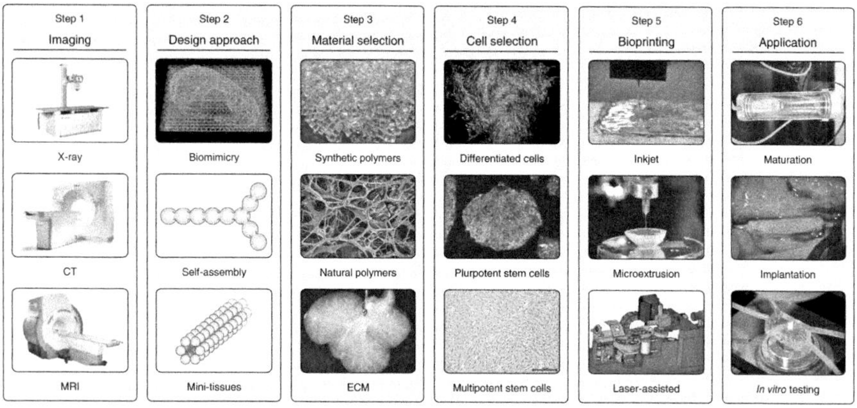

FIGURE 12.1 The process of 3D bioprinting, which begins with taking an image of the area of the body that needs a replacement, then an implant is designed and material to be used for fabrication is selected depending on the purpose. Bioprinting takes place where cells are incorporated into the material that will be used to make the scaffold (Murphy and Atala, 2014).

consists of short, striated, branched cells. Cardiac muscle is only found in the heart and is responsible for the circulation of blood (Marieb, 2006).

Skin is the outer covering of the body and is the largest organ of the body. It acts as a barrier and protects the body from environmental factors such as sun and cold as well as harmful pathogens and germs (Menche, 2012). It is also responsible for regulating body temperature and preventing dehydration, in addition to serving as storage for fat, water, and metabolic products. The skin consists of three layers: the epidermis (keratinized layer), the dermis (collagen-rich layer), and the subcutaneous layer (Menche, 2012; Kundu and Wang, 2013). The skin has limited self-healing capability and that is why extensive damage to the skin can cause loss of skin integrity, which may result in death. Developing a skin replacement is quite tricky; it is a complex structure that houses appendages like hair and hormonal glands in the epidermis (Kundu and Wang, 2013).

Fibrous tissue is mainly composed of fibroblast cells, densely packed collagen fibers, and elastin. Typically found in cartilage and ligament, they provide support, strength, and flexibility. They can be characterized by a thin and dispersed cellularity that is spread over the extracellular matrix (Marieb, 2006). They typically vary in composition and appearance depending on their location and purpose in the body. Cells can be organized either in regular or irregular formations. Fibrous tissue that functions as a strong support and protection, such as ligaments, tends to have cells compacted and more closely stacked to increase strength, whereas tissue that requires flexibility, such as cartilage, has cells that are loosely formed to allow for smooth movement and elasticity (Marieb, 2006).

Other soft tissues include: synovial tissue, which is thin loose tissue that lines the joints of the elbow and the knee. Blood vessels are long elastic tubes that include veins and arteries; they help transfer blood, nutrients, and hormones. There are three types: arteries, capillaries, and veins. They consist of three layers (1) tunica intima, which is the thinnest layer; (2) tunica media, the thickest layer in arteries; and (3) tunica adventitia, which is the thickest layer in veins (Marieb, 2006). Peripheral nerves originate from the spinal cord and cover the whole body; they relay signals to and from the brain (Marieb, 2006).

In the field of soft tissue engineering, as in other engineering applications, the efficiency of the engineered tissue or implant depends on the material used and the design of the implant (Park, 2007). These factors, however, depend upon the location and purpose of the implant within the body. The determining factor on the suitability of the implant will be mainly based upon how it reacts to the physiological conditions in-vivo. The in-vivo clinical performance of the implant represents the biggest hurdle in the process of selecting an appropriate material to be used in the human body.

For the past decade, much of the success of soft tissue engineering is due to advancements in synthetic polymers. Polymers are advantageous because they can be custom made to match the properties of soft tissues. They can be created into a wide range of physical forms, such as gels for filling spaces, solid structures for cosmetics, and networked fabrics for vessels (Park, 2007).

Another important aspect is bio-absorbability, which has garnered attention because of the advantages it presents, such as less complicated revision surgeries, better surgical imaging, biocompatibility, and the non-necessity of removal surgery (Park, 2007).

According to Kim et al. (2008) and Bressan et al. (2011), when developing a soft tissue implant, several factors must be considered:

1. The implant must achieve similar physical properties such as flexibility and texture.
2. Depending on its purpose in the body, the implant shouldn't degrade or change properties after implantation. Otherwise, if it is designed to deteriorate, rates and modes of degradation should follow expected pathway.
3. The implant should be non-carcinogenic, non-toxic, non-immunogenic, and not cause any adverse effects.
4. Implants should be sterilizable and have a low cost of fabrication.
5. Implants must have sufficient porosity and structural integrity for the transport of cells, gases, metabolites, nutrients, and signal molecules within the scaffold and with the local environment.

12.1.1 Types of Materials

Existing approaches to soft tissue replacement and regeneration typically rely on autologous tissue or synthetic implants, or a combination of both. Some aspects that include tissue-implant contracture or rupture and tissue resorption usually limit these approaches (Bressan et al., 2011).

The initial emphasis in engineering soft tissue replacement was focused on developing and using biodegradable synthetic materials, such as poly-L-lactic acid (PLLA), polycaprolactone (PCL), polyglycolic acid (PGA), and other polymers, to aid in ECM production. However, these synthetic materials have achieved limited success due to several reasons, mainly because their lack of strength and their elasticity causes a mismatch in compliance (Tran et al., 2010).

Recently, biodegradable elastomers have received much more attention due to their elastic properties and their exceptional cell and tissue compatibility. Some of these elastomers include poly(glycerol sebacate) (PGS), crosslinked biodegradable photoluminescent polymer (CBPLP), and poly(ovtamethylene citrate) (POC) (Yang et al., 2006; Wang et al., 2002). One key disadvantage of using these polymers is that they lack suitable mechanical strength when they are made into scaffolds (Tran et al., 2010).

Table 12.1 showcases a list of the main materials that are used or were used in fabricating soft tissue replacement and soft tissue engineering.

12.2 TYPES OF IMPLANTS

The substitution of a human body part with a biomaterial requires good communication between the host and implanted system for a successful outcome. Soft-tissue replacement can be classified into (1) space filler; (2) mechanical support; and (3) fluid carrier. Table 12.2 lists the types of organs that are implantable, the methods used for their implant, and the limitations of the implant methods.

12.2.1 Surgical Tapes and Sutures

12.2.1.1 Sutures

Sutures are the one of the most frequently used soft tissue implants. Typically, they are used to close wounds up resulting from injury or surgery. Currently, there are two types that exist and they are classified according to their in-vivo behaviour: biodegradable or non-biodegradable. They can also be classified according to the type of material used in fabricating them, either natural polymers (silk or cotton) or synthetic polymers (polyethylene or PLLA) (Roby and Kennedy, 2004). Absorbable sutures like catgut are made from collagen derived from sheep intestine. It is treated chemically to increase the strength and lifespan of the implant prior to use. Synthetic sutures are typically made from PGA and tend to degrade much slower than natural sutures (Park, 2007).

12.2.1.2 Surgical Tapes

Similar to sutures, surgical tapes are typically used to close surgical cuts to avoid pressure necrosis, the formation of scar tissue, and other problems with stitching. However, some problems faced with using tapes include uneven wound edges and poor adhesion (Park, 2007). They are not as popularly used as sutures or staples.

12.2.1.3 Staples

Staples are commonly used to close surgical incisions from large surgical operations. They are made from metals, such as stainless steel or titanium–nickel alloy. They facilitate a similar response to synthetic polymers from the body and they are typically not used for cosmetic surgeries where cosmetic looks are more important (Park, 2007).

TABLE 12.1
Types of Material Used in Fabricating Soft Tissue Substitutes

Material Composition	Matrix Geometry	Advantages	Disadvantages
PVA	Film coating	Inert hydrophilic matrix	Lacks cell binding motif
PLGA	3D printed flow channels, foams, nanoporous scaffold	Biocompatible Biodegradable Suitable for stem cell differentiation	Acidic degradation product Can result in inflammation
PLLA	Nanoporous scaffold	Suitable for stem cell differentiation	–
PLGA/PLLA coated with PVA	Porous scaffold	Improved seeding and biocompatibility due to hydrophilic coating	–
Polydimethyl-sulfoxide	3D printed flow channels	Oxygen permissible membrane	Highly hydrophobic thus it can absorb biomolecules
Polyurethane	Foam	Can easily be modified chemically or mechanically	Degradation by-products are toxic
Polycaprolactone	Porous scaffold, nanofibers	Inert Biodegradable Biocompatible	Slow degradation Hydrophobic Hard to seed cells
Polyethylene glycol	Hydrogels	Hydrophilic Resistant to protein adsorption Modulation polymer MW	
Polyacrylamide	Grafted polymer chains	Surface properties can be modulated through temperature	Non-biodegradable
Elastin-like polypeptide	Polyelectrolyte multilayer	Scaffold composition can be modified Biodegradable	Not cell adhesive Needs functionalization
Natural Polymers Chitosan	Hydrogels, porous scaffold, nanofibers	Resembles glycosaminoglycans Exhibits cell bioactivity	Low mechanical strength Can be immunogenic
Fibrin Gels	Hydrogels	Hydrolytically degradable	Rapid degradation Low mechanical strength Immunogenic
Heparin	Hydrogel	Natural Biocompatible	Low mechanical strength
Matrigel	Coatings, films, gels	Biocompatible Structurally modified	Composition varies Low mechanical strength
Collagen	Films, gels, foams	Easily binds to cells Low antigenicity	Low mechanical strength Costly

(Continued)

TABLE 12.1 (CONTINUED)
Types of Material Used in Fabricating Soft Tissue Substitutes

Material Composition	Matrix Geometry	Advantages	Disadvantages
Alginate	Porous scaffolds, hydrogels	Hydrophilic Promotes cell growth Easy to fabricate	Low adherence Immunogenic
Hyaluronic Acid	Hydrogels, sponges	Biocompatible	Low mechanical strength High viscosity
Native ECM	Decellularized matrix	Intact functional and structural components Imitates native cell	Survival times vary

Source: Jain et al. (2014).

TABLE 12.2
Limitations of Current Approaches

Targeted Organ	General Practice	Limitation
Peripheral nerve grafts	Autologous grafts Homograft	Low quality Vasospasm (spasm of blood vessels) Restricted length Fast degeneration
Cornea	Synthetic keratoprosthesis	High host rejection rate
Skin	Allograft Xenografts	Inadequate availability Limited permanent revascularization
Ligament	Autograft Allografts Xenografts	Compromises normal healthy tissue and prolonged surgical time Risk of disease transmission and immune rejection
Heart	Heart transplantation left ventricular assist device	Scarcity of donors High cost

12.2.2 Percutaneous Skin Implants

Percutaneous implants are usually used in artificial organs like the kidney and the heart. Some challenges exist when attaining a working sustainable interface between the tissue and the implant. Some of these challenges include (1) that the attachment between the implant and surrounding tissues cannot be maintained overtime due to the continuous turnover of cells; (2) that degradation of epithelium around the implant or growth of implant may occur; or (3) the chance of bacteria infiltrating the opening, resulting in an infection.

Another form of soft tissue substitutes are artificial skin implants, which are usually used to maintain the body temperature or maintain the hydration of burn victims by preventing loss of fluids, electrolytes, and other essential minerals. 1st and 2nd degree burns are typically treated with temporary membranes, whereas 3rd degree burns are treated with autografts (Morgan et al., 2004; Park, 2007).

Many polymers have been used to address skin loss due to burns; some of these include collagen, methyl-2-cyanocrylate, and vinyl chloride acetate copolymer. Other methods include the culturing of cells in-vitro by growing an epidermal layer of cells from the skin of the burned victim. However, early failure in regenerating autogenic skin has slowed down research into regenerating autogenous skin from burned patients, and this presents itself as a field of focus for the future (Morgan et al., 2004; Park, 2007).

12.2.3 Maxillofacial Implants and Space Fillers

Maxillofacial implants can be classified into either intraoral or extraoral. This field of soft tissue substitutes focuses on the science of anatomic, functional reconstruction of components in the maxilla, mandible, and face (Miloro et al., 2003).

Several polymeric options exist to make extraoral implants, the colour and texture of which must match that of the patient, along with being chemically and mechanically viable. Some polymeric materials currently used include poly(methyl methacrylate) (PMMA), silicone, and polyurethane (Miloro et al., 2003; Park, 2007). For intraoral implants, similar requirements apply to those followed in making extraoral implants. Some polymeric materials used intraorally for soft tissue include silicone rubber or PMMA (Park, 2007).

12.2.3.1 Ear Implants

Replacements of a damaged ear are in essence done for mainly cosmetic reasons and fall under maxillofacial remodelling. Damage done to the middle ear would fall under hard tissue replacement because it would involve the bones used for conduction (malleus, incus, stapes). Many different materials can be used to fabricate ear implants including polyethylene, PMMA, silicone rubber, porous polyethylene, and others.

12.2.3.2 Eye Implants

Damage to the cornea or lens is usually addressed using eye implants. Contact lenses are not considered implants or replacements, they are rather considered as a visual aid. The standard care for treating damaged corneas is by administering a transplant from a suitable donor. However, cornea implants exist and they are typically made from acrylics like PMMA (Park, 2007). The main challenge with using cornea implants is that the lifespan of the implant is unreliable due to fixation and infections.

12.2.3.3 Space-Filling Implants

Breast implants are the most popular space-filling implants. In the past, breast augmentation was done with paraffin wax, petroleum jelly, and vegetable oil through a direct injection (Scalfani and Jacono, 2015). Consequences from using these materials such as infection, pain, and loss of original shape resulted in the banning of such practices (Scalfani and Jacono, 2015). This pushed the field into finding new ways that are longer lasting and much safer. Some of the safer and currently common materials used include silicone gel-filled rubber bags; these promote tissue growth from fixation (Park, 2007). Other variations include using saline solution. Saline is considered to be a safer option than silicone (Park, 2007).

12.3 FABRICATION TECHNOLOGIES

Several techniques exist today that are employed to design and fabricate soft tissue replacements. In this review, we will discuss the most significant ones utilized in the fabrication of soft tissue and explain the process in which each technique is carried out.

12.3.1 3D Bioprinting

3D printing was first explained Charles Hull back in the mid-1980s. His technique was described as thin layers of material that are treated with ultraviolet light and printed in layers to form a three-dimensional construct. The development of advanced systems that are solvent-free allowed the direct fabrication of three-dimensional scaffolds made from biological materials (Murphy and Atala, 2014). The recent advancements in materials science, cell biology, and 3D printing allowed for the next step to be taken towards 3D bioprinting as a form of tissue engineering. But the main challenge in 3D printing is to recreate the complex microstructure of extracellular matrix constituents and the different cell types in an adequate resolution in order to support biological functioning (Murphy and Atala, 2014; Rodriguez et al., 2017).

3D bioprinting can be described as a layer-by-layer high precision positioning of living cells, biomaterials, and biochemicals that is used to fabricate three-dimensional structures (Murphy and Atala, 2014). This approach could lead to the regeneration of healthy functional tissues or organs for patients, thereby eliminating the need for tissue grafts and mechanical devices (Pati et al., 2015; Rodriguez et al., 2017). There are three design approaches to 3D bioprinting that were developed to fabricate three-dimensional biological constructs and they include biomimicry, mini-tissue building blocks, and autonomous self-assembly (Murphy and Atala, 2014).

12.3.1.1 Design Approaches

1. Biomimicry refers to biologically focused engineering that aims at identical reproduction of cellular and extracellular components of tissues. It requires a comprehensive understanding of the microenvironment, such as the arrangement of functional and supporting cell types and the composition of the ECM (Marga et al., 2007).
2. Autonomous self-assembly uses embryonic organ development as a guide. This approach depends on the cell to drive histogenesis, and direct the localization, composition, and functional properties of the tissue (Derby, 2012). It requires comprehensive knowledge of embryological developmental mechanisms.

3. The mini-tissue building blocks approach is rooted in the idea that organs and tissues are made up of smaller, functional units or 'mini-tissues' (Kelm et al., 2010). Organs can be fabricated by assembling mini-tissues through rational design and self-assembly (Murphy and Atala, 2014).

The process of bioprinting 3D tissues is initiated by imaging the damaged tissue to provide information on the 3D structural and functional characteristics at the cellular, tissue, and organ levels to guide the design of the bioprinted tissues. After following the design approaches, a choice of material must be selected that is specific to the tissue form and function. Cell sourcing may be autogenic or allogeneic depending on the purpose in the body. Finally, all these components must be coordinated together to integrate with the bioprinting system that will be used whether it is inkjet, microextrusion, or laser-assisted (Murphy and Atala, 2014). A summary of the steps is shown in Figure 12.2.

A study by Pati et al. (2015) was done to address the need of reconstructing soft tissue, with the main focus being on adipose tissue restoration. Researchers utilized an in-house-developed tissue printing technology known as the multi-head tissue/organ building system (MtoBS). It has multiple heads that can dispense different hydrogels or polymers to fabricate a construct. The constructs were designed and printed through a biomimetic approach. Decellularized adipose tissues (DAT) were used to make up the bioink due to its naturally constituted ECM, which provided a

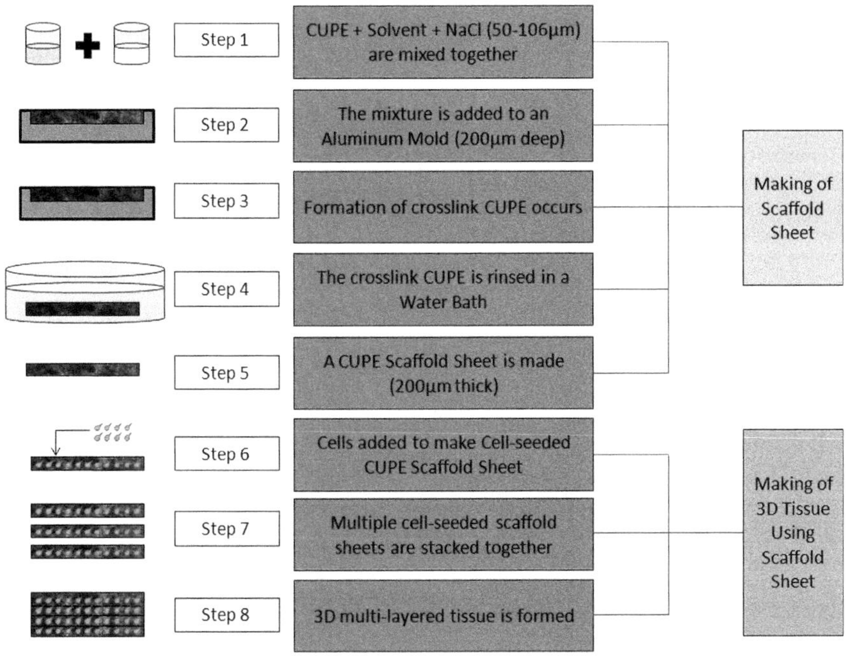

FIGURE 12.2 The following steps showcase the procedure followed using particulate leaching to create porous scaffolds where they are stacked to form a 3D multi-layered tissue.

proper environment for the encapsulated human adipose derived mesenchymal stem cells (hASCs) to regenerate (Pati et al., 2015).

PCL, a biopolymer, was used to fabricate the framework of the constructs that housed the bioink and prevented the structure from collapsing. PCL however has a slow degradation rate and is stiffer than natural adipose tissue. Pati et al. (2015) concluded that their use of hASCs did not result in any deleterious effects and the DAT constructs induced a greater commitment of hASCs to adipogenic lineage compared to non-DAT gel. The animal studies conducted on mice concluded that tissue printing with hASCs encapsulated in DAT gel resulted in soft tissue regeneration. 3D printing has some additional complexities, such as type of material used, type of cell, and growth and signalling molecules that are related with fabricating tissues and implants (Rodriguez et al., 2017).

12.3.2 Injectable Implants

Soft tissue injection refers to a technique developed to augment soft tissue. Injectables, otherwise known as fillers, are materials that are injected into soft tissue through a syringe (Chacon, 2015). It is a process of implanting tissue or biocompatible material to treat wrinkles or soft tissue defects in the body.

It is an older approach compared to the other ones, as incidences of soft tissue augmentation using dermal injections date back to the nineteenth century. The earliest known injectable filler was paraffin wax, which was mainly used for facial augmentations (Chacon, 2015). Its use did not last long before it was abandoned following the realization of its complications such as migration, granuloma, and embolization (Scalfani and Jacono, 2015). Silicone, which was another type of material used in dermal fillers, was abandoned due to similar complications.

The first dermal filler for cosmetic injections to be internationally approved was bovine collagen in the 1990s (Chacon, 2015; Scalfani and Jacono, 2015). In the early 2000s, a class of new agents was being developed known as hyaluronic acid (HA) fillers and has since become the industry standard (Chacon, 2015). Research into enhanced materials continues, and more improved fillers are becoming available such as poly-L-lactic acid and hydroxylapatite.

Soft tissue injections can be categorized based on numerous characteristics that include the site of injection (dermal or subcutaneous), the injection material (autologous, natural, synthetic, or xenogenic), and the lifespan of the implant (resorbable or nonresorbable) (Chachon, 2015; Dermal Fillers, n.d.).

Resorbable (temporary) materials:

- Collagen: a type of protein that makes up a majority of the skin and other tissue in the body. Collagen can be sourced from bovine or human cells. Its impact can last up to 4 months; it is considered to be on the shortest lasting injectable material.
- Calcium hydroxyapatite: a type of mineral that can be found in teeth and in bone. Commonly used as a filler for face or hand wrinkles, where it can last somewhere around 18 months.

- Hyaluronic acid: a polysaccharide that is widely present in skin and cartilage tissue. Its ability to combine with water allows it to have a smooth effect when implanted. Depending upon the purpose of the implant, hyaluronic acid can be modified chemically through crosslinking to make it last longer. The impact of the implant can last up to a year.
- Poly-L-lactic acid: a biodegradable, biocompatible, synthetic material that is long lasting. It is usually administered in several injections over several months. The impact of a PLLA implant can last up to 2 years.

Nonresorbable (permanent) material:

PMMA microspheres: non-biodegradable, biocompatible, and synthetic. Its applications include bone cement and intraocular lenses. The microspheres are small and smooth. As injectable filler, PMMA microspheres are suspended in a gel solution that contains bovine collagen.

In the study by Okabe et al. (2009), they tested injectable implants in rats, which included using a scaffold on its own as well as with growth factors and cells. They concluded that the use of a scaffold on its own cannot induce skin growth and regeneration. The method that achieved the most success involved the use of a scaffold along with mesenchymal-derived stem cells, skin growth factors, and signalling molecules. The study recommended using this approach because it presents the possibility of longer lasting soft tissue augmentation without repeated injections.

12.3.3 Layer-by-Layer Technique with Particulate Leaching

Particulate leaching is a simple process that is characterized by the ability to control the porosity of particles. This is an attractive feature of this approach, which allows the selection of particle size and the amount of salt that is added to control pore size (Sin et al., 2010). Conversely, it can be inconsistent due to the uneven distribution of salt within the polymer solution. This inconsistency is due to the difference in density between salt and the polymer solution along with uncontrolled contact between salt and polymer (Sin et al., 2010; Tran et al., 2010).

The steps taken in this approach are summarized concisely by Figure 12.2.

The procedure begins with making a polymer solution that is appropriate for the purpose of the implant. It is followed by adding salt into the solution and mixing it properly, then it is centrifuged and the solvent is allowed to evaporate. The following step is leaching salt by placing it in a water bath. Finally, the scaffold sheet is lyophilized to remove any remaining water. The scaffold sheet is ready to be cell seeded with the desired cells (Sin et al., 2010).

As mentioned earlier, the simplicity of the particulate leaching approach and the ease of synthesis of these scaffolds with the ability to modify both mechanical and material properties along with potential cost effectiveness makes this approach a widely popular one among researchers (Tran et al., 2010).

Particulate leaching is usually combined with the layer-by-layer technique to build artificial tissue. This is capable of creating stratified tissue by stacking cell sheets; the stacked layers also provide a 3D environment that promotes improved cell proliferation (Dey et al., 2008). Moreover, the layer-by-layer approach can potentially facilitate the compartmentalization of multiple cell types, which can give potential to create complex structures such as blood vessels, skin, liver lobules, and kidney glomeruli (Dey et al., 2008; Sin et al., 2010).

In the study done by Tran et al. (2010), they utilized the mechanism of particulate leaching along with the layer-by-layer technique to create an artificial scaffold made of crosslinked urethane doped polyester and seeded with 3T3 mice fibroblast cells. The tissue was cultured in-vitro and was evaluated for cell proliferation, tensile peak stress, modulus, elongation, and suture retention. The scaffold showcased adequate cell proliferation, proper mechanical properties, and increased suture strength, which potentially make it suitable for immediate implantation in applications for in-vivo tissue engineering (Tran et al., 2010).

12.3.4 Electrospinning

The last major technique to be discussed in this review is electrospinning. In a nutshell, electrospinning is a process that utilizes electric charges to create polymeric nanofibers. The quality of nanofibers is typically impacted by a number of parameters such as the tip-to-target distance, polymer concentration, needle gauge size, solvent, and the applied voltage. Recent advances in this field have enabled the fabrication of nanofibers with diameters that range between 100–500 nm (Sundaramurthi et al., 2014). These nanofibers can be utilized to create nanofibrous scaffolds, which can potentially lead to revolutionary innovations in tissue engineering applications, especially in the field of soft tissue replacement.

The electrospinning set-up (Figure 12.3) is typically comprised of a syringe with an appropriate needle through which the polymeric solution to be electrospun is ejected, as well as a high voltage power supply and a grounded collector, which will collect the nanofibers (Sundaramurthi et al., 2014).

To make these nanofibers, a syringe pump typically forces the polymeric solution in the syringe through the needle, which will form a suspended droplet at the tip (Doshi and Reneker, 1995). When a high voltage charge is applied at the tip, it causes the surface of the polymeric droplet to elongate and form what is commonly known as the Taylor Cone. Once the electrostatic charge overpowers the surface tension of the droplet, the polymeric solution jets towards the collector (Doshi and Reneker, 1995; Sundaramurthi et al., 2014). In essence, nanofibers form following the evaporation of the solvent along with the bending instabilities in the jet that are caused by collisions with air molecules (Doshi and Reneker, 1995; Sundaramurthi et al., 2014).

Electrospun nanofibers are remarkable because they can be customized to achieve the desired pore size and distribution, cell adhesion, proliferation, and surface-area-to-volume ratio (Sundaramurthi et al., 2014). Their flexibility in terms of customization is partly due to the fact that they resemble the native ECM of skin (Doshi and Reneker, 1995). These characterizations of electrospun nanofibers make them suitable as skin substitutes due to their ability to prevent nutrient and fluid loss

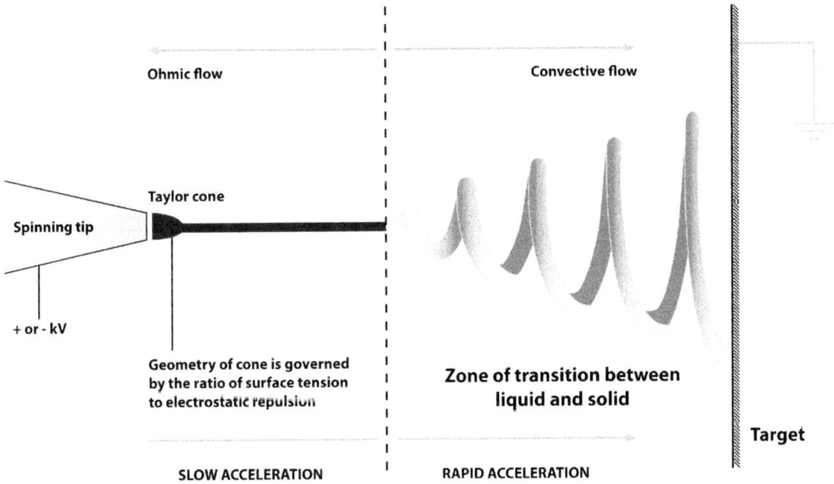

FIGURE 12.3 The set up for an electrospinning approach, consisting of a tip that ejects a polymeric solution through the cone and travels across towards the grounding plate (Gatford, 2008).

from a wounded area, aid in cleanup of exudates, prevent infections, and showcase anti-adhesion features, as well as promoting the proliferation of endogenous cells Sundaramurthi et al., 2014).

In conclusion, electrospun nanofibers can be a helpful option in treating cuts and wounds. They are capable of covering wounds as well as aiding in skin regeneration. They also present a better option since they can allow for improved vascularization compared to current technologies.

12.4 CONCLUSION AND FUTURE PERSPECTIVE

This review discussed the current state of the art of soft tissue replacement as well as the types of material used to fabricate these replacements and implants. This review has also covered the main technological methods used to fabricate soft tissue implants. Obviously, with the continuous need for soft tissue replacements, increased demand for more efficient, biocompatible, rapidly fabricated, and affordable replacement implants will lead to further innovations in the future. One of the main challenges that exist today is to be able to assess and compare the long-term stability and efficacy of soft tissue implants and their potential replacements.

As mentioned in this review, soft tissue replacements must be biocompatible and nom-immunogenic; they should also have suitable tissue integration along with tissue conductive traits. The mechanical characteristics of these replacement implants must be able to guarantee proper clinical handling and ensure physical stability. Future research must focus on developing soft tissue alternatives that can replace autogenic implants that can be rapidly fabricated at affordable costs. Future replacements should also possess tissue-genetic and tissue-inductive characteristics that will help make them as similar to native cells as possible. The incorporation of tissue engineering is needed to improve such properties, such as the inclusion of living biological cells, growth proteins, or signalling molecules. Prospective advances should focus on soft tissue replacements that can be vascularized either pre-implantation vascularization or post-implantation vascularization. Vascularization helps in maintaining cell viability and ensuring proper implant functioning (Zuhr et al., 2014).

Other aspects that deserve attention in future studies include understanding the essential proteins and other important structural components that needed to be present in the ECM to maintain the structural morphology of soft tissue substitutes. Additionally, identifying if any would be lost when using decellularized tissues or when using synthetic polymers. Past studies on soft tissue substitutes seemed to overlook the importance of having proper morphological features (Pati et al., 2015); thus, future studies should focus on developing replacement constructs that possess mechanical properties that are as similar as possible to native tissue.

Additional emphasis in future studies should focus on developing electrospun scaffolds that incorporate wound healing factors along with growth factors to promote skin regeneration.

In essence, future research should focus on regenerating soft tissue instead of replacing it through a multidisciplinary approach that includes input from physicians, biomaterial scientists, as well as engineers to innovate methods that will regenerate soft tissue.

REFERENCES

Alkhouli, N. (2013). The mechanical properties of human adipose tissues and their relationships to the structure and composition of the extracellular matrix. *American Journal of Physiology, 305*(12), E1427–1435.

Bressan, E., Favero, V., Gardin, C., Ferroni, L., Iacobellis, L., Favero, L., and Zavan, B. (2011). Biopolymers for hard and soft engineered tissues: Application in odontoiatric and plastic surgery field. *Polymers*, 3(1), 509–526.

Chacon, A. (2015). Fillers in dermatology: from past to present. *Cutis*, 96(5), E17–19.

Dermal Fillers (Soft Tissue Fillers). (n.d.). Retrieved from U.S. Food and Drug Administration website: https://www.fda.gov/MedicalDevices/ProductsandMedicalProcedures/CosmeticDevices/ucm619837.htm

Derby, B. (2012). Printing and prototyping of tissues and scaffolds. *Science*, 338(6109), 921–926.

Dey, J., Xu, H., Shen, J., Thevenot, P., Gondi, S.R., Nguyen, K.T. and Yang, J. (2008). Development of biodegradable crosslinked urethane-doped polyester elastomers. *Biomaterials*, 29(35), 4637–4649.

Doshi, J., and Reneker, D. H. (1995). Electrospinning process and applications of electrospun fibers. *J. Electrostat*, 35(2–3), 151–160.

Gatford, J. (2008). *Electrospinning Diagram*. [Online Image]. Retrieved from https://commons.wikimedia.org/wiki/File:Electrospinning_Diagram.jpg

Jain, E., Damania, A., and Kumar, A. (2014). Biomaterials for liver tissue engineering. *Hepatology International,* 8(2), 185–197.

Kelm, J. M., Lorber, V., Snedeker, J. G., Schmidt, D., Broggini-Tenzer, A., Weisstanner, M., and Hoerstrup, S. P. (2010). A novel concept for scaffold-free vessel tissue engineering: self-assembly of microtissue building blocks. *Journal of Biotechnology*, 148(1), 46–55.

Kim, I. Y., Seo, S. J., Moon, H. S., Yoo, M. K., Park, I. Y., Kim, B. C., and Cho, C. S. (2008). Chitosan and its derivatives for tissue engineering applications. *Biotechnology Advances*, 26(1), 1–21.

Kundu, R., and Wang, X. (2013). Silk fibroin biomaterial for tissue regeneration. *Advanced Drug Delivery Reviews*, 65(4), 457–470.

Marga, F., Neagu, A., Kosztin, I., and Forgacs, G. (2007). Developmental biology and tissue engineering. *Birth Defects Research C Embryo Today*, 81(4), 320–328.

Marieb, E. N. (2006). *Essentials of Human Anatomy and Physiology.* (8th Edition). San Francisco, CA: Pearson Benjamin Cummings.

Menche, N. (Ed.) (2012). *Biologie Anatomie Physiologie.* Urban & Fischer/Elsevier.

Miloro, M., Ghali, G. E., Larsen, P., and Waite, P. (Eds.) (2003). *Peterson's Principles of Oral and Maxillofacial Surgery.* B.C. Decker.

Morgan, J. R., Sheridan, R. L., Tompkins, R. G., Yarmush, M. L., and Burke, J. F. (2004). Burn dressings and skin substitutes. In B. Ratner, A. Hoffman, F. Schoen, and J, Lemons (Eds.), *Biomaterials Science: An Introduction to Materials in Medicine.* Elsevier Academic.

Murphy, S. V., and Atala, A. (2014). 3D Bioprinting of tissues and organs. *Nature Biotechnology*, 32(8), 773–785.

Okabe, K., Yamada, Y., Ito, K., Kohgo, T., Yoshimi, R., and Ueda, M. (2009). Injectable soft-tissue augmentation by tissue engineering and regenerative medicine with human mesenchymal stromal cells, platelet-rich plasma and hyaluronic acid scaffolds. *Cytotherapy*, 11(3), 307–316.

Park, J. B. (2007). Soft tissue replacement I: Sutures, skin, and maxillofacial implants. *Biomaterials*, 291–329. Springer.

Pati, F., Ha, D. H., Jang, J., Han, H. H., Rhie, J. W., and Cho, D. W. (2015). Biomimetic 3D tissue printing for soft tissue regeneration. *Biomaterials*, 62, 164–175.

Roby M. S., and Kennedy J. (2004). Sutures. In B. Ratner, A. Hoffman, F. Schoen, and J, Lemons (Eds.), *Biomaterials Science: An Introduction to Materials in Medicine.* Elsevier Academic.

Rodriguez, M. J., Brown, J., Giordano, J., Lin, S. J., Omenetto, F. G., and Kaplan, D. L. (2017). Silk based bioinks for soft tissue reconstruction using 3-dimensional (3D) Printing with in vitro and in vivo assessments. *Biomaterials* 117, 105–115.

Sin, D., Miao, X., Liu, G., Wei, F., Chadwick, G., Yan, C., and Friis, T. (2010). Polyurethane (PU) scaffolds prepared by solvent casting/particulate leaching (SCPL) combined with centrifugation. *Materials Science and Engineering: C*, 30(1), 78–85.

Sundaramurthi, D., Krishnan, U. M., and Sethuraman, S. (2014). Electrospun nanofibers as scaffolds for skin tissue engineering. *Polymer Reviews*, 54(2), 348–376.

Tran, R. T., Thevenot, P., Zhang, Y., Gyawali, D., Tang, L., and Yang, J. (2010). Scaffold Sheet Design Strategy for Soft Tissue Engineering. *Materials*, 3(2), 1375–389.

Wang, Y. D., Ameer, G. A., Sheppard, B. J., and Langer, R. A tough biodegradable elastomer. *Nature Biotechnology*, 20(6), 602–606.

Yang, J., Webb, A. R., Pickerill, S. J., Hageman, G., and Ameer, G. A. (2006). Synthesis and evaluation of poly(diol citrate) biodegradable elastomers. *Biomaterials*, 27(9), 1889–1898.

Zuhr, O., Baumer, D., and Hurzelr, M. (2014). The addition of soft tissue replacement grafts in plastic periodontal and implant surgery: Critical elements in design and execution. *Clinical Periodontology*, 41(15), S123–142.

13 Biomaterials in 3D Printing/Bio-printing Techniques

13.1 INTRODUCTION

Three dimensional (3D) printing is a method that is speedily developing and being integrated into manufacturing and our day-to-day lives. Several people have known of its emergence into the commercial world, though it has been labeled by different designations, such as additive manufacturing (AM), rapid prototyping (RP), layered manufacturing (LM), and solid freeform fabrication (SFF). Theoretically, AM are approaches where 3D designs can be built directly from a computer-aided design (CAD) file without any part-specific tools or dies. In 3D printing, CAD models of fragments to be constructed are initially sliced in an authentic environment to generate a stack of two-dimensional (2D) parts. A 3D printing apparatus then forms the portions one layer at a time based on the 2D slice information, stacking and assembling successive layers to create the ultimate 3D entity. CAD modelling has provided us with the capability to generate, modify, and, if desired, analyse, designs in an authentic world. With the arrival of 3D printing, such virtual strategies can now be turned into physical 3D entities that can assist as prototypes or be directly handled as functional parts for a range of uses. The creation of modern 3D printing can be dated back to the 1980s when Hull contrived the stereolithographic (SLA) apparatus (Figure 13.1), the first 3D printing technology (Hull, 1986) Stereolithography is an additive fabrication method utilizing a vat of liquid UV-curable photopolymer "resin" and a UV laser to form parts one layer at a time. On each layer, the laser beam traces a cross-sectional pattern on the surface of the liquid resin. Contact with the UV laser cures or hardens the shape traced on the resin, as well as adhering it to the layer below.

In the area of regenerative engineering, there are growing efforts to develop the capability of printing tissues and organoid structures, not only for implantable therapeutics but also for use as in-vitro prototypes for learning tissue and organ improvement and disease, as well as medical drug toxicity analysis. 3D printing allows for the ability to spatially configure cells and biomaterials in order to further closely review the structural, physical, chemical, and biological intricacy of tissues and organs, which has not been possible with other scaffold production technologies. Despite these advantages, a remarkable quantity of research is still needed to discover the minimum design requirements and mechanisms that lead to optimal tissue growth and function, as well as the appropriate size for vascularized multi-tissue

by Yaser Dahman and Thivjhan Kanaganavagam

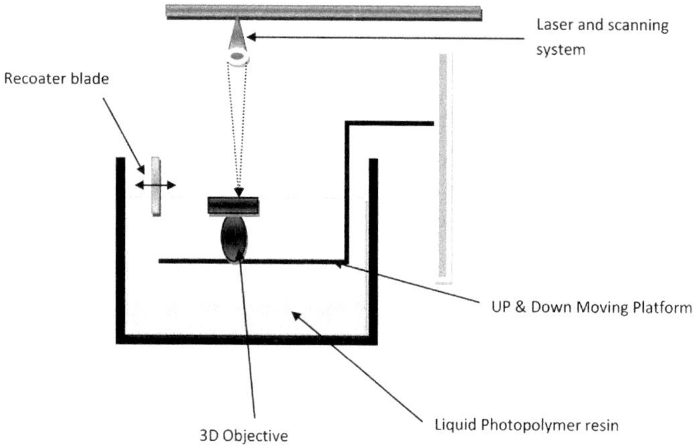

FIGURE 13.1 Stereolithographic (SLA) apparatus (Hull, 1986).

structures. Despite the recent diversification, our palette of 3D printable biomaterials remains significantly restricted. Increasing the biomaterial ink palette will allow the development of a variety of engineered environs with dissimilar and tunable physical, biochemical, and electrical properties necessary for different tissue categories (Figure 13.2), from bone to brain and the whole gamut in-between. This viewpoint offers an overall guide for 3D biomaterial ink improvement and structure estimates, with the purpose of generating further mimetic environs for complex tissue and organ engineering.

13.2 3D PRINTING TECHNOLOGIES

AM is an industrial process that deposits materials layer-by-layer (LBL) to build a touchable manufactured product. The most common, and currently the most popular, is 3D printing. AM is claimed to have triggered a third industrial revolution because the technology presents new and growing technical, economic and social impacts (Economist, 2012). Mainly, increased accessibility to 3D printing capabilities has allowed mass customization to become more widespread in manufacturing, such as in healthcare and consumer markets. Since the advent of mass production in the early twentieth century, consumers' demands have been met by producing large numbers of production in considerably less time than in the past.

The latest variety of 3D printing platforms are diverse, ranging from (1) inkjet deposition of solutions and cell the suspensions; (2) UV or photo facilitated stereolithography of photosensitive polymer solutions (Melchels et al., 2010); (3) selective laser sintering (SLS) of polymeric (Jeong and Hollister Scott, 2010) and metallic fine particles (Duan and Wang, 2011); (4) fused deposition modelling (FDM) of hot liquid extruded synthetic thermoplastics in addition composites (Lee et al., 2010); to (5) straight extrusion of solutions, gels, and colloidal as well as non-colloidal suspensions (Michna et al., 2005). Raised temperatures, strong light energies, solvent pools, or powder or fine particle beds restrict FDM and SLS technologies to

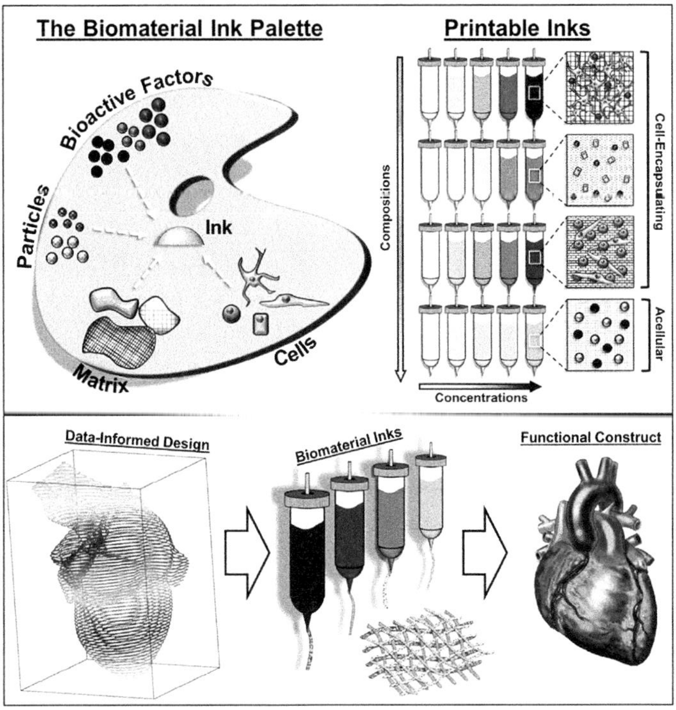

FIGURE 13.2 The biomaterial ink palette must be expanded to include components relevant to a variety of target tissues and organs. These components may include bioactive factors, discrete nanomaterials and matrix materials, and/or cells. By utilizing these ingredients and designing an expanded range of acellular and cell-encapsulating inks to meet the requirements outlined in Figure 13.2, we can use data-informed tissue and organ design to 3D print a variety of complex tissue and organ structures (Jakus, Rutz, and Shah, 2016).

a selection of appropriate materials, and consequently, these 3D printing standards are not perfect for 3D biomaterial printing. Inkjet and extrusion founded printing are commonplace with the largest selection of materials, but so far inkjet techniques still have notable disadvantages. This consists of their incapability of successfully printing viscous suspensions or solutions, and as a result, biomaterial inks having higher-fraction polymer or extracellular matrix (ECM) components are not capable of being directly inkjet printed; however, low viscosity and a low number of cross-links can be printed into dense polymer solutions (Cui and Boland, 2009) to generate healthy structures. Moreover, highly concentrated cell inks or inks containing a high density of rigid nanolevel or microlevel particles, such as drug-eluting microspheres or bioactive inorganic salts (e.g. calcium phosphates (Ca^{++})), are not well-suited to inkjet techniques because of the tendency of particles to aggregate under specified pressure, clogging the providing tool (Murphy and Atala, 2014). Extrusion type 3D printing, which have the capability of extruding viscous cell- and element-laden inks, has previously been positively used to 3D print synthetic, natural, and composite materials (e.g. hydrogels, polymers, ceramics) for soft tissues, hard tissues, and vascularized tissues with or without the accumulation of cells (Malda et al., 2013). Because of material flexibility, extrusion type 3D printing will be the emphasised in this chapter.

13.3 BIOMATERIAL INK

The expression "3D biomaterial printing" refers to the physical extrusion of an "ink" that promptly stabilizes upon deposition through at all ranges of mechanisms and is repeated LBL to create a construct intended to cooperate with biology (from cells to organisms). The biomaterial inks might or might not have active cells, hereafter referred to as cell-encapsulating inks, or bioinks, and acellular inks, respectively.

To be successfully 3D printed, biomaterial inks must show two important features. The primary essential ink characteristic is that they must be extrudable through the outlets ("nozzles") to permit the design of suitably shaped tissue and organ structures, where the scale of the preferred requirement is mainly dependent on the nozzle diameter. For instance, the glomeruli of kidneys exist are nearly 200 μm in width (Maezawa et al., 2013), and would necessitate sub 100 μm orifices to design the required structural features. On the other hand, the human liver lobule, one more functional organ division, is as big as 1.5 mm in diameter and could hypothetically be made-up using orifices greater than 100 μm (Crawford et al., 1998). Multi-level tissue constructions, for instance blood vessel networks, which take structures that range in size from micrometers (capillaries) to a few centimetres (the human aorta) (Miller et al., 2012), could necessitate numerous orifices of changeable diameters. The creation of bulky vasculatures, as well as comparatively homogeneous but inadequately vascularized tissues such as cartilage, skin, and bone, could be well-suited to larger nozzles (approximately 200 to 1000 μm). It is important to note that when specifying the fitting structural, chemical, and biological prompts, numerous cells hold the capability to self-consolidate into mimetic tissue arrangements. For instance, endothelial and mural cells can accumulate into similar porous structures (Takebe et al., 2013) and hepatocytes can make bile canaliculi (Abu-Absi, 2002). Hence, each aspect of a tissue may not be capable of being printed, but 3D printed scaffolds may instead be capable of better guiding cellular groups into supplementary well-ordered assemblies characteristic of natural tissue. To what extent cells and material requirements may be systematized inside the 3D construct will be a significant question to answer in the future of 3D biomaterial printing.

The second key feature relates to structural designation and reliability: on extrusion of the biomaterial ink against a substrate, it must perfectly keep the customer specified design through the whole construction. This means that the material must be mechanically weak enough to be extruded, but that its stiffness is adequate to be self-supporting upon installation, maintaining the printed construction through several layers. These two ink features can co-exist by one of two mechanisms, (1) fast LBL gelation or solidification through crosslinking or (2) straight extrusion of high viscosity liquid (e.g. pastes) or, to some extent, crosslinked soft gels. Fast gelation or hardening upon deposition allows even comparatively low viscosity biomaterial inks to be used to generate concrete structures, self-supporting configurations with distinct architectures. For instance, molten inks composed of biocompatible polymers and possible inorganic or organic bioactive materials mixed in organic diluents can extrude and result in a concrete structure, self-supporting fibers over rapid vaporization of the diluents, and precipitation of the polymer matrix on contact with air (Leigh et al., 2012). Supplementary solidification appliances could consist of thermal, ionic, or photon crosslinking; however, the rates of these responses need to be enormously speedy to allow the extruded material to hold its shape during deposition (Khalil and Sun, 2009). Increasing the viscosity or density of solutions, which can also be handled alone or in combination with LBL gelation or solidifying, can be done by increasing the polymer portion (Gaetani et al., 2012) or incorporating an inherently dense component (e.g. hyaluronic acid $(C_{14}H_{21}NO_{11})_n$) (Schuurman et al., 2013). There are known examples of published gel phase inks, which are prepared by adding a low dense of a cross-linker without the side effect of fast solidification of LBL (Skardal et al., 2010). Gel phase inks may need a supplementary post-printing equilibrium phase of further crosslinking, though they might only be lightly crosslinked to allow extrusion. Further post-printing treatment stages may consist of those general in traditional tissue engineering scaffold practice: washing or lyophilizing (to eliminate cytotoxic crosslinkers or organic solvents) (Yoon et al., 2003), sintering (Deville et al., 2006), acclimatizing (i.e. mechanical loading, incubation in media) (Dewez et al., 1999), addition (i.e. coating, infusion, adsorption) of extra components or bioactive materials (i.e. development factors, bioactive peptides, minor molecules) (Wu et al., 2006).

13.3.1 Present Biomaterial Inks and Their Restrictions

Biomaterial inks must meet traditional biomaterial concerns (biocompatibility, biodegradability, etc.) in addition to 3D printing concerns. Therefore, most biomaterial printing research has worked on developing 3D printing procedures before creating biomaterials, which include hydrogels and polyesters, in addition to polymer–ceramic composites (Figure 13.3). Hydrogels have been printed over a range of mechanisms: generally, by LBL crosslinking and as solidified solutions (usually higher-fraction polymers) and less frequently as minor accompaniments of crosslinkers. Agar/agarose (Landers et al., 2002), alginate (Fedorovich et al., 2008), and polyethylene glycol (PEG) derivatives (Maher et al., 2011) hydrogel inks take printed with great cell compatibility, however when untouched, are "blank-slate" materials that lack cell bonding and cell degradation sites. These materials would necessitate the accumulation of such sites by incorporating orders such as RGD (Arginylglycylaspartic acid)- and MMP (Matrix metalloproteinases)-sensitive sequences to impart marginal bioactivity, requiring fewer proper bioactivity to replicate exact cellular and tissue task (Lutolf et al., 2003). Other printed natural protein hydrogels contain gelatin (Wang et al., 2017) and atelocollagen/collagen (Smith et al., 2007). Even though these materials have inherent bioactivity (cell-binding and cell-degradable sites), they suffer from poor mechanical properties, especially in the cases of fibrinogen and collagen. Thermoplastics are one of the most common and notable types of 3D printable resources. These materials take in biocompatible polymers, polylactic acid (PLA), poly-lactico-glycolic-acid (PLGA), and polycaprolactone (PCL). Filaments or granular powders of these substances are liquefied at temperatures exceeding 60°C to permit extrusion through a small orifice. Upon deposition onto a substrate or an earlier deposited part of the material, the polymer hardens at or near room temperature. Even though comparatively easy to 3D print, these polymers and related printing methods suffer from a number of drawbacks that have stopped common clinical usage. These thermoplastics do not intensely sustain cell bonding or tissue-specific reactions on their own and frequently must be exchanged post-printing with biological aspects or protein coatings (Serra et al., 2013). Another substitute methodology for increasing bioactivity is by adding high-temperature resilient bioactive elements, such as calcium phosphates (e.g. tricalcium phosphate or hydroxyapatite) for osteochondral applications (Park et al., 2011). On the other hand, due to the fact that the extrusion depends on the property of the polymer's melt flow, the composite structure must constantly be majority-polymer, and this limits the density of bioactive particles.

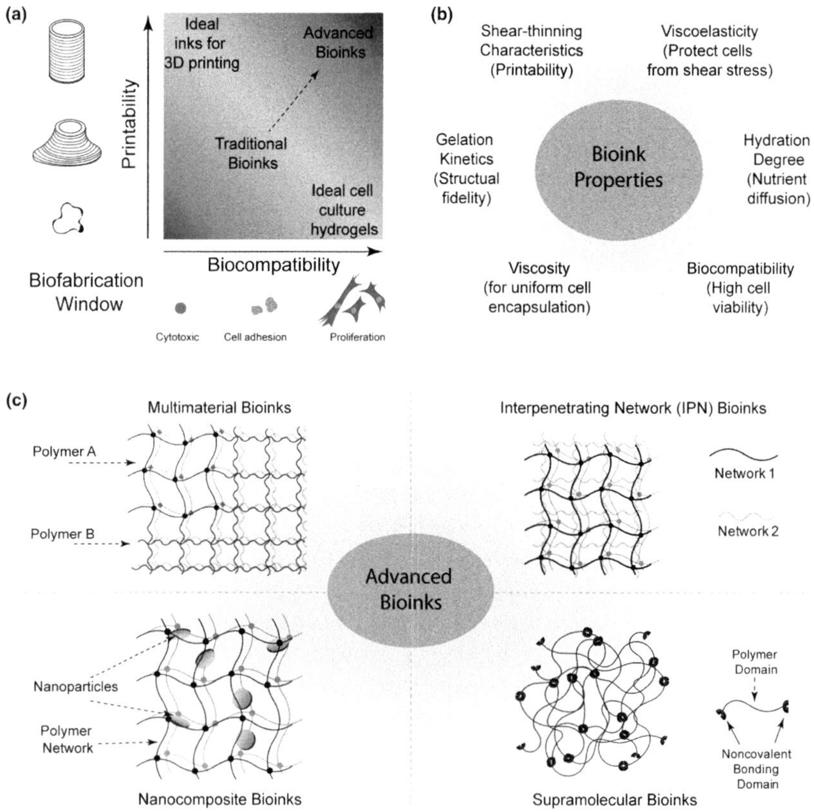

FIGURE 13.3 Advanced bioinks for 3D printing. (a) Biofabrication window for rational design of bioinks requires compromise between printability and biocompatibility. (b) Ideal bioink characteristics require interplay between different materials' properties. (c) Advanced bioinks can be classified into four major categories (Chimene et al., 2016).

13.4 INK-PRINTING AND POST-PROCESSING

While all 3D biomaterial printed constructs eventually have to bear the feasibility, proliferation and occupation of the cells, biocompatibility may look to be an understandable condition but has dissimilar stringency levels depending on the content of the ink as well as of the application of the printed structure. To better appreciate the necessities of cell compatibility, we divide biomaterial inks into two leading classes: acellular and cell-encapsulating, both of which will play important roles in the future of biomaterial 3D printing.

13.4.1 Acellular Inks

Acellular inks and the consequential theories offer plentiful rewards encouraging to commercialization and clinical transformation. Acellular biomaterial inks do not essentially need to be cell-compatible prior to, during, or after being 3D printed, as long as they are changed post-printing to create cell compatibility prior to application. For instance, inks or created structures can hold organic solvents and cytotoxic reagents as long as they are successfully detached leading to the demand of cell printing or implantation. Additionally, high temperatures (>60°C) applied for extrusion and post-printing treating stages, for instance, sintering or heat, gas, or plasma sterilization, could also be employed. In the future, acellular inks could signify a greater diversity of materials with distinct, functional properties from cell-encapsulating inks, owing to the fact that considerably more material and processing options are permissible. A promising illustration of this is the organic solvent-based 3D printing of graphene, which made possible the neurogenic variation of mesenchymal stem cells without the necessity for supplementary biochemical neurogenic aspects or electrical stimulation (Jakus et al., 2015).

13.4.2 Cell-Encapsulating Inks

Cell-encapsulating bioinks, though, must satisfy a more rigorous set of cell compatibility necessities. Bioinks are composed of living cells, which are mixed with polymer aqueous solutions and chemically or physically crosslinked to form hydrogels. Hydrogels offer the encapsulated cells a high aquatic content (95 to 99%) environment that permits the passive dispersal of nutrients and waste to maintain cell viability (Seliktar, 2012).

The bioink acts not only as a structural medium for spatially modelling cells in 3D but also as an incubator, supporting their viability before, during, and after printing, over which time the cells experience substantial stresses. These stresses may not only influence cell viability and health, but also their whole actions, functions, and efficacy. Stresses consist of the physical mixing necessary to incorporate cells into the bioink matrix, shear forces arising during extrusion over a lower diameter nozzle, and pre-printing or post-printing methods such as contact with chemicals or temperature modifications often essential for ink gelation (Fedorovich et al., 2008). It would be noted that separate cell types, stretching from immortalized cell ranks, to stem cells, to multi-cell spheroids, can differ in sensitivity and thus, react negatively to the printing method (i.e. post-printing possibility may change depending on cell type) (Williams et al., 2005).

13.4.3 Ink Necessities for Printing

To overcome poor printing ability, for instance those of small polymer segment hydrogels, materials can be co-printed with support materials. There are two kinds of support inks, (1) "fugitive" and (2) "permanent". If the ink is selectively eliminated at a future time, this is referred to as fugitive ink. The process is implemented in the existence of cells, and thus both the fugitive ink and the process of eliminating it should also be "cell friendly" (Kolesky and Truby, 2014). On the other hand, support inks can be eliminated within the structure, for instance the co-printing of hydrogels with hot flux thermoplastic polymers. For this reason, the cell-compatibility as well as tissue necessities for the final application must also be contemplated (Shim et al., 2012). For instance, the restrictions of high toughness as well as acidic degradation products of stiff thermoplastic polymers might not be perfect for soft tissue production. Moreover, to tolerate inks, dense and/or hydrophobic plotting may be used to offer support to the printed material (Maher et al., 2009). On the other hand, these support mediums must regularly be eliminated prior to the usage of the printed object, confusing the construction method. More complex 3D printed tissue and organ systems will expectedly necessitate a multiplicity of biomaterial inks, all of which must be connectable with each other, which taking into consideration both cell compatibility and printability. For instance, multi-material printing of cellular and acellular inks requires that the printing and post-processing stages of the acellular ink are cell friendly. Thus, the use of organic solvents or high temperatures is not suitable, as it would compromise cell feasibility inside the printed pattern. Despite the practical challenges in emerging multi-material systems, there have previously been examples of multi-material 3D biomaterial printing (Shim et al., 2012). On the other hand, the most promising approach to multi-biomaterial 3D printing is to improved sessions of biomaterial inks that are not only separate in composition and biological functionality but can also be personalized to have very comparable or even the same printing and post-processing requirements (Rutz et al., 2015). In such cases, skills in chemistry, materials science, and engineering will play essential parts in supplementing the essential biological expertise.

13.4.4 Hydrogel-Based Bioinks

In general, the compulsory properties of hydrogel-based bioinks are:

- Comparatively greater viscosity to afford identical cell suspension and primary structural reliability
- High shear-thinning manners to reduce cell injury
- Quick gelation in constructing a 3D tissue structure

Particularly, gelation appliances of hydrogels are carefully constructed on the printing procedure. Numerous gelation procedures have been working on growing the stability of 3D hydrogel-based bioinks for cell printing. One method is the practice of thermosensitive hydrogels, like gelatin and Pluronic F127, to sustain the pattern until a crosslinkable bioink is induced for extensive stages of hydrogel stability at physiological settings. The following sections will discuss the presently uses of hydrogels as bioink materials used for cell printing. Figure 13.4 shows a diagram of variables that are hazardous to 3D bioprinting strategy.

13.4.5 Alginate

Alginate is a naturally derived anionic polysaccharide (brown algae) showing gelation in the occurrence of bivalent ions, for instance, Ca^{++} (Jin et al., 2010). These hydrogels have assisted as a cell transport material for numerous tissue engineering practices, due to the ease of their creation and their comparatively high cell compatibility; however, the major disadvantage is the absence of degradation in mammalian enzymatic activity, which restricts tissue regeneration when implanted. Likewise, there is insufficient cell supplementation to alginate chains

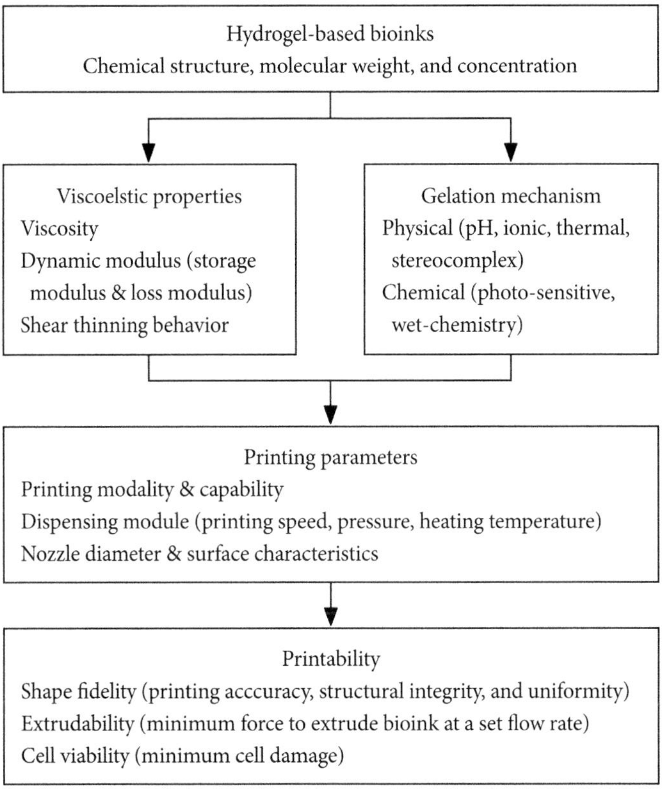

FIGURE 13.4 Schematic diagram of variables critical to 3D bioprinting strategy. The hydrogel-based bioinks determine the viscosity, gelation mechanism, and printing parameters, eventually, bioprinted tissue constructs. (Kim, Yoo, and Lee, 2016).

without chemical alteration (Nair and Laurencin, 2007). In the initial phase of cell printing, a jetting printing arrangement was altered for 3D printing by the printing of a Ca^{++} (salt paper) solution into a basin of cardiac cells combined with alginate solution (Xu et al., 2009). The assembling of the calcium chloride solution caused gelation to create a hollow shell structure in the desired pattern, with each shell having a mean outer diameter of 25 μm. An elevator system stirred the gelled construct down to expose fresh alginate solution to the printed solution to permit the creation of a 3D structure in the figure of a two-chambered heart-like structure.

13.4.6 Hyaluronic Acid

Hyaluronic acid (HA) is a glycosaminoglycan substance in most advanced of tissues in the group, specifically the skin, vitreous humor, and synovial fluid. The high molecular weight and huge volume of branching of HA allows for intermolecular hydrogen connection and higher viscosity. Similar to further polysaccharides, HA provides the cell sustainability, but has short tie motifs in cell connection. HA has been utilized in the accumulation of PEG-based arms for crosslinking by UV originated acrylate polymerization (Skardal et al., 2010). 4-armed PEG connectors were applied for chemical alteration to established TetraPAc crosslinker molecules that crosslinkers were reacted with thiolated HA, and thiolated gelatin (Gtn-DTPH), to generate a crosslinkable printable bioink to advance the cell attachment. The crosslinked hydrogel was allotted as cylindrical filaments, loaded to create a tubular shape, and sheltered with agarose to preserve the construction and alignment of the filaments. The outcomes specified the conservation of cell viability, stability, and structural alignment with a lumen for nearly 4 weeks (Figure 13.2 d–f).

13.4.7 Collagen

Collagen is the most plentiful protein in the human body. We have numerous types of collagen in our body. Collagen type 1 is the focus of this segment as it is the most frequently used for cell printing protein. Under suitable temperature and pH, a pure collagen solution undergoes gelation to form a gel with properties dependent on its solution density. Cells give collagen over integrin tie and enzymatically reduce collagenous fibers allowing cell movement and extracellular matrix (ECM) renovation. Plentiful reactive moieties allow for chemical alteration and crosslinking of chemical and physical properties (Nair and Laurencin, 2007). Unlike other hydrogels, collagen-based bioinks must be handled with care to avoid premature setting, regularly preserved at lower temperatures of 277–293 K. Roth et al. (2007), showed that 1% solution of collagen could be written with an altered inkjet printer into the chosen patterns on agarose layered glass coverslips. The collagen solution for this methodology was kept to some extent acidic to avoid clogging. After printing, the collagen hydrogel was dehydrated and reconstituted in advance cells which could then be cultured on the patterns of ararose layered glass.

13.4.8 Gelatin

Gelatin is a thermally altered collagen which forms a temperatire reversible hydrogel with a strength relying on its density. Laser-induced forward transfer (LIFT) methodology was applied to print ranges of precipitation droplets with embryonic stem cells (ESCs) (Raof et al., 2011). The ribbon should be coated with 20vol.% gelatin solution, then an ESC suspension of $2–5 \times 10^6$ cells was employed on the gelatin coating surface. Extra fluid was subtracted from the surface such that the ESCs were moderately incorporated and faced down over the acceptance substrate during printing. The acceptance substrate was coated with a nearly 10% gelatin solution to permit printing through the droplets. The printed designs displayed proliferation and embryonic body materialization after one week in culture, signifying that the printing of ESCs was capable of preserving their phenotype and vitality as guaranteed by immunostaining for OCT4, Myf 5, and PDX 1. Chemical alteration of gelatin can be done to improve crosslinking (increase strength) and bioactivity. The usage of gelatin methacrylate (GelMa) for printing a matrix architecture having cells and vasculature has been demonstrated (Kolesky et al., 2014).

13.5 PRESENT USES FOR BIOPRINTING

A vast number of studies on unnatural or natural manmade tissues and organs using 3D bioprinting methods have been undertaken not only for short-term respite and maintenance functions, which have been achieved previously, but also for the determination of tissue reparation and renewal (Table 13.1). A growth factor that can augment cell differentiation can be printed together with scaffold or biomaterials and cells can be printed straight.

13.5.1 Bone

Of the regeneration of hard tissues, the most predictable and quickly developing area of usage of 3D printers is with bone. Bone consists of a simpler structure than other tissues and the failure zone is frequently non-uniform in structure; a lot of studies have been directed towards creating bone using 3D printers (Bose, Roy, and Bandyopadhyay, 2012). Since bone usually has to survive in high loads, scaffolds prepared through ceramic and biodegradable polymers have been created for bone regeneration purposes. A research group led by W. D. Kim fabricated a biodegradable PCL scaffold using an extrusion technique for bone tissue fabrication (Lee et al., 2016).

13.5.2 Cartilage

To regenerate cartilage tissue fabrication, a scaffold is desirable to propagate chondrocytes or stem cells. For this reason, polymer mixtures or natural polymers are made into a scaffold whose construction uses a 3D bio-printer extrusion technique. Some of the illustrative prototypes are regenHU (Villaz-St-Pierre, Switzerland) and EnvisionTEC (Gladbeck, Germany) (Hutmacher, 2000).

TABLE 13.1
Overview of 3D Bioprinting Technologies for Generation of Tissue and Organ

Bioprinting Technique	Building Materials	Cell Types	Applications	Printing Conditions
Powder bed fusion	HA	MC3T3-E1	Bone	Sintering: 2 h Temperature: 1,300°C
Extrusion	PCL	MC3T3-E1	Bone	Heating dispenser: 80°C Air pressure: 300 kPa
Extrusion	PEGT/PBT	Bovine articular chondrocytes Human articular chondrocytes	Cartilage	Temperature: 180, 200°C Applide force: 1.5, 2.0 kN
Extrusion	Nanofibrillated cellulose (NFC) Alginate	Human nasoseptal chondrocytes	Cartilage	Printer head: microvalve dispenser Valve opening time: 400–1,200 μs Dispensing pressure: 20–60 kPa
Extrusion	Collagen hydrogel	Fibrochondrocyte	Cartilage	Baseplate temperature: 37°C
Extrusion	Hydrogel	Multicellular spheroid: human umbilical vein smooth muscle cells (HUVSMCs) Human skin fibroblasts (HSFs)	Blood vessel	Scaffold-free tissue fabrication technology
Electrospinning	Collagen type I Elastin Poly(D,L-lactide-coglycolide) (PLGA)	Bovine endothelial cells Bovine smooth muscle cells	Blood vessel	Voltage potential: 25 kV Distance: 15 cm Rotation rate: 50 rpm
Electrospinning, extrusion	PCL Chitosan (CTS)	–	Blood vessel	–
Inkjet	–	Fibroblasts Keratinocytes	Skin	HP deskjet 640c series inkjet printer modified
Extrusion	Alginate hydrogel	Chondrocyte	Ear	–
Inkjet	Collagen Polyglycolic acid	Urothelial Muscle cells	Bladder	–
SLS (selective laser sintering)	PCL HAp	–	Trachea (personalized medical devices)	–

Source: Murphy and Atala (2013).

13.5.3 Blood Vessels

Blood vessels essentially possess high elasticity and stability that can tolerate frequent expansion and shrinkage. Polymer compounds have created artificial blood vessels. However, due to difficulties such as thrombosis and stenosis, straining on the usage of biodegradable polymers has been shown (Stitzel et al., 2006). A scaffold fabrication through electrospinning using numerous approaches has the advantage of having the characteristics of ECM analogues by reason of its porosity; on the other hand, its pore size is minute so the connectivity is short (Naito et al., 2014).

13.5.4 Skin

Skin tissue hurt due to burns or wounds can be treated via an autograft of the patient's specific skin, a homograft or allograft, which consists of the transfer of donor/patient skin, or a heterograft or xenograft, which consists of replacement with animal skin. On the other hand, the above approaches have disadvantages due to immune elimination. To overcome this restriction, research has led to the creation of artificial skin (Martínez-Santamaría et al., 2012), in addition to skin regeneration using 3D bioprinters as an alternative to artificial skin.

13.5.5 Ear

A non-natural ear was made-up by a combined research group from Princeton University and Johns Hopkins University via a syringe extrusion 3D printer. The figure of the ear was printed by chondrocyte holding alginate hydrogel. The device used was a Fab@Home 3D printer. Once printing finished, a coil antenna was fixed to receive wireless signals via silver nanoparticles (AgNP). Sound waves were observed directly through the antenna (Mannoor et al., 2013).

13.5.6 Liver

A non-natural liver was created by the Organovo Company using an improvement of a 3D printer containing liver cell cartridge. To do this, they used an extrusion method with Novogen equipment. The artificial liver contained a number of cell categories as well as human liver cells. It was synthesized in 20 layers and had similar cell density as that of the original liver tissue. The new liver tissue worked similar to the real liver for 40 days, forming albumin (depending on the ink containing cells), transferrin, and fibrinogen. This type of manufactured liver tissue can be likely used for medical research, such as the analysis of different drugs (Struecker, Raschzok, and Sauer, 2014).

13.5.7 Trachea

The University of Michigan invented a lung splint via bio-absorbable powder material through a powder bed fusion (PBF) technique and transferred it to the bronchi of an 18-month old child who was in distress due to shortness of respiration. An ideal structure that was coordinated to the child's airway structure was made up, and a splint was synthesized using 3D printing equipment. The splint was constructed by a 3D printer (EOS P 100) via the SLS technique using 96% PCL and 4% HA (Morrison et al., 2015).

13.6 CONCLUSIONS AND FUTURE DIRECTIONS

Three-dimensional or additive manufacturing bioprinting technologies hold countless promise to overcome the present restrictions in tissue engineering and regenerative treatment. Presently, there has been great work on improving novel printing techniques and hydrogel-based bioinks to achieve better resolution of the product. Advanced printing techniques could offer progressively more complex designs with anatomical and functional likenesses to native tissues. The present jet-based technique using cell-laden hydrogel bioinks has achieved comparatively greater resolution of roughly 20–100 μm; however, this method is restricted to constructing huge tissue constructs with structural reliability. The extrusion-based technique has printed cell-laden hydrogel bioinks measuring approximately 50–400 μm in layer thickness, yet the shear stress to the hydrogel bioinks through the nozzle sharply rises when the nozzle diameter drops, causing high cell injury. Alternatively, a new hydrogel bioink scheme for cell printing needs to be developed for improving printability with good resolution capability. However, the accessibility of presently available hydrogels that can suit as cell printing bioinks which also offer tunable mechanical properties, cell-matrix interface, and insignificant cytotoxicity is narrow. The benefits of the bioprinting of appropriate cell-compatible hydrogel-based bioinks are serious for the long-term achievement of cell printing for tissue regeneration.

3D bioprinting methods are capable of fabricating 3D freeform structures holding multiple cell kinds, biomaterials, and bioactive particles, bringing about sophisticated constructions that have the chance to heal injured or unhealthy human tissues and organs. Thus, 3D bioprinting methods hold high potential in tissue engineering as well as regenerative medicine. There is much effort towards progressing these technologies on the way to successful clinical translation.

REFERENCES

Abu-Absi, S. (2002). Structural polarity and functional bile canaliculi in rat hepatocyte spheroids. *Experimental Cell Research*, 274(1), 56–67.

Advancing the Field of 3D Biomaterial Printing. (2016). Retrieved from http://www.bioinksolutions.com/bbs/board.php?bo_table=news&wr_id=23.

Crawford, A. R., Lin, X. Z., and Crawford, J. M. (1998). The normal adult human liver biopsy: a quantitative reference standard. *Hepatology*, 28(2), 323–331.

Cui, X., and Boland, T. (2009). Human microvasculature fabrication using thermal inkjet printing technology. *Biomaterials*, 30(31), 6221–6227.

Deville, S., Saiz, E., Nalla, R. K., and Tomsia, A. P. (2006;2017;). Freezing as a path to build complex composites. *Science*, 311(5760), 515–518. doi:10.1126/science.1120937

Dewez, J., Doren, A., Schneider, Y., and Rouxhet, P. G. (1999). Competitive adsorption of proteins: Key of the relationship between substratum surface properties and adhesion of epithelial cells. *Biomaterials*, 20(6), 547–559. doi:10.1016/S0142-9612(98)00207-5

Duan, B. and Wang, M. (2011). SLS and its application in biomedical engineering. *MRS Bulletin*, 36(12), 998–1005.

Economist. "The Third Industrial Revolution." (2012). Retrieved from https://www.economist.com/leaders/2012/04/21/the-third-industrial-revolution

Fedorovich, N. E., de Wijn, J. R., Verbout, A. J., Alblas, J., and Dhert, W. J. A. (2008). Three-dimensional fiber deposition of cell-laden, viable, patterned constructs for bone tissue printing. *Tissue Engineering. Part A*, 14(1), 127–133. doi:10.1089/ten.a.2007.0158

Gaetani, R., Doevendans, P. A., Metz, C. H. G., Alblas, J., Messina, E., Giacomello, A., and Sluijter, J. P. G. (2011;2012;). Cardiac tissue engineering using tissue printing technology and human cardiac progenitor cells. *Biomaterials*, 33(6), 1782–1790. doi:10.1016/j.biomaterials.2011.11.003

Hull, C. W. (1986). Apparatus for production of 3D objects by stereolithography 4, 575, 330.

Jakus, A. E., Rutz, A. L., and Shah, R. N. (2016). Advancing the field of 3D biomaterial printing. *Biomedical Materials (Bristol, England)*, 11(1), 014102. doi:10.1088/1748-6041/11/1/014102

Jakus, A. E., Secor, E. B., Rutz, A. L., Jordan, S. W., Hersam, M. C., and Shah, R. N. (2015). Three-dimensional printing of high-content graphene scaffolds for electronic and biomedical applications. *ACS Nano*, 9(4), 4636–4648. doi:10.1021/acsnano.5b01179

Jeong, C. G., and Hollister, S. J. (2010). A comparison of the influence of material on in vitro cartilage tissue engineering with PCL, PGS, and POC 3D scaffold architecture. *Biomaterials*, 31(15), 4304–4312.

Jin, R., Moreira Teixeira, L. S., Dijkstra, P. J., van Blitterswijk, C. A., Karperien, M., and Feijen, J. (2010). Enzymatically-crosslinked injectable hydrogels based on biomimetic dextran -hyaluronic acid conjugates for cartilage tissue engineering. *Biomaterials*, 31(11), 3103–3113. doi:10.1016/j.biomaterials.2010.01.013

Khalil, S., and Sun, W. (2009). Bioprinting endothelial cells with alginate for 3D tissue constructs. *Journal of Biomechanical Engineering*, 131(11), 111002. doi:10.1115/1.3128729

Kim, J. H., Yoo, J. J., and Lee, S. J. (2016). Three-dimensional cell-based bioprinting for soft tissue regeneration. *Tissue Engineering and Regenerative Medicine*, 13(6), 647–662.

Kolesky, D. B., Truby, R. L., Gladman, A. S., Busbee, T. A., Homan, K. A., and Lewis, J. A. (2014). 3D bioprinting of vascularized, heterogeneous Cell-Laden tissue constructs. *Advanced Materials*, 26(19), 3124–3130. doi:10.1002/adma.201305506

Landers, R., Pfister, A., Hübner, U., John, H., Schmelzeisen, R., and Mülhaupt, R. (2002). Fabrication of soft tissue engineering scaffolds by means of rapid prototyping techniques. *Journal of Materials Science*, 37(15), 3107–3116. doi:10.1023/A:1016189724389

Lee, C. H., Cook, J. L., Mendelson, A., Moioli, E. K., Yao, H., and Mao, J. J. (2010). Regeneration of the articular surface of the rabbit synovial joint by cell homing: a proof of concept study. *Lancet*, 376(6739), 440–448.

Lee, D., Lee, J. H., Lee, S. J., Yoon, T. R., Kim, W. D., Kim, H. K., and Kwon, I. K. (2016). Surface modification of 3D-printed porous scaffolds via mussel-inspired polydopamine and effective immobilization of rhBMP-2 to promote osteogenic differentiation for bone tissue engineering. *Acta Biomaterialia*, 40, 182–191. doi:10.1016/j.actbio.2016.02.006

Leigh, S., Bradley, R., Purssell, C., Billson, D., and Hutchins, D. (2012). A simple, low-cost conductive composite material for 3D printing of electronic sensors. *PLoS One*, 7(11) doi:10.1371/journal.pone.0049365

Lutolf, M. P., Raeber, G. P., Zisch, A. H., Tirelli, N., and Hubbell, J. A. (2003). Cell-responsive synthetic hydrogels. *Advanced Materials*, 15(11), 888–892. doi:10.1002/adma.200304621

Maezawa, Y., Cina, D., and Quaggin, S. E. (2013). *Glomerular Cell Biology Seldin and Geibisch's*. Elsevier.

Maher, P. S., Keatch, R. P., Donnelly, K., Mackay, R. E., and Paxton, J. Z. (2009). Construction of 3D biological matrices using rapid prototyping technology. *Rapid Prototyping Journal*, 15(3), 204–210. doi:10.1108/13552540910960307

Maher, P. S., Keatch, R. P., Donnelly, K., and Vorstius, J. B. (2011). Bio-mechanical analysis for characterising a commercial 3D printed composite. *International Journal of Medical Engineering and Informatics*, 3(3), 275. doi:10.1504/IJMEI.2011.042873

Malda, J., Visser, J., Melchels, F. P., Jüngst, T., Hennink, W. E., Dhert, W. J. A., and Hutmacher, D. W. (2013). Engineering hydrogels for biofabrication. *Advanced Materials*, 25(36), 5011–5028.

Mannoor, M. S., Jiang, Z., James, T., Kong, Y. L., Malatesta, K. A., Soboyejo, W. O. and McAlpine, M. C. (2013). 3D printed bionic ears. *Nano Letters*, 13(6), 2634–2639.

Martínez-Santamaría, L., Guerrero-Aspizua, S., and Del Río, M. (2012). Skin bioengineering: preclinical and clinical applications. *Actas Dermo-Sifiliográficas*, 103(1), 5–11.

Melchels, F. P., Domingos, M. A., Klein, T. J., Malda, J., Bartolo, P. J., and Hutmacher, D. W. (2012). Additive manufacturing of tissues and organs. *Progress in Polymer Science*, 37(8), 1079–1104.

Melchels, F. P. W., Feijen, J., and Grijpma, D. W. (2010). A review on stereolithography and its applications in biomedical engineering. *Biomaterials*, 31(24), 6121–6130.

Michna, S., Wu, W., and Lewis, J. A. (2005). Concentrated hydroxyapatite inks for direct-write assembly of 3D periodic scaffolds. *Biomaterials*, 26(28), 5632–5639.

Miller, J. S., Stevens K. R., Yang, M. T., Baker B. M., Nguyen, H. T., Cohen, D. M. and Chen, C. S. (2012). Rapid casting of patterned vascular networks for perfusable engineered 3D tissues. *Nature Materials*, 11(9), 768–774.

Morrison, R. J., Hollister, S. J., Niedner, M. F., Mahani, M. G., Park, A. H., Mehta, D. K. and Green G. E. (2015). Mitigation of tracheobronchomalacia with 3D-printed personalized medical devices in pediatric patients. *Science Translational Medicine*, 7(285), 285ra64.

Murphy, S. V., and Atala, A. (2014). 3D bioprinting of tissues and organs. *Nature Biotechnology*, 32(8), 773–785.

Nair, L. S., and Laurencin, C. T. (2007). Biodegradable polymers as biomaterials. *Progress in Polymer Science*, 32(8), 762–798. doi:10.1016/j.progpolymsci.2007.05.017

Naito, Y., Rocco, K., Kurobe, H., Maxfield, M., Breuer, C., and Shinoka, T. (2014). Tissue engineering in the vasculature. *The Anatomical Record*, 297(1), 83–97. doi:10.1002/ar.22838

Park, S. A., Lee, S. H., and Kim, W. D. (2011). Fabrication of porous polycaprolactone/hydroxyapatite (PCL/HA) blend scaffolds using a 3D plotting system for bone tissue engineering. *Bioprocess and Biosystems Engineering*, 34(4), 505–513. doi:10.1007/s00449-010-0499-2

Raof, N. A., Schiele, N. R., Xie, Y., Chrisey, D. B., and Corr, D. T. (2011). The maintenance of pluripotency following laser direct-write of mouse embryonic stem cells. *Biomaterials*, 32(7), 1802–1808. doi:10.1016/j.biomaterials.2010.11.015

Roth, E. A., Xu, T., Das, M., Gregory, C., Hickman, J. J., and Boland, T. (2004). Inkjet printing for high-throughput cell patterning. *Biomaterials*, 25(17), 3707–3715. doi:10.1016/j.biomaterials.2003.10.052

Rutz, A. L., Hyland, K. E., Jakus, A. E., Burghardt, W. R., and Shah, R. N. (2015). A multimaterial bioink method for 3D printing tunable, cell-compatible hydrogels. *Advanced Materials*, 27(9), 1607–1614. doi:10.1002/adma.201405076

Schuurman, W., Levett, P. A., Pot, M. W., van Weeren, P. R., Dhert, W. J., Hutmacher, D. W., Melchels, F. P., Klein, T. J., and Malda, J. (2013), Gelatin-methacrylamide hydrogels as potential biomaterials for fabrication of tissue-engineered cartilage constructs. *Macromolecular Bioscience*, 13(5), 551–561. doi:10.1002/mabi.201200471

Seliktar, D. (2012). Designing cell-compatible hydrogels for biomedical applications. *Science*, 336(6085), 1124–1128. doi:10.1126/science.1214804

Serra, T., Mateos-Timoneda, M. A., Planell, J. A., and Navarro, M. (2013). 3D printed PLA-based scaffolds: A versatile tool in regenerative medicine. *Organogenesis*, 9(4), 239–244. doi:10.4161/org.26048

Shim, J. H., Lee, H. C., Kim, S., Shin, Y. M., Kim, K. M., Lim, Y., and Suh, D. J. (2012). Which response criteria best help predict survival of patients with hepatocellular carcinoma following chemoembolization? A validation study of old and new models. *Radiology*, 262(2), 708–718. doi:10.1148/radiol.11110282

Skardal, A., Zhang, J., and Prestwich, G. D. (2010). Bioprinting vessel-like constructs using hyaluronan hydrogels crosslinked with tetrahedral polyethylene glycol tetracrylates *Biomaterials*, 31(24), 6173–6181.

Smith, C. A., Lau, K. M., Rahmani, Z., Dho, S. E., Brothers, G., She, Y. M., Berry, D. M., Bonneil, E., Thibault, P., Schweisguth, F., Le Borgne, R. and McGlade, C. J. (2007). aPKC-mediated phosphorylation regulates asymmetric membrane localization of the cell fate determinant Numb. *The EMBO Journal*, 26(2), 468–480. doi:10.1038/sj.emboj.7601495

Stitzel, J., Liu, J., Lee, S. J., Komura, M., Berry, J., Soker, S., Atala, A. (2006). Controlled fabrication of a biological vascular substitute. *Biomaterials*, 27(7), 1088–1094. doi:10.1016/j.biomaterials.2005.07.048

Struecker, B., Raschzok, N., and Sauer, I. M. (2014). Liver support strategies: Cutting-edge technologies. *Nature Reviews. Gastroenterology & Hepatology*, 11(3), 166–176. doi:10.1038/nrgastro.2013.204

Wang, X., Ao, Q., Tian, X., Fan, J., Tong, H., Hou, W., and Bai, S. (2017). Gelatin-based hydrogels for organ 3D bioprinting. *Polymers*, 9(12), 401. doi:10.3390/polym9090401

Wu, Y., Shaw, S., Lin, H., Lee, T., and Yang, C. (2006). Bone tissue engineering evaluation based on rat calvaria stromal cells cultured on modified PLGA scaffolds. *Biomaterials*, 27(6), 896–904. doi:10.1016/j.biomaterials.2005.07.002

Xu, T., Baicu, C., Aho, M., Zile, M., and Boland, T. (2009). Fabrication and characterization of bio-engineered cardiac pseudo tissues. *Biofabrication*, 1(3), 035001. doi:10.1088/1758-5082/1/3/035001

14 Application of "Smart Polymers" as Biomaterials

14.1 INTRODUCTION TO SMART POLYMERS

14.1.1 Introduction

Smart polymers or stimuli-responsive polymers are the new generation of polymers; they have the ability to modify their chemical and physical structure as the result of changes in their environment. They can respond to changes in factors such as temperature, pH, light, magnetic field, shape, etc., that, depending on the physical structure of the material, could respond by swelling, changing in phase, etc. (Aguilar et al., 2007).

Linear smart polymers will change from monophasic to biphasic in order to produce the reversible sol-gel hydrogels and crosslinked polymers will change their structure so that the network changes from a collapsed structure to an expanded structure. All these changes help us to design and produce smart devices with multiple functions, for instance, minimally invasive injectable systems (Arora et al., 2010; Tran et al., 2013).

In addition, polymer engineering has a lot of possibilities for producing different types of smart polymers for different applications, because polymers can be active either by pre-polymerization or by post-polymerization (Guillerm et al., 2012; Arnold et al., 2012) systems.

14.1.2 Types of Smart Polymers

There are different types of stimuli-responsive polymers that will be covered in the following chapter, and with their ability to change their structure to respond the changes in environment, stimuli-responsive polymers are becoming a very important field in engineering, biomedical, automotive, and even nuclear energy. However, among the temperature-responsive polymers, those that contain lower critical solution and upper critical solution temperature (LCST and UCST) (Figure 14.1) have gained huge attention due to their potential in wide applications (Yang et al., 2013; Bajpai et al., 2010).

pH responsive polymers, photo-responsive polymers, hydrogel polymers, magnetic responsive polymers and enzyme responsive polymers are different type of smart polymers and we will discuss each of them in this chapter.

by Yaser Dahman and Amirhossein Kouhpour

FIGURE 14.1 Sol-to-gel transition followed by a gel-to-sol transition (Yoshida et al., 2017).

14.1.3 Application of Smart Polymers

Responsive polymers are a highly important field in tissue engineering because they modulate cell behaviour when in contact with an external factor. Addressing some applications of smart polymers, we can point at smart biomineralization (Huang et al., 2008), vascular graft tissue engineering (Fioretta et al., 2012), drug delivery (Moroni et al., 2008), the development of new medical devices such as minimally invasive surgery medical devices by using shape-responsive polymers (Yakacki and Gall et al., 2010), the fabrication of new devices for cancer diagnosis, the development of hyperthermia treatment and magnetic resonance imaging diagnosis by using magnetic nanoparticles (Karimi, Karimi and Shokrollahi, 2013), the making of biosensors based on smart polymers, bioseparation and biotechnological applications based on smart polymers (Galaev and Mattiasson, 2007), and textiles using smart polymers in their formulations (Hu et al., 2012), and these applications are just some examples of the usage of smart polymers in biomedical and tissue engineering.

14.2 DIFFERENT TYPES OF SMART POLYMERS

14.2.1 Temperature-Sensitive Polymers

Smart polymers that can respond to changes in environmental conditions are an interesting and promising class of materials, and temperature-sensitive polymers, as mentioned before, have gained a lot of attention due to variation in temperature.

There are three important classes of temperature-sensitive polymers:

- Shape-memory materials
- Liquid crystalline materials
- Responsive polymer solutions

The first class of TRP's are thermoplastic elastomers that contain a solid phase with high temperature and a melting phase with mid temperature which is the reason for their temperature behaviour (Lendlein and Kelch, 2002; Liu et al., 2007).

Liquid crystalline materials have three phases: a liquid crystalline phase, a glassy state phase and an isotropic rubbery phase (Donald et al., 2005; Weiss and Ober, 1990). Polymers with crystalline blocks in their backbone have a fully reversible mechanism for the main chain in their crystalline phase. When a polymer with crystalline blocks adsorb enough heat, the polymer contracts to a random coil state and changes to an isotropic phase. This method has been used as the switching mechanism for artificial muscles (Li and Keller, 2006).

Stimuli-responsive polymer solutions, which are an important type of polymers to understand, are the systems that pass through a liquid–liquid phase conversion and respond to variations in temperature and change from a homogeneous phase to a concentrated and dilute polymer phase. This phase change usually appears as a change from a clear solution to a cloudy solution for polymers with low concentration. The presence of the cloudy phase is because of the structure of the high concentration polymer droplets and the index gaps of the other two phases.

It is important to mention that the phase transition due to increase in temperature is referred to as the LCST, and the phase transition as a result of decreasing temperature is referred to as UCST.

In addition, the thermos-responsive polymer phase transition in water solution is the most interesting field because of its potential in biomedical applications.

The basic principle of temperature-responsive polymers in water solution is:

The main polymer phase transitions that can occur in water solutions are LCST and UCST. However, a third type of transition has been found and that is when the LCST and UCST exist together, and that is called closed loop coexistence.

As shown in Figure 14.2, LCST stands for the lowest temperature of the binodal curve and UCST stands for the highest temperature of the binodal curve. Closed loop coexistence has been seen in some polymers, for instance in poly ethylene glycol.

It is important to mention that most polymers have LCST transition and on the contrary, UCST and closed loop coexistence can rarely be found in polymers.

Polymers with LCST behaviour:

Polymers that pass through the LCST transition are soluble in aqueous environments with low temperatures, and the phase transition occurs by elevating temperature. By increasing the temperature, water molecules leave the partially dehydrated polymer chain and return to the bulk water, which causes the polymer to collapse and turn into a complete polymer phase. This transition is completely reversible by the hydration process, which gives access to reversible temperature (Figure 14.3).

In total, LCST totally depends on the polymer structure and more hydrated polymers have higher LCST, while less hydrated polymers have lower LCST. Also, it is important to note that by increasing the molecular weight LCST will decrease due to decreasing hydration and on the other hand, by adding more hydrophilic monomers, LCST will increase.

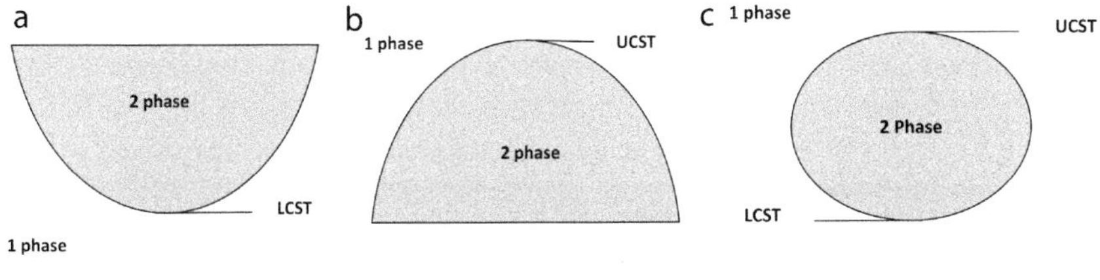

FIGURE 14.2 Polymer phase diagrams for (a) LCST behaviour, (b) UCST behaviour, and (c) closed-loop coexistence (Hoogenboom, 2014).

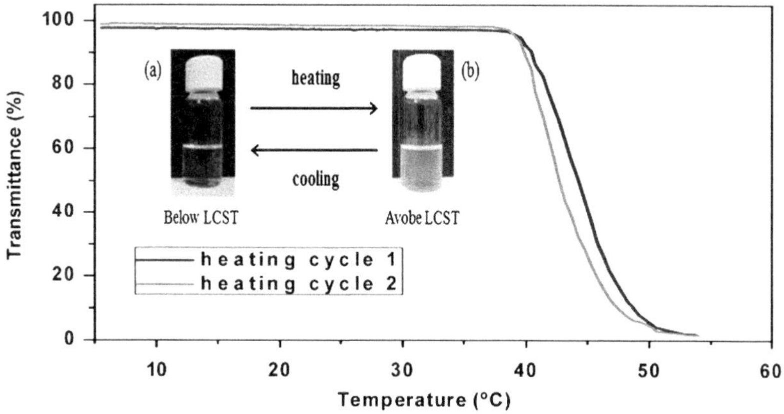

FIGURE 14.3 Temperature behaviour of a chemical species (a) below LCST and (b) above LCST. The blue curve shows the heating process and the green curve shows the cooling process (Montolio et al., 2014).

Polymers with UCST behaviour:

LCST is an entropic process and UCST is an enthalpic process and there is an important difference between these two behaviours. The UCST transition in aqueous solution relies on associative interactions; however, the power of supramolecular interactions has a direct effect on UCST transition, which may cause unwanted effects. For example, hydrophobic chains decrease the solubility of polymer in aqueous solutions and will decrease the LCST; however, they may cause the creation of a hydrophobic environment and enhance the hydrogen links' strength and result in a higher UCST transition. However, this description of UCST behaviour in water would not explain the UCST transition.

The UCST transition in highly heated water is not related to the changes in polymer bonds, instead it relates to changes in solvent properties. By elevating the temperature of an aqueous solution, the polarity of water decreases, making it an efficient environment to dissolve the polymer bondings. Because of this occurrence, the majority of polymers will go under the UCST transition in an alcohol–water mixture.

In recent decades, temperature-responsive polymers with LCST and UCST behaviour in water solution such as poly(acrylamide)s and poly(vinyl amide)s have been introduced and produced with different synthetic methods (Aoshima and Kanaoka, 2008; Aseyev et al., 2006). An example for poly(acrylamide)s is poly(N-isopropylacrylamide) (PNIPAM) and for poly(vinyl amide)s is poly(N-vinyl caprolactam) (PVCL) and despite their chemical similarity (Aseyev et al., 2006), PNIPAM has been the best candidate for biomaterial applications and has been used in most applications, rather than PVCL.

Applications for thermoresponsive polymers:

The LCST transition in polymers has been used to develop polymeric temperature sensors for accurate local temperature to determine living cells. In addition, by combining LCST and pH-responsive solvatochromic dye, a dual sensor has been developed which allows determination of the solution's temperature and pH by using UV-measurement (Pietsch et al., 2009).

Furthermore, by modifying the concentration of the polymer during LCST dehydration and aggregation of polymer bonding, we can monitor the osmotic strength and influence a reversible transfer of water through a membrane (Noh et al., 2012).

Future trends:

The future outlook of thermo-responsive polymers is expected to focus on polymers with UCST and LCST behaviour, especially LCST in aqueous solution, as well as their applications. In addition, the combination of temperature-responsive polymers with other responsive polymers, which are called multi-responsive polymers, will be expected.

14.2.2 *pH*-Responsive Polymers

pH-sensitive polymers are polyelectrolytes that have a weak acidic structure and which, with change in the pH of the environment, release or add protons to their structure. The ionization of these acidic or basic groups of the polyelectrolytes are the same as acidic or basic groups of monoacids, but it's more difficult due to the electrostatic effects of other ionized groups, which lead to a different dissociation constant (Ka) from similar monoacids. Also, by changing the electrolyte concentration or charges in the polymer backbone, the physical properties (chain conformation, solubility, etc.) could change.

This transition between a highly coiled polymer and an expanded structure could be the result of any change in electrostatic repulsion (pH, ionic strength, etc.)

There are two different ways to adjust the pH range where the phase change happens:

Choosing the ionizable fraction between the polyacid and the polybase (acceptable pH range)

Inclusion of a hydrophobic fraction in the polymer backbone (Na et al., 2004)

In addition, controlled radical polymerization (CRP) is a route to synthesize pH-responsive polymers (Gregory and Stenzel, 2012; Mu et al., 2011).

Definition and properties of pH-responsive polymers:

> Polyelectrolytes can separate to give polymeric ions in a proper ionized solvent and expand the structure of polymer. But with an unsuitable solvent that prevents ionization, the polymeric structure will remain folded (Figure 14.4). The behaviour of a pH-responsive polymer relies on the interaction between electrostatic repulsion and hydrophobic surface energy. The reason that pH-responsive polymers respond to different environmental pH is because pH has a significant effect on the degree of ionization of a poor polyelectrolyte.

Every pH-responsive polymer has acidic (anionic) or basic (cationic) groups that can receive or release protons by changes in the pH of the environment. Acidic polymers will be swollen and ionized by increasing the pH of the environment and basic group polymers will be swollen and ionized by decreasing the pH. An example of

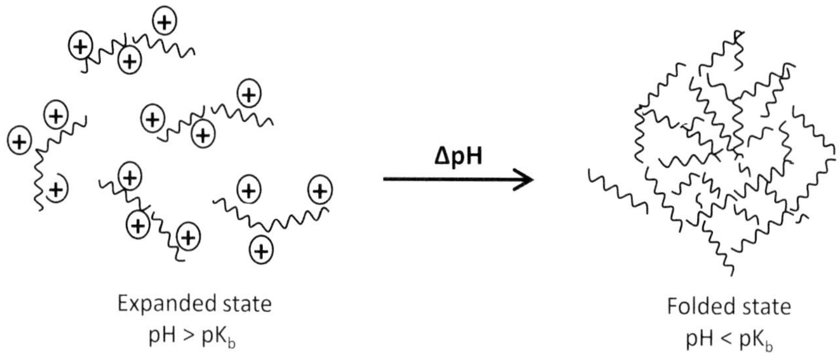

FIGURE 14.4 Polybasic states depending of the ionization of the ionic chain groups of the pH-responsive polymer (Reyes-Ortega, 2014).

a pH-responsive polymer is poly(carboxylic acid); in a low pH environment, hydrophobic interaction has the most effect on carboxyl groups, but by increasing the pH, the carboxyl group will turn into carboxyl ions and swell, resulting in an increase in polymer density (Kang and Bae, 2002).

In addition, besides the synthetic polymers that play a key role in biomaterial applications, natural biodegradable polymers such as alginate, chitosan, hyaluronic acid, etc., are very interesting because of their good biocompatibility and ease of modification.

Different methods for preparing a pH-responsive polymer:

> pH-responsive polymers can be prepared in different ways and of those, emulsion polymerization is a popular way to prepare vinyl-base pH-responsive polymers (Rao and Geckeler, 2011). This method usually consists of a monomer, water, an initiator and a surfactant. However, this method has a disadvantage, which is that removing the surfactant might be difficult. Mini-emulsion polymerization is another route to produce synthetic pH-responsive polymers, which consist of hydrogen molecules, a co-stabilizer, a monomer mixture, an initiator and a surfactant. The differences between emulsion polymerization and mini-emulsion polymerization are the use of low molecular mass as co-stabilizer and the use of a high shear device. Also, a new way to prepare pH-responsive polymer is microemulsion polymerization, and unlike emulsion polymerization, which has three reaction rates, microemulsion polymerization has two reaction rates. However, regardless of microemulsion polymerization's potential, this method of polymerization is not commonly used because it needs huge amount of surfactant in the formulation.

Another method of polymerization is group transfer polymerization (GTP), which is the usual way for producing methacrylate, and reversible addition-fragmentation chain transfer polymerization (RAFT) for the synthesis of macromolecular architectures that usually contain a monomer, a radical initiator, and a chain transfer agent.

Atom transfer radical polymerization (ATRP) is another method for polymerization and is the most common CRP system for monitoring the polymerization of vinyl and acrylic monomers (Matyjaszewki, 2009).

Applications of pH-responsive polymers:

> Every tissue and every organ in human body contains different pH ranges, and pH-responsive polymers can respond to different pH ranges and change their structure and this ability makes pH-responsive polymers very important in drug delivery, chromatography, and many other biomedical applications.

To describe more about their application, the human body has a pH ranging from 2 in the stomach to 8.2 in the duodenum. Due to this range of pH in living organs and tissues, it is possible to generate a pH-responsive carrier that can keep its structure in low-pH level or more acidic environments such as the gastrointestinal tract release its contents in a higher-pH level or more alkaline environment such as the duodenum, and this condition is important for drug delivery.

Another application of pH-responsive polymers is the invention of an insulin delivery system for diabetic patients (Hu and Liu, 2010). Delivering insulin is different from other drugs, because insulin must deliver a certain amount at the needed time, which makes it difficult to deliver. pH-responsive polymers with the use of the glucose oxidase enzyme can produce a glucose responsive polymer and can provide a mechanism for responding to reductions of glucose in the bloodstream and by releasing insulin can confront the reduction and maintain the concentration of insulin within the regular range.

Furthermore, the recent discovery of Park et al. (2012) has shown that pH-responsive polymers have great potential as non-viral gene carriers.

Future trends:

> The future development of pH-responsive polymers is focused on finding different analytes, while at the same time, maximizing their responses and finding ways to use these materials in medical devices to gain the most benefit in in-vitro or in-vivo circumstances.

In addition, multi-stimuli-responsive materials are another path to gaining the most benefits from the combination of smart polymers, and by finding the appropriate copolymerization or crosslinking, smart polymers can be used for specific applications in biomedical applications. However, most of the developments until now are experimental and future trends should focus on ways to use these materials for clinical therapies.

14.2.3 Photo-Responsive Polymers

As the term represents, photo-responsive polymers are a type of polymer that can respond to light. Some molecular characteristics that can be light-regulated are conformation, polarity, amphiphilicity, etc., and these polymers' responses are reflected in microscopic changes such as in shape, adhesion, solubility, wettability, etc.

Photo-responsive polymers have some advantages compared to other types of responsive polymers such as:

Can be triggered from outside the body
Easy to control in order to have the desirable response
Have accurate positional resolution of the response.

Photo-responsive polymers can respond to light because of a chromophore group, which is a photo-responsive group in their structure. Reversibility of an interaction is very important in biomedical applications and for photo-responsive materials, the type of chromophore dictates the reversibility or irreversibility of the response.

Chromophores and their responses:

> As mentioned, there are two types of chromophores available in photo-responsive polymers, reversible and irreversible. Reversible chromophores are the groups that go through a reversible isomerization at the time of light radiation with particular light properties and this reversibility is a result of photochromic interconversion between two phases. Some examples of reversible photo-response groups are azobenzene (Figure 14.5), spiropyran (Figure 14.6), and diarylethene (Figure 14.7) and some examples of irreversible photo-responsive groups are pyrenylmethyl, o-nitrobenzyl and 2-napthoquinone-3-methide.

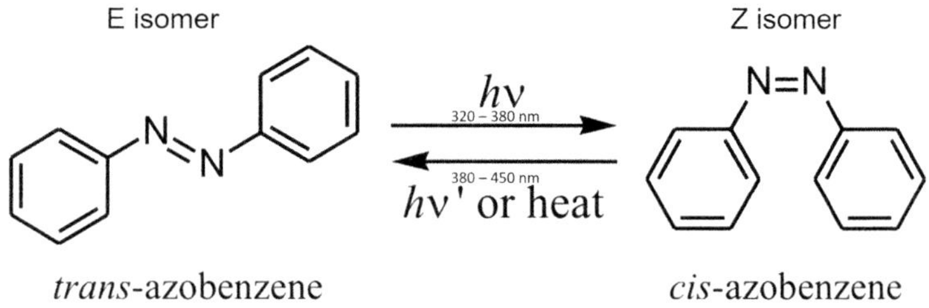

FIGURE 14.5 Isomerization of azobenzene (Su-no-G, 2007).

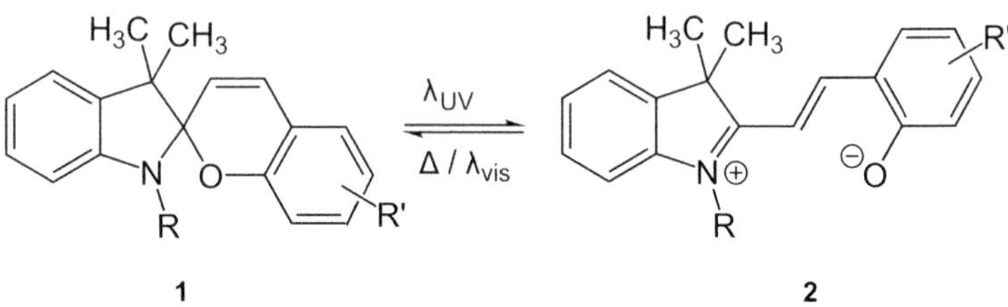

FIGURE 14.6 Isomerization of (1) Spiropyran to (2) Merocyanine (Yikrazuul, 2014).

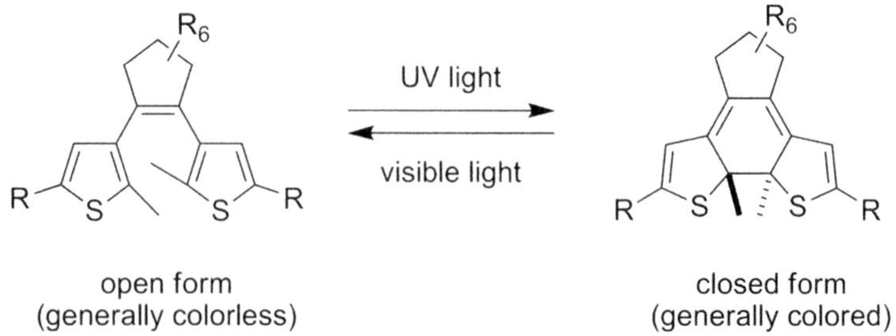

FIGURE 14.7 Dithienylethene acting as a molecular switch (Hbf878, 2018).

Application:

Photo-responsive polymers have been used in different applications as well as other stimuli-responsive polymers and have demonstrated useful feedback. Some applications are:

Drug delivery:

Polymeric micelles that were prepared for photo-responsive groups have been used as a carrier in drug delivery applications. Photo-responsive polymers can disassemble their structure (Zhao et al., 2009, 2012) by being in touch with a specific light wave, and by changing the light properties, micelles can be controlled (Figure 14.8). Light waves can change the hydrophobicity of the micelles and improve their solubility.

Responsive hydrogels:

Responsive hydrogels have been used in various applications and in most applications, the molecular structure alters and can change the degree of swelling in the hydrogels (Figure 14.9).

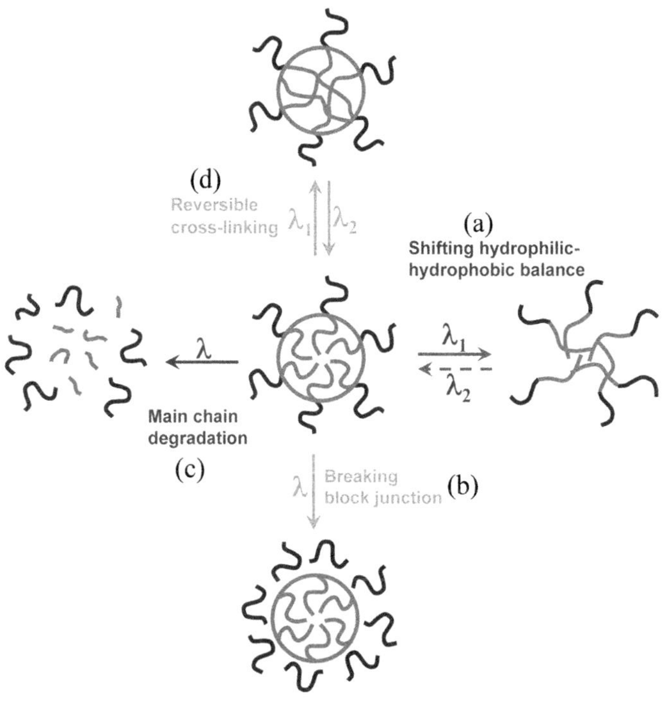

FIGURE 14.8 Schematic illustration of various types of light-responsive block copolymer micelles (a to d) (Zhao, 2012).

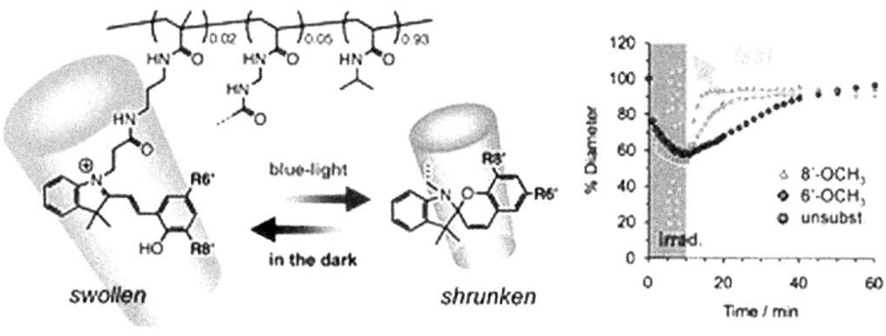

FIGURE 14.9 Swelling and shrinking of a chemical in the presence of light (Satoh et al., 2011).

Photodegradable materials:

Photodegradable materials are materials that have photolabile groups in their structure and go through chain breakage by exposure to light. The o-nitrobenzyl chromophore is an example of these materials, and various applications of it with different derivatives have been reported.

Future trends:

The future prospects of photo-responsive materials are work on materials that can respond to light with long waves to reach maximum support in and compatibility with the body. Although different types of chromophores have been used in-vitro, they haven't reached the level of satisfaction for use in in-vivo applications and need more development.

14.2.4 Magnetic-Responsive Polymers

Molecular magnetic material is a young field of stimuli-responsive polymers and there are two types of magnetic polymers (Miller and Epstein, 1998; Sato et al., 1996):

Molecular magnetic polymers that show weak responses at low temperature
Polymer composites that contain magnetic particles.

These composites have a rigid polymer type which contains magnetic particles and a highly elastic type which is divided into magnetic gels and magnetic elastomers. The magnetic elastomers have low flexibility and their elastic properties and shapes will remain constant in the presence of a magnetic field. Elastomers and magnetic gels have been recently developed and provide a new generation of composites containing nanosized magnetic particles mixed into elastic polymers which can easily change their physical and chemical properties and enhance their structure. In the presence of an external magnetic field, magnetic-responsive polymers adjust their polymer chain and this adjustment appears as changes in shape and deformation. However, by removing the magnetic field this deformation returns to the original shape, and this feature could be very helpful in different applications.

Different methods for preparing magnetic-responsive polymers:

> Magnetic-responsive polymers can be produced in many ways, such as magnetic filled particles, core-shell particles, etc. (Figure 14.10). The core-shell particle has magnetic particles surrounded by a polymer layer. The interesting part of preparing magnetic-responsive polymers is that there is no need to use a specific type of magnetic particles of polymers, but an important characteristic that is required is adsorptive interaction between the polymer chain and the magnetic particles.

However, an important challenge in preparing magnetic-responsive polymers is incompatibility when mixing magnetic particles with polymers, and to prevent this, strong adsorption interaction or surfactant is needed to stabilize the structure.

Another way of preparation is to equally distribute magnetic filler in a polymer, and by adding an external magnetic field, the orientation of magnetic particles will change, and magnetic interaction will happen.

In addition, there are two types of interactions for polymer gel with exposure to the magnetic field: field–particle interactions and particle–particle interactions. Polymer gels can also elongate and bend rapidly without trouble when they are in a magnetic field (Fehér et al., 2001; Zrinyi et al., 2001, 2002).

Applications:

> Even though magnetic-responsive polymers are a new field in stimuli-responsive materials, just like other smart polymers, they have been applied in different applications, for instance in controlled drug delivery, biosensors, MRI, cell separation, etc. (Figure 14.11).

Future trends:

> The future of magnetic gels with the ability to control shape changes is in artificial systems and devices. By improving technology, having a small and powerful magnetic field inside the polymer surrounding or the magnetic gel to eliminate the external magnetic field, could lead to the production of artificial tissues and muscles.

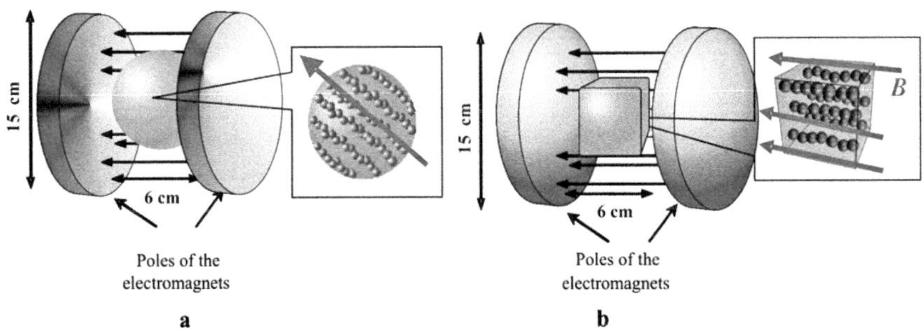

FIGURE 14.10 Preparation of anisotropic filler loaded PDMS gels under uniform magnetic field. (a) Spherical gel beads, (b) cube-shaped gels (Hajsz et al., 2006).

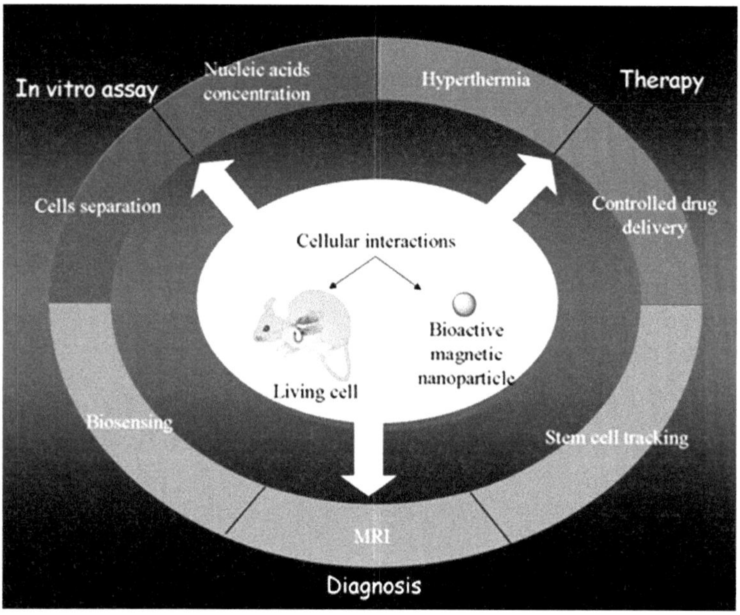

FIGURE 14.11 Biomedical applications of magnetic nanoparticles (Medeiros and Santos, 2010).

14.2.5 Enzyme-Responsive Polymers

Enzyme-responsive polymers (ERM) are another new generation of stimuli-responsive polymers that introduced in 2006 and used in various applications. These materials can respond to specific biological molecules, which is suitable to apply in biomedical application. However, there is a drawback for this material and that is that ERM cannot function in environments where no enzyme activity is detected or a better description is they cannot apply in environments wherein enzyme activity changes over time. This drawback could limit their application in many situations. However, engineers have already been used these materials in biomedical applications such as drug delivery, injectable scaffolding, etc.

Different types of enzyme-responsive polymers:

> During past decade, different types of ERMs have been designed and used and base on their structures, they can be categorized as hydrogels, supramolecular, particles, and self-immolatives.

Polymer hydrogels are the first type of ERMs and contain of hydrophilic polymers and water (Wichterle and Lim, 1960). Crosslinked polymers that are used in hydrogels can be artificial or natural and the main applications of hydrogels are in temporary scaffolding (Mano, 2008) and drug delivery. In addition, different hydrogels have been designed such as enzyme crosslinked hydrogels, enzyme degradation hydrogels (Kumashiro et al., 2002), and enzyme-induced morphological hydrogels

(Thornton et al., 2005, 2008; McDonald et al., 2009) that can respond in different ways in the presence of enzymes.

Supramolecular polymers are polymers that due to phase segregation or the formation of secondary structures can self-assemble the structure and expand (Lehn, 2002) and reverse this change. Like hydrogels, supramolecular polymers have three types: self-assembly, disassembly and enzyme-induced morphological supramolecular.

Polymeric particles usually act as carriers for drugs in order to release them in a desired environment and allows a controlled drug delivery (Delcea et al., 2011; Ghadiali and Stevens, 2008). The drugs are contained in polymeric particles composed of a structure containing polymer and polymeric capsules in which the polymeric capsules surround a solvent. This allows a controlled release of a drug, making polymeric particle carriers very useful for small drug deliveries.

Self-immolative polymers are ester-based polymers which can disassemble polymers to monomeric units that make this type of ERM a useful material for drug delivery (Blencowe et al., 2011).

Applications:

> ERM's have been introduced in the past decade and even though this is a new field of stimuli-responsive materials, because of the water in their structure, their having a 3D structure, the ability to remodel the structure, degradation, reversibility and many other important characteristics, these materials have been widely used for cell support (Ehrbar et al., 2007; Guan et al., 2008), injectable scaffolds (Jin et al., 2010), and drug delivery (Itoh et al., 2006).

Future trends:

> Because of the young age of these materials, their applications are mostly limited to forming or degrading these materials with enzyme reactions. However, recent studies have shown that this material has great potential other than forming or degrading. For example, recent developments are mostly focused on combining ERM with other stimuli-responsive polymers to produce a complex material to respond to different circumstances and changes.

In conclusion, enzyme-responsive materials have shown great potential for biomedical applications either in-vitro or in-vivo, but to achieve their maximum usage and to commercialize this material, more investigation and effort are required.

14.3 COMPARING SMART POLYMERS IN SPECIFIC APPLICATIONS

In this section, we want to compare the types of smart polymers that have been discussed in a specific application, in particular drug delivery, and see the pros and cons of each application.

The first smart polymer that we discussed was a temperature-responsive polymer, and according to the research that took place in Tokyo Women's Medical University in 2006, they tried to design temperature-responsive micelles by using PLA in order to get immediate drug release and sharp thermo-responsiveness. After the test was completed, they successfully synthesized hydroxy-terminated P

(IPAAm-co-DMAAm) [Poly(N-isopropylacrylamide-co-N,N-dimethylacrylamide)] by 40°C phase-transition temperature, and this transition is useful for temperature-induced drug release in living organs. In addition, by changing the molecular weight of polymer blocks, they controlled the micelle concentration for drug delivery targeting. Also, they determined the phase-transition temperature by copolymer ratio of N-isopropylacrylamide and N-dimethylacrylamide units.

Other smart polymers that have been introduced include pH-responsive polymers, and according to the research that has been done by Heebeom Koo and coworkers in 2010, they tried to diagnose and treat tumours by using pH-responsive polymers. After they injected tumour-bearing mice with protoporphyrin loaded into pH-responsive polymeric micelles, they found that the drug delivery efficiency was about 80%. In addition, after injection strong fluorescence signals were observed at the tumour location, which means that pH-responsive polymers could also help for diagnosis. After a day, they observed that the tumour started to bleed and was damaged, which shows that more antitumour drugs were delivered at the tumour site using pH-responsive polymers.

Photo-responsive polymers are another type of smart polymers that have been introduced, and according to the paper prepared by Chaojuan Chen and coworkers in 2011, they used biocompatible micelles made out of amphiphilic hyper branched polyphosphate polymer to apply in controlled release of Courmarin 102. The results showed a successful encapsulation of Courmarin 102 in micelles and successful control of the release of the drug under 365 nm UV light and high biocompatibility without any toxicity, which illustrates the potential of photo-responsive polymers as smart carrier for anticancer drugs.

In other research established by Moom Sinn Aw and coworkers in 2012, the usage of magnetic-responsive polymers was tested by using titanium nanotubes loaded with magnetic nanoparticles at the bottom and drug encapsulated polymer micelles at the top, with the concept of using magnetic field to cause movement in magnetic particles in order to release the drug encapsulated in polymer micelles. The results revealed that complete and quick release of the drug was achieved at the desired time and at the targeted location after applying the magnetic field.

The last type of stimuli-polymers that will be discussed are ERM. Research done by Yao Wang and coworkers in 2016 reported the usage of poly(amino amide) containing bioadhesive and enzyme-responsive components. Their goal was to use this material as a carrier for drug delivery for cancer therapy, and after experimental tests they found impressive results such as complete biocompatibility without toxicity, high drug loading capacity, and constant drug release in the enzymatic environment.

Nowadays, it is important to understand the usage of smart polymers and find the importance of these materials in global marketing. So far, we mainly focused on biomedical applications. However, the applications of stimuli-responsive polymers are not just limited to biomedical applications, and these materials have been used in several other applications such as textile, automotive industry, nuclear energy, etc. Figure 14.12 shows the usage of these materials in different applications in 2015.

In addition, according to a Business wire report in November 2016, the worldwide marketing of smart polymers in 2015 was valued at USD 881.66 million,

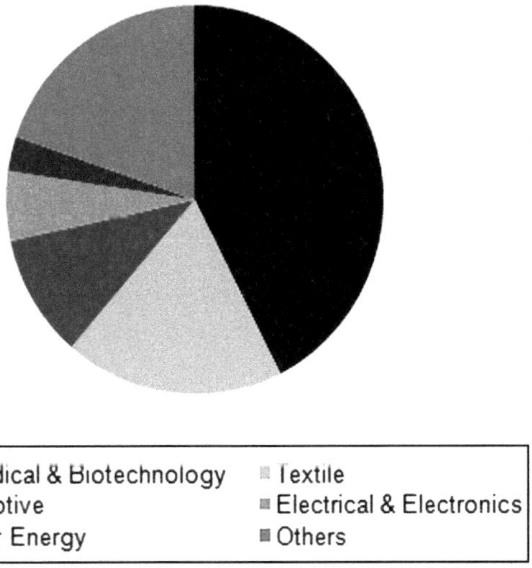

FIGURE 14.12 Smart polymers market share by application (Grand View Research, Inc., 2016).

and they predict that this amount in 2021 will reach approximately USD 3,000 million which shows the growth of smart polymers in different application in near future.

14.4 FUTURE PERSPECTIVE OF SMART MATERIALS

The future of smart polymers is very promising and these materials with their unique characteristics and responses will be helpful in various applications such as detecting and treating cancer, in biosensors, tissue engineering, detecting radioactive radiation, automotive, etc. Also, after gaining an understanding of each type of smart polymers responses and behaviour, a new field of smart polymers emerges and can be explored: the combination of smart polymers to find a suitable material in more different and more complex environment.

In conclusion, the future of smart polymers is bright; however, it is only by taking into consideration the future needs of people and the different technological developments and usage of smart polymers in technology that the true potential of smart polymers is unleashed.

REFERENCES

Aguilar, M. R., Elvira, C., Gallardo, A., Vázquez, B., and San Román, J. (2007). Smart polymers and their applications as biomaterials. In Ashammakhi N., Reis R. L. and Chiellini E., (Eds.) *Topics in Tissue Engineering* Vol. 3 [ebook]. Biomaterials and Tissue Engineering Group.

Aoshima, S., and Kanaoka, S. (2008). Synthesis of stimuli-responsive polymers by living polymerization: Poly(N-isopropylacrylamide) and poly(vinyl ether)s. *Advances in Polymer Science*, 210, 169–208.

Arnold, R. M., Huddleston, N. E., and Locklin, J. (2012). Utilizing click chemistry to design functional interfaces through post-polymerization modification. *Journal of Materials Chemistry*, 22, 19357–19365.

Arora, G., Singh, I., Nagpal, M., and Arora, S. (2011). Recent advances in stimuli induced pulsatile drug delivery system: A review. *Research Journal of Pharmacy and Technology*, 4(5), 691–703.

Aseyev, V. O., Tenhu, H., and Winnik, F. M. (2006). Temperature dependence of the colloidal stability of neutral amphiphilic polymers in water. *Advances in Polymer Science*, 196, 1–85.

Aw, M. S., Addai-Mensah, J., and Losic, D. (2012). Magnetic-responsive delivery of drug-carriers using titania nanotube arrays. *Journal of Materials Chemistry*, 22(14), 6561–6563.

Blencowe, C. A., Russell, A. T., Greco, F., Hayes, W., and Thornthwaite, D. W. (2011). Self-immolative linkers in polymeric delivery systems. *Polymer Chemistry*, 2, 773–790.

Chen, C., Liu, G., Liu, X., Pang, S., Zhu, C., Lv, L., and Ji, J. (2011). Photo-responsive, biocompatible polymeric micelles self-assembled from hyperbranched polyphosphate-based polymers. *Polymer Chemistry*, 2(6), 1389–1397.

Delcea, M., Mohwald, H., and Skirtach, A. G. (2011). Stimuli-responsive LbL capsules and nanoshells for drug delivery. *Advanced Drug Delivery Reviews*, 63(9), 730–747.

Ehrbar, M., Rizzi, S. C., Schoenmakers, R. G., San Miguel, B., Hubbell, J. A., Weber, F. E., and Lutolf, M. P. (2007). Biomolecular hydrogels formed and degraded via site-specic enzymatic reactions. *Biomacromolecules*, 8(10), 3000–3007.

Fehér G. F., Szalma J., and Zrínyi M. (2001). *Colloids and Surfaces A: Physicochemical and Engineering Aspects*, 183:505–515.

Fioretta, E. S., Fledderus, J. O., Burakowska-Meise, E. A., Baaijens, F. P. T., Verhaar, M. C., and Bouten, C. V. C. (2012). Polymer-based scaffold designs for in situ vascular tissue engineering: controlling recruitment and differentiation behaviour of endothelial colony forming cells. *Macromolecular Bioscience*, 12(5), 577–590.

Galaev, I., and Mattiasson, B. (Eds.) (2007). *Smart Polymers. Applications in Biotechnology and Biomedicine*. CRC Press.

Ghadiali, J. E., and Stevens, M. M. (2008). Enzyme-responsive nanoparticle systems. *Advanced Materials*, 20(22), 4359–4363.

Grand View Research, Inc. (2016). Smart Polymers Market Analysis By Stimulus Type (Physical, Chemical, Biological), By Application (Biomedical & Biotechnology, Automotive, Textile, Electrical & Electronics, Textile, Nuclear Energy) And Segment Forecasts, 2018–2025. Retrieved from https://www.grandviewresearch.com/industry-analysis/smart-polymers-market

Gregory, A., and Stenzel, M. H. (2012). Complex polymer architectures via RAFT polymerization: From fundamental process to extending the scope using click chemistry and nature's building blocks. *Progress in Polymer Science*, 37(1), 38–105.

Guan, J., Fujimoto, K. L., and Wagner, W. R. (2008). Elastase-sensitive elastomeric scaffolds with variable anisotropy for soft tissue engineering. *Pharmaceutical Research*, 25(10), 2400–2412.

Guillerm, B., Monge, S., Lapinte, V., and Robin, J. J. (2012). How to modulate the chemical structure of polyoxazolines by appropriate functionalization. *Macromolecular Rapid Communications*, 33(19), 1600–1612.

Hajsz, T., Csetneki, I., Filipcsei, G., and Zrinyi, M. (2006). Swelling kinetics of anisotropic filler loaded PDMS networks. *Physical Chemistry Chemical Physics*, 8(8), 977–984.

Hbf878. (2018). *Dithienylethene*. [Online Image]. Retrieved from https://commons.wikimedia.org/wiki/File:Dithienylethene.svg

Hoogenboom, R. (2014). Temperature-responsive polymers: Properties, synthesis and applications. *Smart Polymers and Their Applications* (pp. 15–44) doi:10.1533/9780857097026.1.15

Hu, J., and Liu, S. (2010). Responsive polymers for detection and sensing applications: Current status and future developments. *Macromolecules*, 43(20), 8315–8330.

Hu, J., Meng, H., Li, G., and Ibekwe, S. I. (2012a). A review of stimuli-responsive polymers for smart textile applications. *Smart Materials and Structures*, 21(5), 053001 .

Hu, J., Zhang, G., and Liu, S. (2012b). Enzyme-responsive polymeric assemblies, nano-particles and hydrogels. *Chemical Society Reviews*, 41, 5933–5949.

Huang, Z., Sargeant, T. D., Hulvat, J. F., Mata, A., Bringas Jr, P., Koh, C. Y., and Snead, M. L. (2008). Bioactive nanofibers instruct cells to proliferate and differentiate during enamel regeneration. *Journal of Bone and Mineral Research*, 23(12), 1995–2006.

Itoh, Y., Matsusaki, M., Kida, T., and Akashi, M. (2006). Enzyme-responsive release of encapsulated proteins from biodegradable hollow capsules. *Biomacromolecules*, 7(10), 2715–2718.

Jin, R., Teixeira, L. S. M., Dijkstra, P. J., van Blitterswijk, C. A., Karperien, M., and Feijen, J. (2010). Enzymatically-crosslinked injectable hydrogels based on biomimetic dextran-hyaluronic acid conjugates for cartilage tissue engineering. *Biomaterials*, 31(11), 3103–3113.

Kang, S. I., and Bae, Y. H. (2002). pH-Induced solubility transition of sulfonamide-based polymers. *Journal of Controlled Release*, 80(1–3), 145–155.

Karimi, Z., Karimi, L., and Shokrollahi, H. (2013) Nano-magnetic particles used in biomedicine: Core and coating materials. *Materials Science and Engineering C*, 33(5), 2465–2475.

Koo, H., Lee, H., Lee, S., Min, K. H., Kim, M. S., Lee, D. S., and Jeong, S. Y. (2010). In vivo tumor diagnosis and photodynamic therapy via tumoral pH-responsive polymeric micelles. *Chemical Communications*, 46(31), 5668–5670. Chicago.

Kumashiro, Y., Lee, T. K., Ooya, T., and Yui, N. (2002). Enzymatic degradation of semi-IPN hydrogels based on N-isopropylacrylamide and dextran at a specific temperature range. *Macromolecular Rapid Communications*, 23(7), 407–410.

Lehn, J. M. (2002). Supramolecular polymer chemistry—scope and perspectives. *Polymer International*, 51(10), 825–839.

Lendlein, A., and Kelch, S. (2002). Shape-memory polymers. *Angewandte Chemie International Edition*, 41(12), 2034–2057.

Li, M. H., and Keller, P. (2006). Artificial muscles based on liquid crystal elastomers. *Philosophical Transactions of the Royal Society A: Mathematical, Physical and Engineering*, 364(1847), 2763–2777.

Liu, C., Qin, H., and Mather, P. T. (2007). Review of progress in shape-memory polymers. *Journal of Materials Chemistry* 17, 1543–1558.

Matyjaszewski, K. (Ed.) (2009). Controlled radical polymerization: State of the art in 2008. In *Controlled/Living Radical Polymerization: Progress in ATRP*, vol. 1023. American Chemical Society.

McDonald, T. O., Qu, H. L., Saunders, B. R., and Ulijn, R. V. (2009). Branched peptide actuators for enzyme responsive hydrogel particles. *Soft Matter*, 5(8), 1728–1734.

Medeiros, S. F., Santos, A. M., Fessi, H., and Elaissari, A. (2011). Stimuli-responsive magnetic particles for biomedical applications. *International Journal of Pharmaceutics*, 403(1–2), 139–161.

Miller, J. S., and Epstein, A. J. (1998). Tetracyanoethylene-based organic magnets. *Chemical Communications*, 13, 1319–1325.

Montolio, S., Gonzáez, L., Altava, B., Tenhu, H., Burguete, M. I., García-Verdugo, E., and Luis, S. V. (2014). LCST-type polymers based on chiral-polymeric ionic liquids. *Chemical Communications*, 50(73), 10683–10686.

Moroni, L., De Wijn, J. R., and Van Blitterswijk, C. A. (2008). Integrating novel technologies to fabricate smart scaffolds. *Journal of Biomaterials Science, Polymer Edition*, 19(5), 543–572.

Mu, B., Liu, P., Tang, Z., Du, P., and Dong, Y. (2011). Temperature and pH dual-responsive cross-linked polymeric nanocapsules with controllable structures via surface-initiated atom transfer radical polymerization from templates. *Nanomedicine: Nanotechnology, Biology and Medicine*, 7(6), 789–796.

Na, K., Lee, K. H., and Bae, Y. H. (2004). pH-sensitivity and pH-dependent interior structural change of self-assembled hydrogel nanoparticles of pullulan acetate/oligo-sulfonamide conjugate. *Journal of Controlled Release*, 97(3), 513–525

Noh, M., Mok, Z., Lee, S., and Kim, H. (2012). Novel lower critical solution temperature phase transition materials effectively control osmosis by mild temperature changes, *Chemical Communications*, 48, 3845–3847.

Park, H. J., Yang, F., and Cho, S. W. (2012). Nonviral delivery of genetic medicine for therapeutic angiogenesis. *Advanced Drug Delivery Reviews*, 64(1), 40–52.

Pasparakis, G., Manouras, T., Argitis, P., and Vamvakaki, M. (2012). Photodegradable polymers for biotechnological applications. *Macromolecular Rapid Communications*, 33(3), 183–198.

Pietsch, C., Hoogenboom, R., and Schubert, U. S. (2009). Soluble polymeric dual sensor for temperature and pH value. *Angewandte Chemie International Edition*, 48(31), 5653–5656.

Rao, J. P., and Geckeler, K. E. (2011). Polymer nanoparticles: Preparation techniques and size-control parameters. *Progress in Polymer Science*, 36(7), 887–913.

Reyes-Ortega, F. (2014). pH-responsive polymers: Properties, synthesis and applications. *Smart Polymers and Their Applications* (pp. 45–92) doi:10.1533/9780857097026.1.45

Satoh, T., Sumaru, K., Takagi, T., and Kanamori, T. (2011). Fast-reversible light-driven hydrogels consisting of spirobenzopyran-functionalized poly(N-isopropylacrylamide). *Soft Matter*, 7(18), 8030–8034.

Schumers, J.M., Fustin, C.A., and Gohy, J.F. (2010). Macromol. Rapid Commun., 31, 1588–1607.

Su-no-G. (2007). *Azobenzene Isomerization*. [Online Image]. Retrieved from https://commons.wikimedia.org/wiki/File:Azobenzene_isomerization.svg

Thornton, P. D., Mart, R. J., Webb, S. J., and Ulijn, R. V. (2008). Enzyme-responsive hydrogel particles for the controlled release of proteins: designing peptide actuators to match payload. *Soft Matter*, 4(4), 821–827.

Thornton, P. D., McConnell, G., and Ulijn, R. V. (2005). Enzyme responsive polymer hydrogel beads. *Chemical Communications*, 47, 5913–5915.\

Tran, H. V., Piro, B., Reisberg, S., Tran, L. D., Duc, H. T., and Pham, M. C. (2013). Label-free and reagentless electrochemical detection of microRNAs using a conducting polymer nanostructured by carbon nanotubes: Application to prostate cancer biomarker miR-141. *Biosensors and Bioelectronics*, 49, 164–169.

Wang, Y., Luo, Y., Zhao, Q., Wang, Z., Xu, Z., and Jia, X. (2016). An enzyme-responsive nanogel carrier based on PAMAM dendrimers for drug delivery. *ACS Applied Materials & Interfaces*, 8(31), 19899–19906.

Wichterle, O., and Lim, D. (1960). Hydrophilic gels for biological use. *Nature*, 185, 117–118.

Yakacki, C. M., Lyons, M. B., Rech, B., Gall, K., and Shandas, R. (2008). Cytotoxicity and thermomechanical behavior of biomedical shape-memory polymer networks post-sterilization. *Biomedical Materials*, 3(1), 015010.

Yikrazuul. (2014). *Spiropyran to Merocyanine*. [Online Image]. Retrieved from https://commons.wikimedia.org/wiki/File:Spiropyran_to_Merocyanine.svg

Yoshida, Y., Kawahara, K., Mitsumune, S., Kuzuya, A., and Ohya, Y. (2017). Injectable and biodegradable temperature-responsive mixed polymer systems providing variable gel-forming pH regions. *Journal of Biomaterials Science, Polymer Edition*, 28(10–12), 1158–1171.

Zhao, H., Sterner, E. S., Coughlin, E. B., and Theato, P. (2012). o-Nitrobenzyl alcohol derivatives: Opportunities in polymer and materials science. *Macromolecules*, 45(4), 1723–1736.

Zhao, Y. (2009). Photocontrollable block copolymer micelles: What can we control? *Journal of Materials Chemistry*, 19(28), 4887–4895.

Zhao, Y. (2012). Light-responsive block copolymer micelles. *Macromolecules*, 45(9), 3647–3657.

Zrinyi, S., Szabó, D., Filipcsei, G., and Fehér, J. (2001). Electric and magnetic field-sensitive smart polymer gels, In Y. Osada, and A. Khokhlov (Eds.), *Polymer Gels and Networks*. Marcel Dekker, Inc.

Index